Life Sciences in Comics
The Complete Series, Units I through Unit V

Randy Yang

© 2013 Randy Yang
All Rights Reserved.

No part of this publication may be reproduced, stored in a retrieval system, or transmitted, in any form or by any means, electronic, mechanical, photocopying, recording, or otherwise, without the written permission of the author.

First published by Dog Ear Publishing
4010 W. 86th Street, Ste H
Indianapolis, IN 46268
www.dogearpublishing.net

ISBN: 978-1-4575-2178-2

This book is printed on acid-free paper.

Printed in the United States of America

Table of Content

(*State Science Standards are Listed in Parenthesis)

Unit I: Ecology   Page
Chapter 1:   Material and Energy Cycles (* Standard #6.5.b) — 2
Chapter 2:   An Organism's Role in an Ecosystem — 12
             (* Standard #6.5.c)
Chapter 3:   How the Environment Affects an Organism — 19
             (* Standard #6.5.e)
Chapter 4:   Biodiversity (* Standard #Bio/LS.6.a) — 28
Chapter 5:   Changes in an Ecosystem (* Standard #Bio/LS.6.b) — 32
Chapter 6:   Population Size (* Standard #Bio/LS.6.c) — 47
Chapter 7:   The Biosphere and Other Spheres) — 57
             (* Standard #Bio/LS.6.d)
Chapter 8:   Producers and Decomposers (* Standard #Bio/LS.6.e) — 61
Chapter 9:   Energy-Flow in an Ecosystem (* Standard #Bio/LS.6.f) — 67

Unit 2: Cell Biology
Chapter 10: The Nucleus Stores Genetic Information — 76
             (* Standard #7.1.c)
Chapter 11: Cellular Respiration and Photosynthesis — 84
             (* Standard #7.1.d)
Chapter 12: Chloroplasts and Photosynthesis — 106
             (* Standard #Bio/LS.1.f)
Chapter 13: Cell Division (* Standard #7.1.e) — 112
Chapter 14: Elements in Living Things (* Standard #8.6.b) — 125
Chapter 15: Carbon Compounds (* Standard #8.6.c) — 132
Chapter 16: Cell Membrane and Diffusion — 142
             (* Standard #Bio/LS.1.a)
Chapter 17: Virus, Prokaryote, and Eukaryote — 158
             (* Standard #Bio/LS.1.c)

Unit 3: Genetics   Page

Chapter 18:	Sexual Reproduction (* Standard #7.2.a)	162
Chapter 19:	Dominant and Recessive Traits (* Standard #7.2.c) (* Standard #7.2.d)	165
Chapter 20:	What Are "Genes" Made Of? (* Standard #7.2.e.)	179
Chapter 21:	Meiosis and Gametes (* Standard #Bio/LS.2.b)	184
Chapter 22:	Sexual Reproduction's Advantage (* Standard #Bio/LS.2.d) (* Standard #Bio/LS.2.e)	192
Chapter 23:	Sex Chromosomes (* Standard #Bio/LS.2.f)	195
Chapter 24:	Predicting One's Offspring (* Standard #Bio/LS.3.a)	199
Chapter 25:	DNA and RNA (* Standard #Bio/LS.5.a)	209

Unit 4: Evolution

Chapter 26:	Genetics and Evolution (* Standard #7.3.a) (* Standard #Bio/LS. 7.c) (* Standard #Bio/LS.7.d.)	216
Chapter 27:	Charles Darwin (* Standard #7.3.b)	242
Chapter 28:	Fossils and the Evidence for Evolution (* Standard #7.3.c) (* Standard #Bio/LS.8.e)	251
Chapter 29:	Natural Selection and Phenotype (* Standard #Bio/LS.7.a) (* Standard #Bio/LS.7.b)	271
Chapter 30:	Three Types of Natural Selections (* Standard #Bio/LS.8.a)	278
Chapter 31:	Diversity and Survival (* Standard #Bio/LS.8.b)	282

Unit 5: Anatomy and Physiology

Chapter 32:	Cells, Tissues, Organs, and Systems (* Standard #7.5.a)	288
Chapter 33:	Bones and Muscles (* Standard #7.5.c)	304
Chapter 34:	Blood Circulation (* Standard #7.6.j)	311
Chapter 35:	Material Exchange by Body Systems (* Standard #Bio/LS.9.a)	321
Chapter 36:	The Nervous System (* Standard #Bio/LS.9.b)	338
Chapter 37:	The Immune Response (* Standard #Bio/LS.10.b)	344
Chapter 38:	Vaccines (* Standard #Bio/LS.10.c)	351
Chapter 39:	The War against Pathogens (* Standard #Bio/LS.10.d)	355

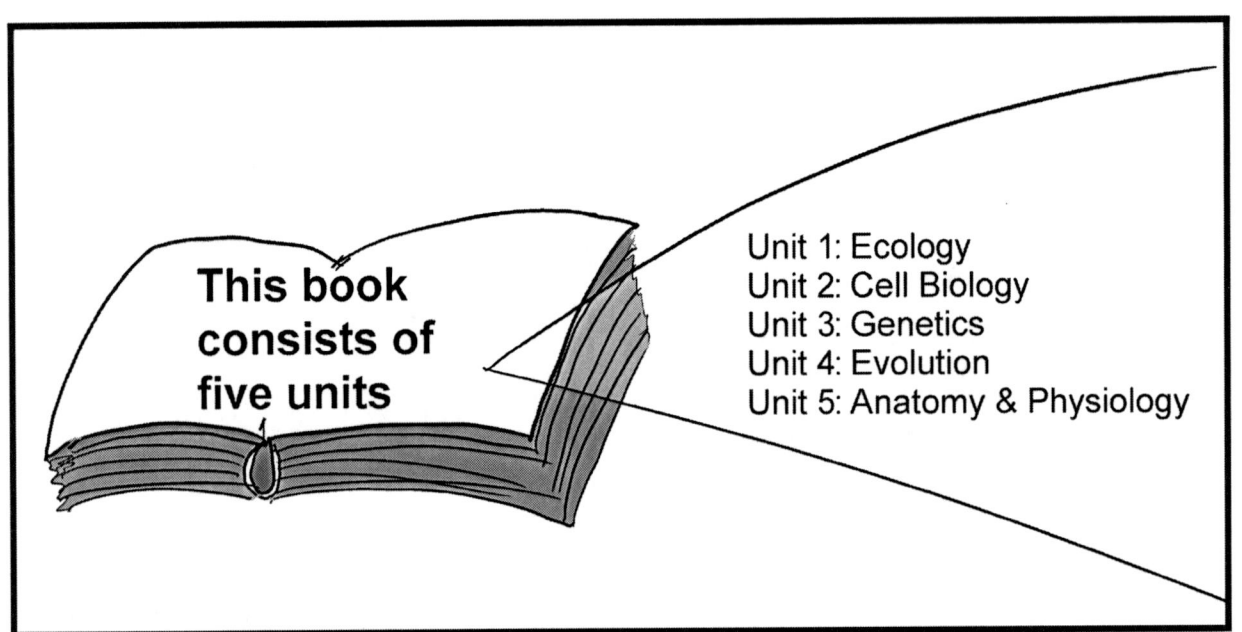

Each unit contains different chapters.

Unit 1: Chapter 1~9

Unit 2: Chapter 10~17

Unit 3: Chapter 18~25

Unit 4: Chapter 26~31

Unit 5: Chapter 32~39

At the begnning of each chapter, you will see the following information:

"Here is the science *standard (9th Grade) this chapter covers.

(* 9th grade science standard in the state of California)"

"This is the name of the chapter."

Chapter 38: Vaccine

Science Standard Bio/LS . 10 . c : Students know how vaccination protects an individual from infectious diseases..

NEW VOCABS

* **Antibody:** A substance produced by your white blood cells.

* **Antigens:** A substance that covers the surface of pathogens, and it can cause an immunological response.

* **Pathogen:** A micro-organism that can cause sickness and is contagious.

* **Vaccine:** A substance that increases your immunity to a particular disease.

"The new vocabularies you will encounter in this chapter are listed here, along with their definitions."

Intro-Page C

Unit 1: Ecology

Chapter 1: Material and Energy Cycles

Science Standard: 6.5.b
Students know matter is transferred over time from one organism to others in the food web and between organisms and the physical environment.

NEW VOCABS

* **Ammonia:** Its chemical formula is NH_3. It is a nutrient useful to plants.

* **Biome:** There are various types of biomes on earth. A biome is the embodiment of all the regions that feature a similar climate, ecosystems, and geographic location (latitude).

* **Biosphere:** The parts of the earth where living organisms exist, from the high sky to the deep underground.

* **Combustion:** The burning of matter such as wood, fossil fuel…etc.

* **Community:** Multiple populations of different species living in the same area.

* **Condensation:** The process in which water vapor (gas) becomes droplets of water (liquid).

* **Decomposition:** A process during which bacteria break down an organism's [dead] body.

* **Denitrification:** This process is done by bacteria, and it converts nitrates and nitrites into nitrogen gas.

Unit 1: Chapter 1

* **Ecology:** The study of how organisms interact with each other and with the environment.

* **Ecosystem:** A community of organisms + nonliving things in a region.

* **Evaporation:** The process during which liquid water turns into water vapor (gas).

* **Excretion:** The act of an organism expelling wastes (such as feces/urine) out of its body.

* **Food chain:** A list of "who-eats-whom" in an ecosystem.

* **Food web:** A diagram that features 2 or more food chains in an ecosystem and illustrates every organism's relationship with each other.

* **Geologic uplifting:** Geologic activities, such as volcanic eruption, causing under-water landmasses to be lifted above the sea level.

* **Material Cycles:** The movement of materials (such as water, carbon, nitrogen, and phosphorus) in an ecosystem, cycling back and forth between organisms and the environments.

* **Nitrates and nitrites:** Nitrates and nitrite's chemical formula are, respectively $-NO_3$ and $-NO_2$. These chemicals are important nutrients to plants.

* **Nitrification:** This process is done by bacteria. It turns ammonia into nitrates and nitrite.

* **Nitrogen fixation:** This process is done by bacteria. It turns nitrogen gas into ammonia.

* **Nitrogen gas:** Its chemical formula is N_2 and exists abundantly in the atmosphere.

* **Organism:** A living thing.

* **Population:** A group of the same species of organisms living together.

* **Precipitation:** Raining (water droplets falling from the sky).

* **Primary consumer:** Also known as "herbivores". An organism who eats producers.

* **Producer:** Any organism that can make its own energy. In most ecosystem, "producers" are plants and algae. Another name for producer is "autotroph".

* **Secondary consumer:** Any organism who eats primary consumers.

* **Sedimentation:** A process in which substances settle to the bottom of a body of water.

* **Seepage:** A process in which water seeps through the soil, washing certain soil-bound substances into a body of water.

* **Tertiary consumer:** Any organism who eats secondary consumers.

* **Trophic level:** It means each of the positions in a food chain.

* **Weathering:** A process in which water/wind break down small bits of rock, which are added to the soil.

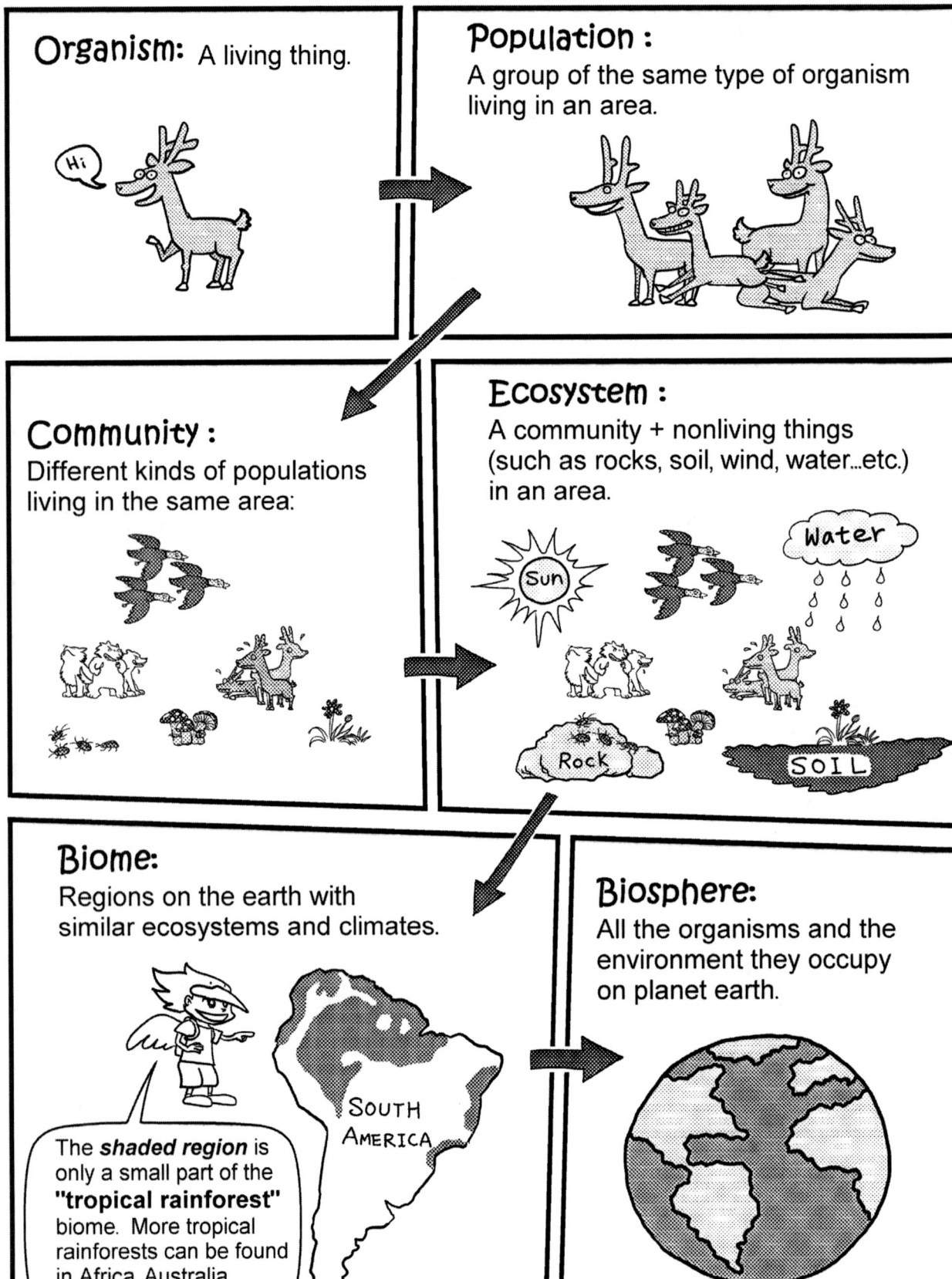

In an ecosystem, materials cycle back and forth:

Water Cycle

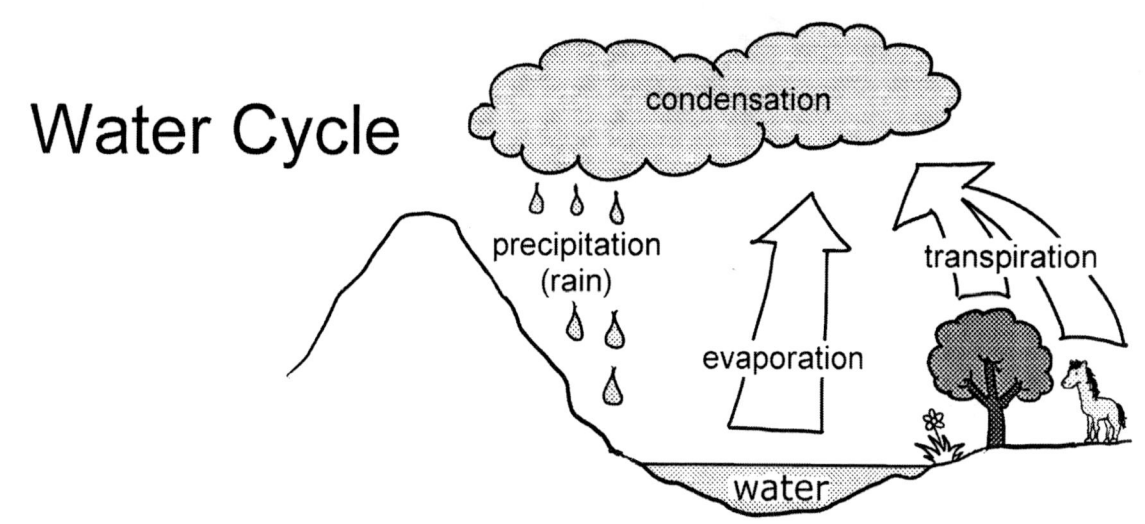

* **Condensation:** water vapor (gas) turns into water droplets (liquid).
* **Evaporation:** liquid water turns into water vapor (gas).
* **Precipitation (rain):** Water vapor (gas) solidifies into drops of water (liquid).
* **Transpiration:** Water vapor escapes from living organisms into the environment.

Carbon Cycle

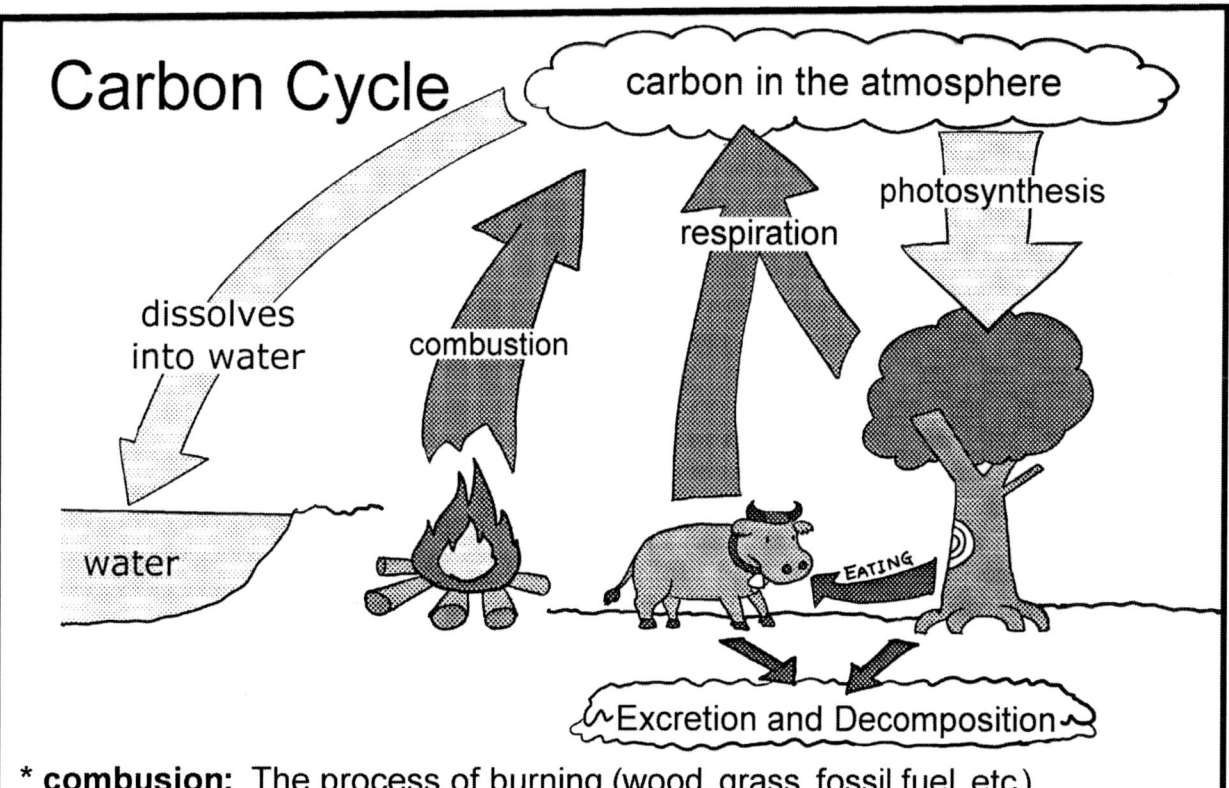

* **combusion:** The process of burning (wood, grass, fossil fuel...etc.)
* **excretion and decomposition:** Pooping, peeing, <u>or</u> dying onto the ground.
* **photosynthesis:** A process that involves plants taking in carbon dioxide.
* **respiration:** The act of breathing in O_2 and breathing out CO_2.

Nitrogen Cycle

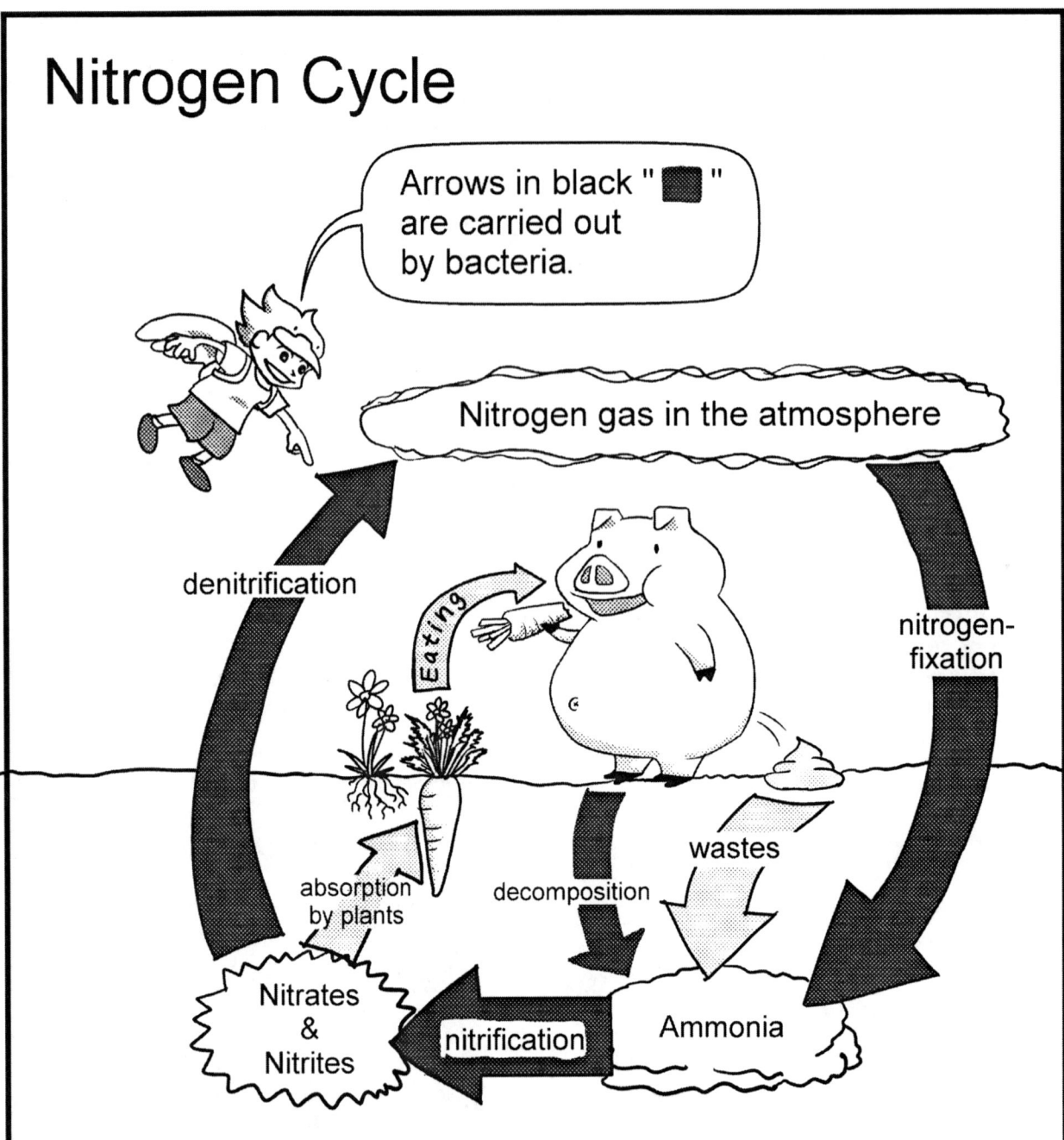

* **Ammonia:** A nitrogen-rich chemical that occurs naturally in animal feces/urines.
* **Decomposition:** Decomposition happens when an organism's dead body is broken down by bacteria.
* **Denitrification:** It's done by bacteria. Nitrates and nitrites are turned into nitrogen gas.
* **Nitrates & Nitrites:** These 2 nitrogen-rich chemicals are nutrients for plants.
* **Nitrification:** It's done by bacteria. Ammonia is turned into nitrates and nitrites.
* **Nitrogen fixation:** It's done by bacteria. Nitrogen gas is turned into ammonia.
* **Wastes:** Feces (poop), urine (pee) are both considered wastes.

Unit 1: Chapter 1

Phosphorus Cycle

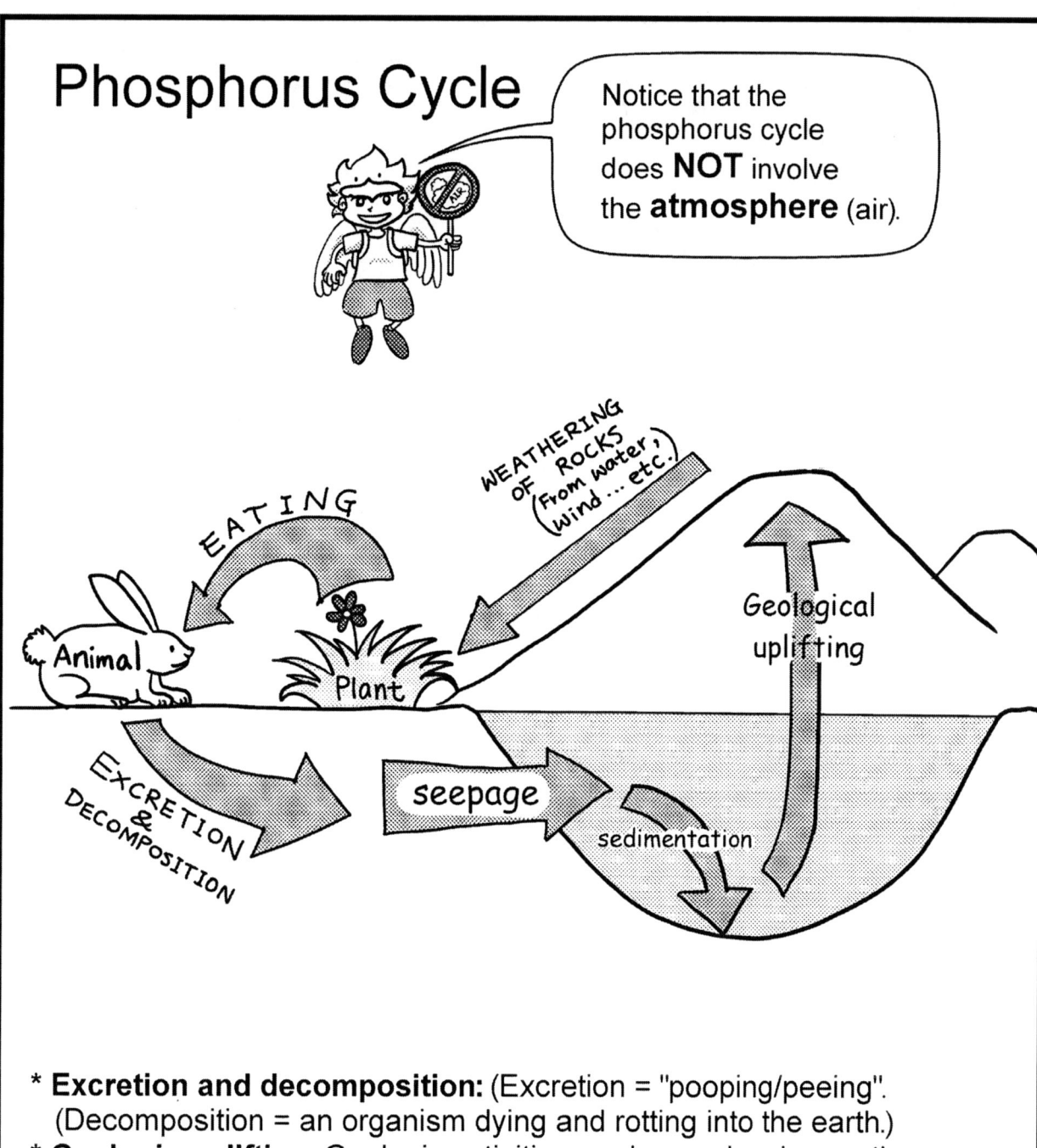

* **Excretion and decomposition:** (Excretion = "pooping/peeing". (Decomposition = an organism dying and rotting into the earth.)
* **Geologic uplifting:** Geologic activities, such as volcanic eruption, causing under-water landmasses to be lifted up above sea level.
* **Sedimentation:** Substances settling to the bottom of a body of water.
* **Seepage:** This happens when water seeps through the soil, washing certain soil-bound substances into a body of water.
* **Weathering** (of rocks from water/wind): Water or wind can break down bits and pieces of rock and add them to the soil.

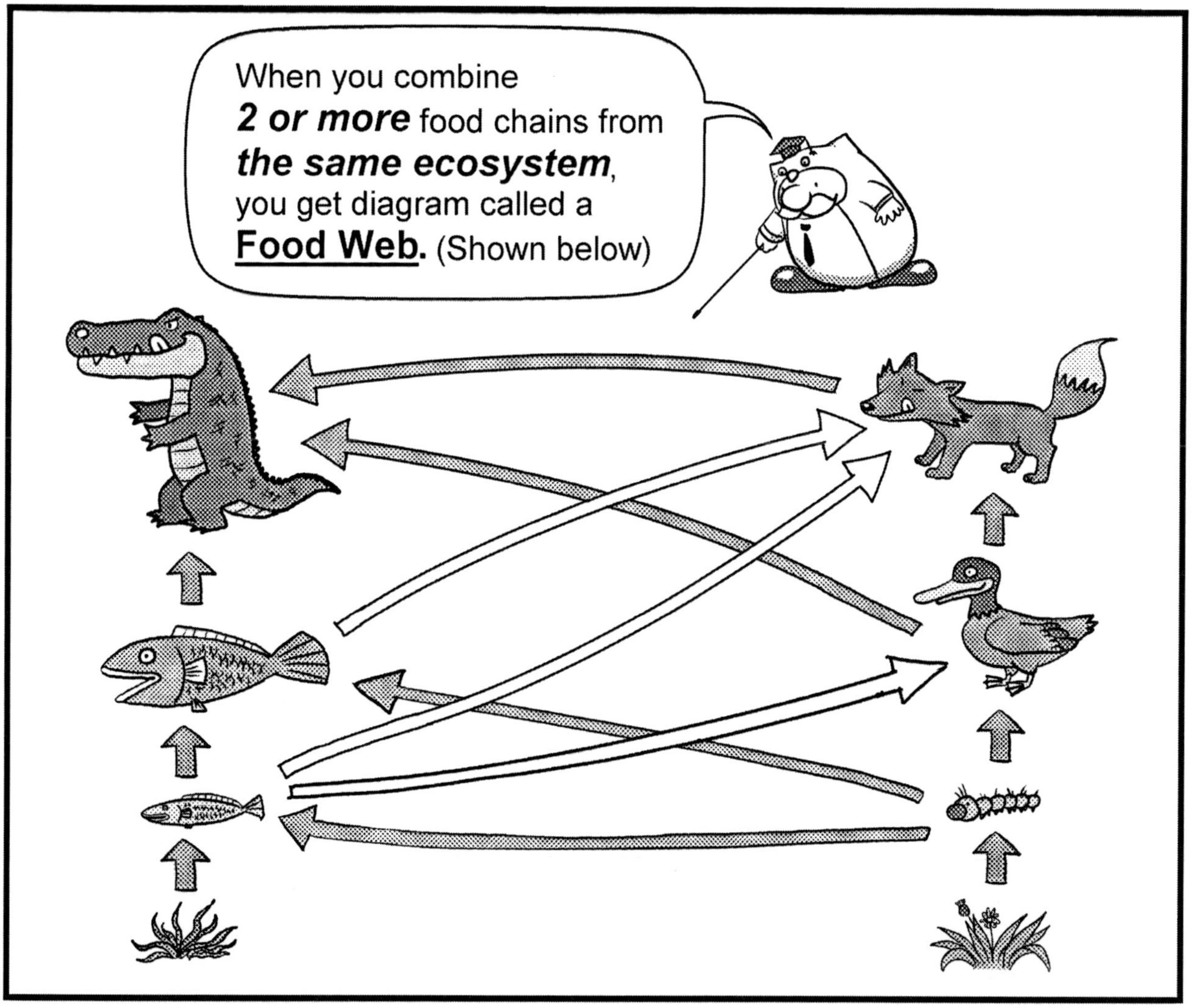

Chapter 2: An Organism's Role in an Ecosystem

Science Standard: 6.5.c
Students know populations of organisms can be categorized by the functions they serve in an ecosystem.

NEW VOCABS

* **Autotroph:** Also known as producers. Organisms that can make their own food—either by photosynthesis or chemosynthesis.

* **Carnivore:** Organisms that eat other animals for food.

* **Chemosynthesis:** A process in which an organism makes food by extracting energy from certain chemicals in the environment. (Note: Unlike photosynthesis, chemosynthesis does not require light.)

* **Commensalism:** A type of symbiotic relationship in which 1 side benefits. The other side receives neither harm or benefit.

* **Competition:** 2 organisms competing for the same resource (food, water, space…etc.)

* **Decomposer:** Organisms that chemically break down substances for food. Examples of decomposers are bacteria and fungus.

* **Detritivore:** Animals that feed on small scraps/crumbs of food left by other larger organisms.

page 12

Unit 1: Chapter 2

* **Generalist:** A term used to describe an organism that can survive on a wide variety of food/environment/resources. A generalist is the opposite of a "specialist."

* **Herbivore:** Animals that eat plants/algae for food.

* **Heterotroph:** Also known as consumers. Organisms that must eat other organisms to survive.

* **Host:** In parasitism, the host is the organism that is being taken advantage of by the parasite.

* **Mutualism:** A type of symbiosis, in which both sides of benefit from the relationship.

* **Omnivore:** An animal that can eat animals/plants/algae/fungus for food.

* **Parasite:** In parasitism, a parasite is the organism that takes advantage of the host.

* **Parasitism:** A type of symbiosis, in which 1 organism benefits, while the other organism suffers.

* **Predation:** A relationship in which 1 organism eats another organism.

* **Producer:** See "Autotroph."

* **Specialist:** A term used to describe an organism that can only survive on a specific type of food/environment/resource. A specialist is the opposite of a "generalist."

* **Symbiosis:** A relationship in which 2 different species of organisms live closely together and have biological interactions.

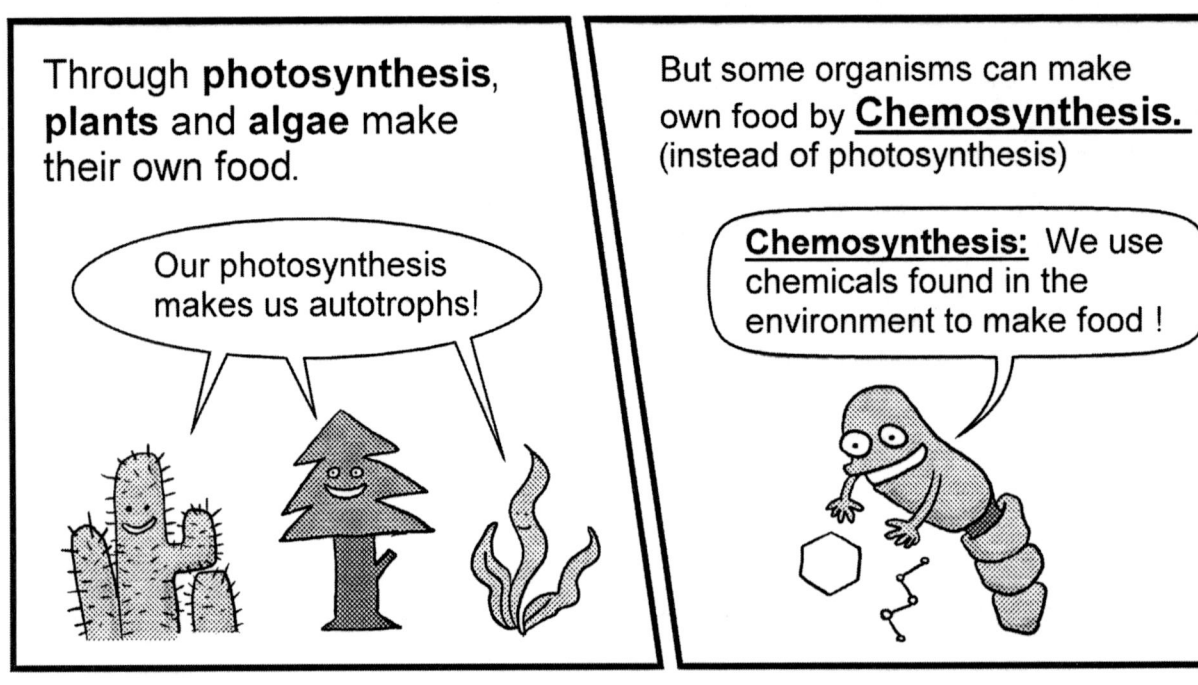

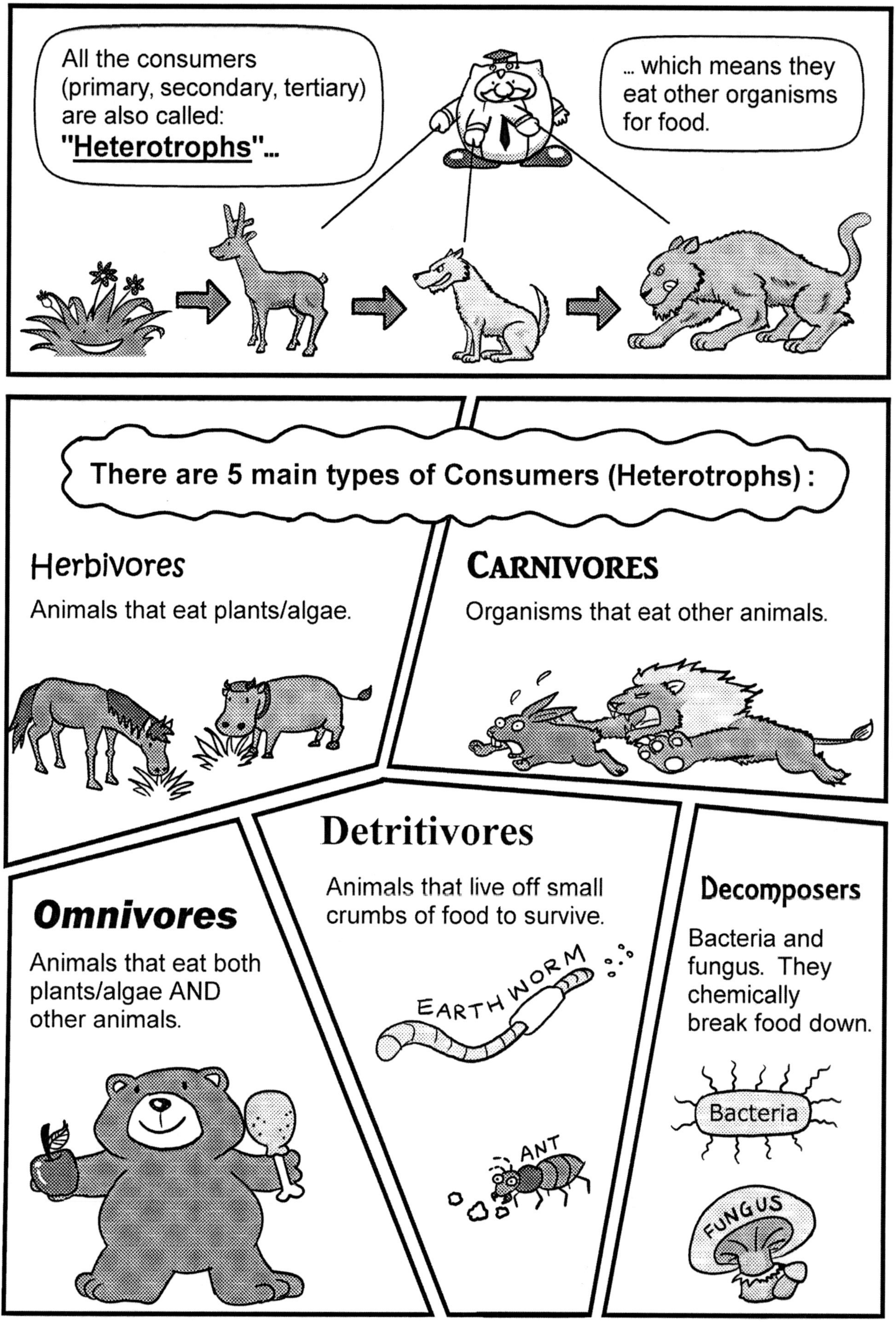

Decomposers and Producers work *together* to support all the organisms in an ecosystem!

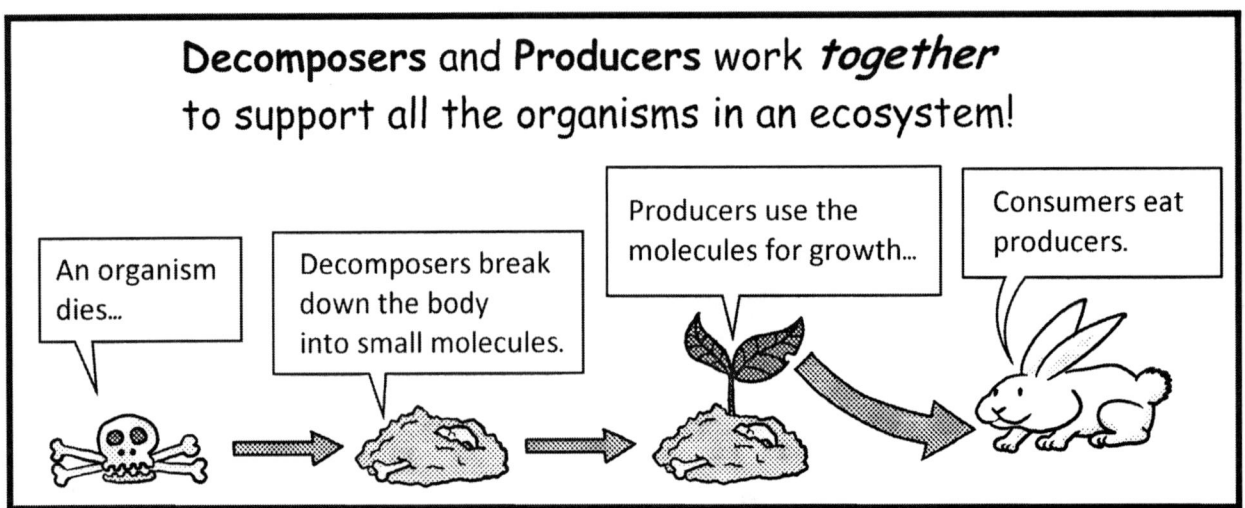

An organism dies... → Decomposers break down the body into small molecules. → Producers use the molecules for growth... → Consumers eat producers.

New Topic: Relationship between Organisms

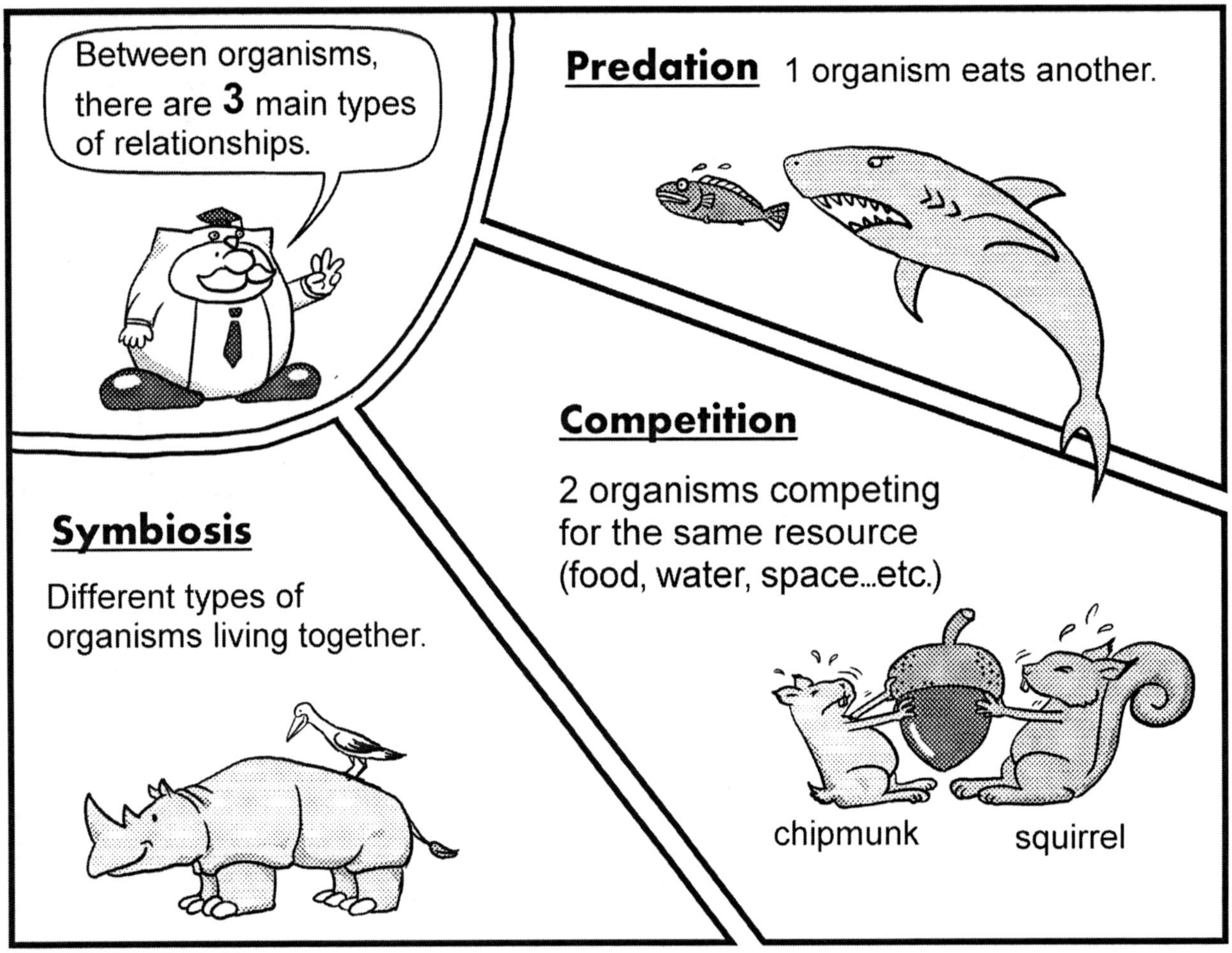

Between organisms, there are **3** main types of relationships.

Predation — 1 organism eats another.

Competition — 2 organisms competing for the same resource (food, water, space...etc.)

chipmunk   squirrel

Symbiosis — Different types of organisms living together.

Unit 1: Chapter 2

Chapter 3: How the Environment Affects an Organism

Science Standard: 6.5.e
Students know the number and types of organisms an ecosystem can support depends on the resources available and on abiotic factors, such as quantities of light and water, a range of temperatures, and soil composition.

NEW VOCABS

* **Abiotic factors:** The "non-living" parts of an ecosystem. For example, the climate, soil, wind, water, temperature...etc.

* **Biotic factors:** The "living" parts of an ecosystem--the organisms.

* **Climate:** The long-term weather pattern of a region. The climate of a region tends to stay the same for a very long time.

* **Competitive exclusion principle:** This principal states that when two species, living in the same region, happen to have the same niche, they will not be able to coexist. One of them will eventually dominate over the other. The weaker species will be forced to change its niche, move out, or face extinction.

* **Habitat:** The place where an organism lives, complete with all the biotic and abiotic factors.

* **Microclimate:** A small region's climate that is different from the climate of its surrounding area. Common causes of microclimate are geography, human influence, or the presence of plant life.

* **Niche:** A description of everything (including habitat) a species needs to survive, stay healthy, and reproduce.

Unit 1: Chapter 3

* **Polar:** The coldest climate zone on earth. Polar zones are located at the northern and southern extremities of the earth.

* **Rain shadow effect:** A phenomenon in which the wind-facing side of the mountain receives more rain, whereas the side of the mountain away from the wind is relatively dry.

* **Temperate:** It is the "warm climate" zone. It is the region between the earth's tropical and polar zones.

* **Tropical:** The hottest climate zone on earth. Tropical zones are at the equator (central region) of the earth.

* **Weather:** A region's condition of sun, wind, rain (…etc.) on a particular day. A region's weather could be different every single day! (Note: Compare this with the definition of "climate".)

"Let's learn some new vocabularies!"

WORDS of the Day

ECOSYSTEM: (Review)
A community of organisms + nonliving things in a region.

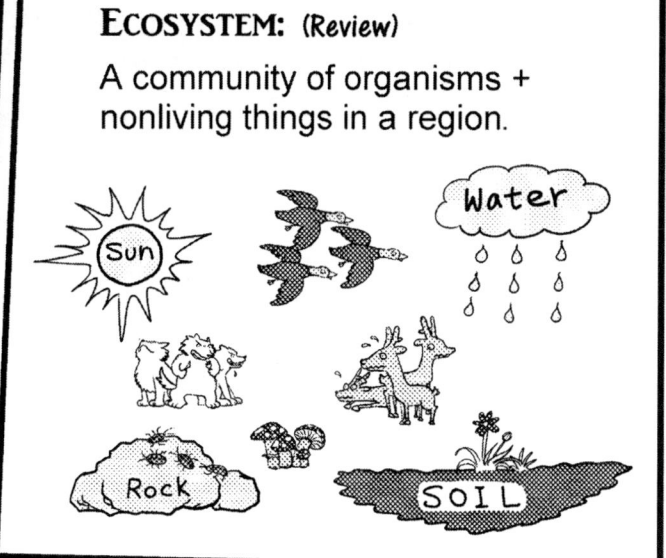

BIOTIC FACTORS
The "living" part of an ecosystem. Basically, the organisms.

ABIOTIC FACTORS
The "nonliving" part of an ecosystem. (Climate, water, soil…etc.)

WEATHER
The sun, rain, wind, etc. condition on a particular day.
The weather is subject to change every day.

"It is rainy, but tomorrow might be sunny"

CLIMATE
The long-term weather pattern of a particular region.
The climate of a region tends to stay the same for thousands of years.

"The Nouth Pole will **ALWAYS** be cold."

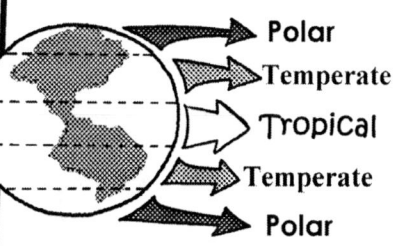

Due to different amount of sun exposure, There are 3 different **climate zones** on earth:

* <u>Tropical</u> (hot)
* <u>Temperate</u> (warm)
* <u>Polar</u> (cold)

→ Polar
→ Temperate
→ Tropical
→ Temperate
→ Polar

page 21

Unit 1: Chapter 3

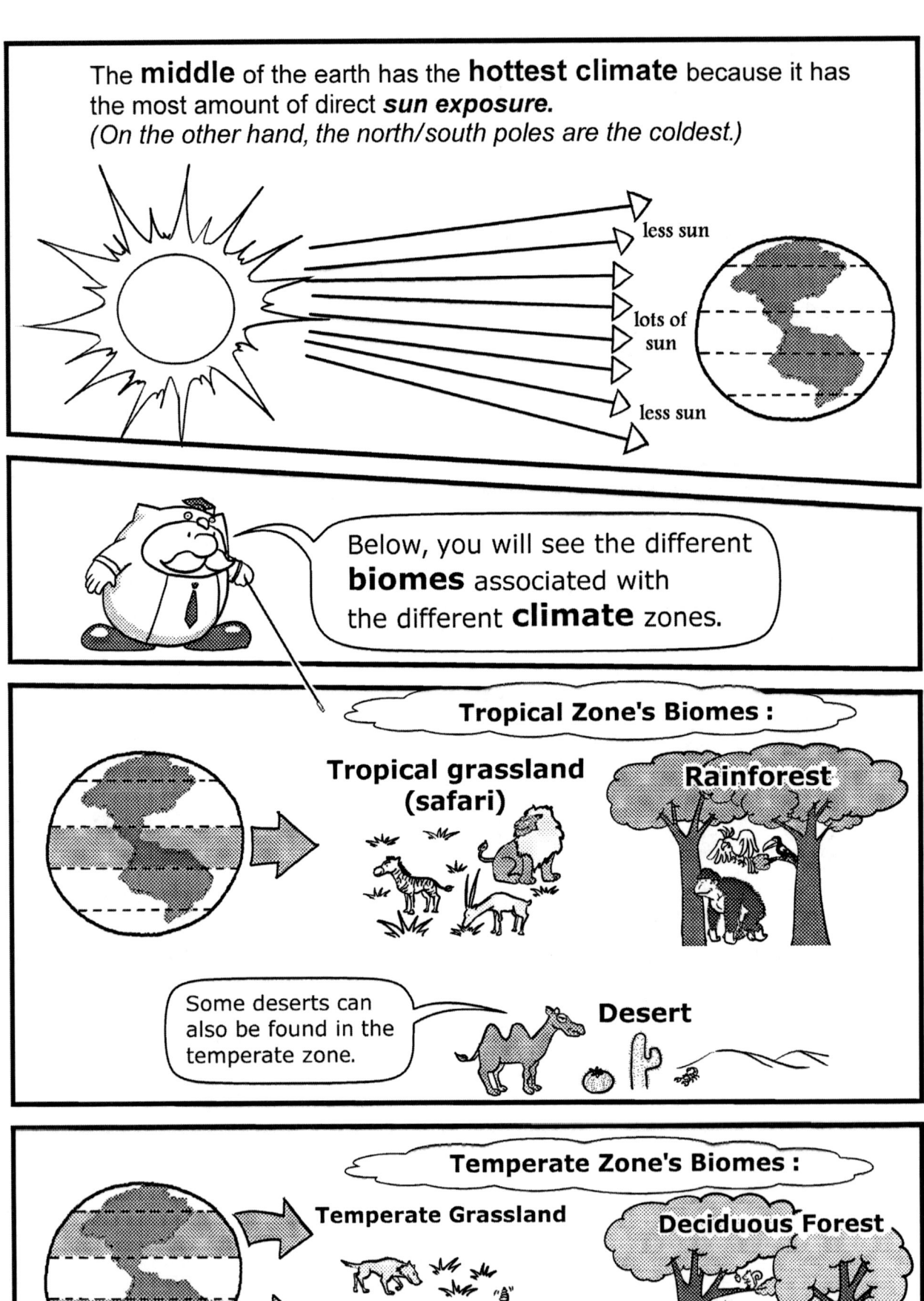

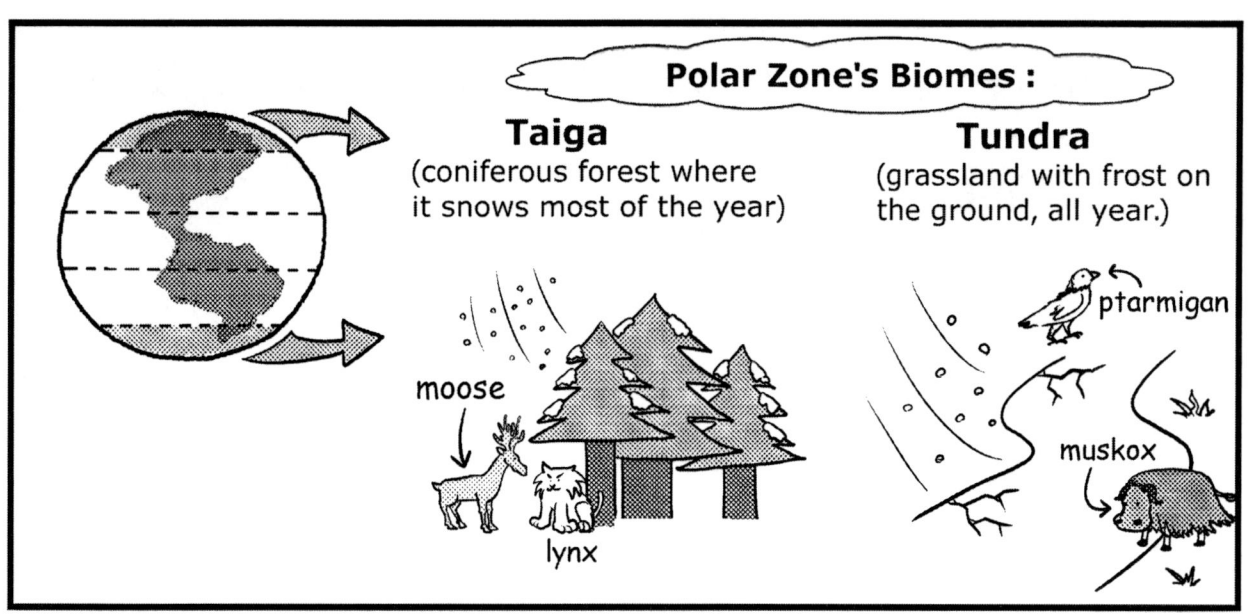

Polar Zone's Biomes:

Taiga
(coniferous forest where it snows most of the year)

Tundra
(grassland with frost on the ground, all year.)

New Concept: "Microclimate":

Microclimate's definition:
A small region's climate that is different from the climate of its surrounding area.

Things that cause microclimates:

Tall buildings can cause a **warmer** climate by **trapping** heat.

Tall plants (such as trees) can increase the moisture in the air.

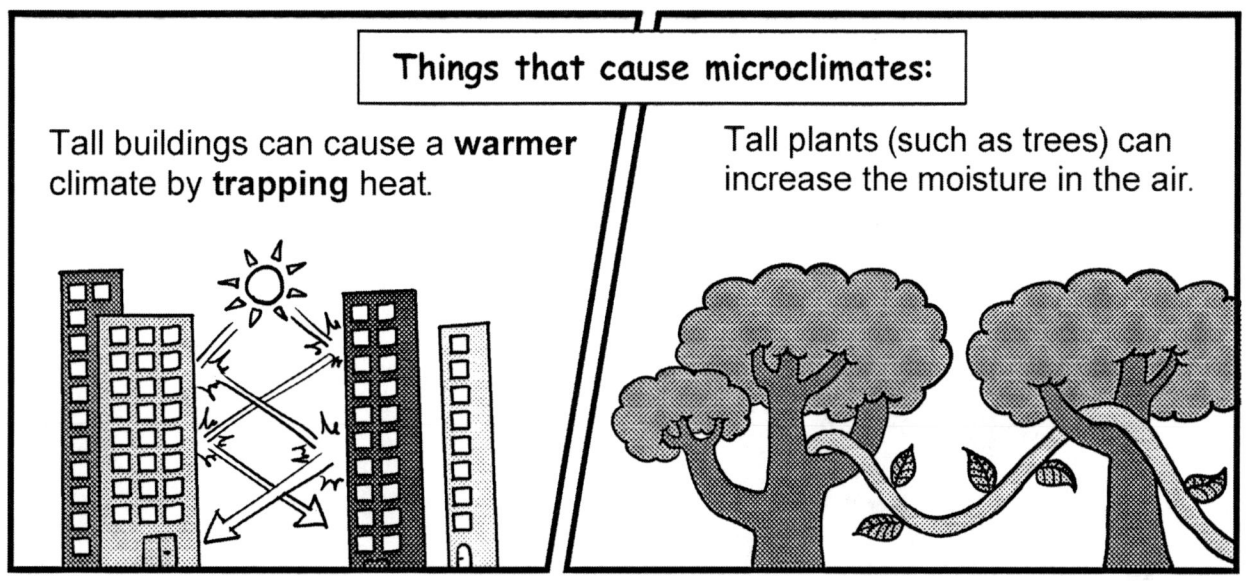

Unit 1: Chapter 3

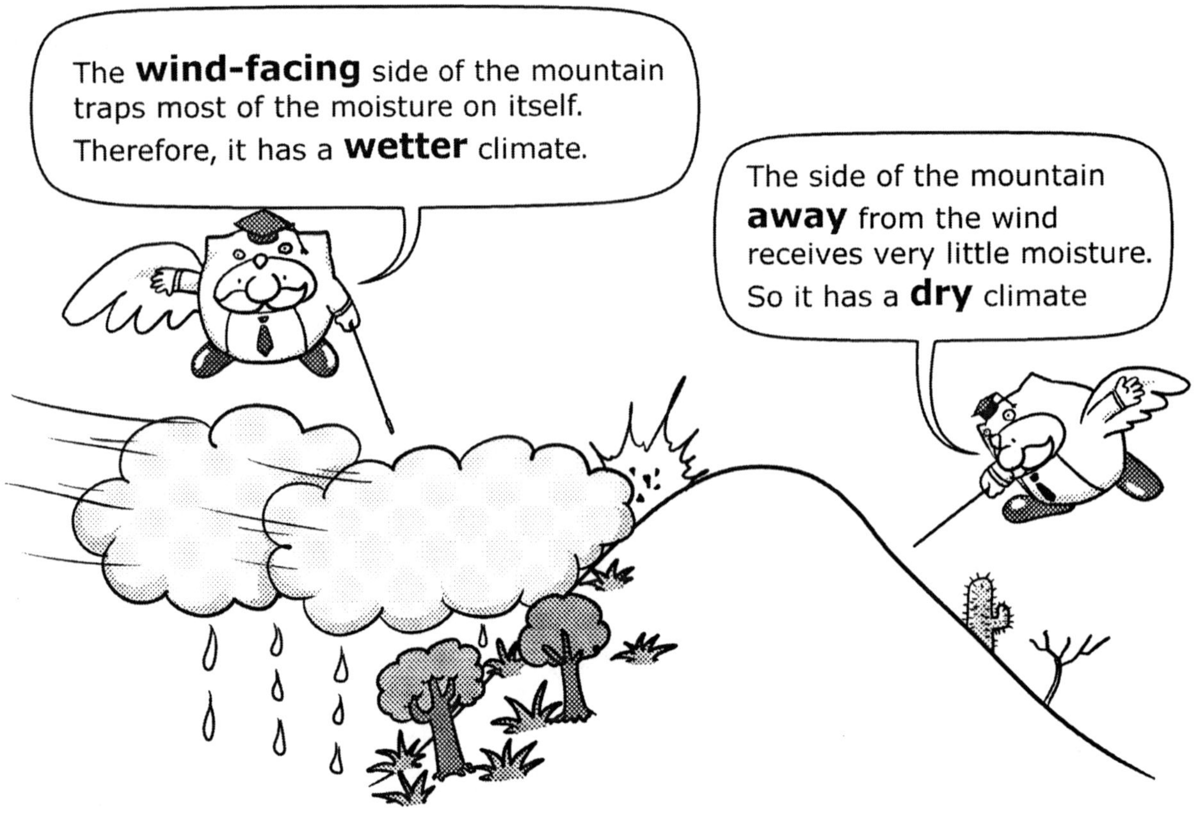

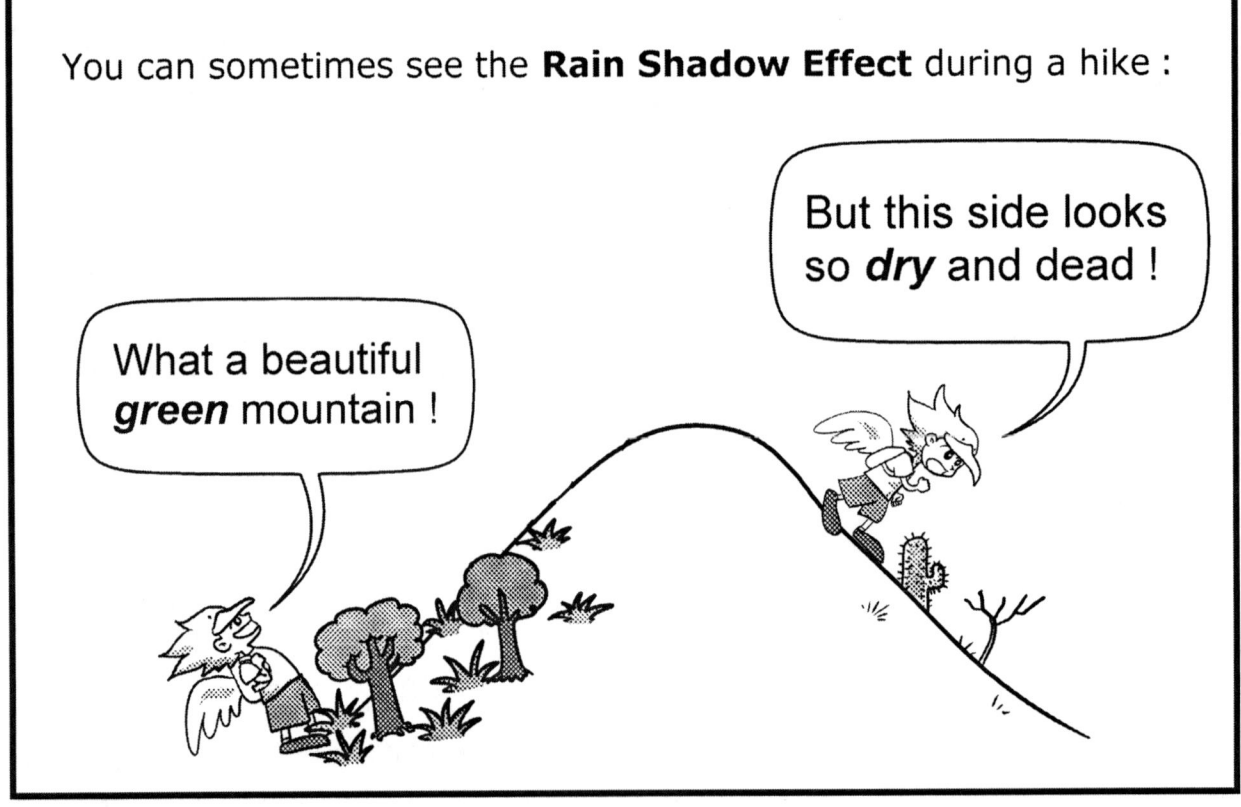

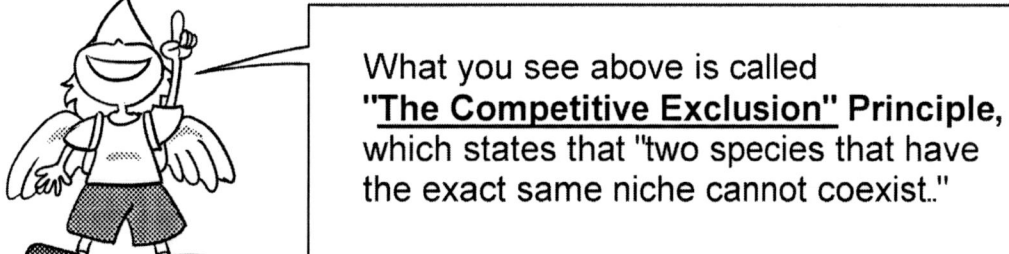

Chapter 4: Biodiversity

Science Standard Bio/LS . 6 . a :
Students know biodiversity is the sum total of different kinds of organisms and is affected by alterations of habitats.

NEW VOCABS

* **Biodiversity:** The amount of varieties of living organisms on earth (or in an ecosystem). Biodiversity is a combination of species diversity and genetic diversity.

* **Ecological equivalents:** This word is used to describe 2 different species having similar roles in their respective ecosystems.

* **Genetic diversity:** The amount of varieties of genetic characteristics in a single species.

* **Keystone species:** Any species that has an usually big influence on its ecosystem. The extinction (or near extinction) of a keystone species can severely harm an ecosystem.

* **Species diversity:** The amount of different species on earth (or in an ecosystem).

Biodiversity:
The amount of variety of living organisms on earth.

Biodiversity consists mainly of 2 parts: **species diversity** and **genetic diversity**

Species Diversity definition:
How many *species are present.
(* "Species" means: organisms of the same type/kind.)

"This place has a lot of **species diversity!**"

Genetic Diversity definition:
How many varieties of genetic characteristics exist in a species.

"This group of rabbits has a lot of **genetic diversity!**"

High level of *biodiversity* makess an ecosystem *healthier* and more *stable.*

page 29

Unit 1: Chapter 4

Biodiversity and ecosystem *depend* on each other.

...this is because the ecosystem gives the species *a place to live.* And the species return the favor by bringing life into the ecosystem..

New Vocabularies

Vocab: <u>Keystone Species</u>

Keystone species is a species that has an unusually big *influence* on an ecosystem.

Example:

Mountain lion is a keystone species because it controls the herbivores' population.

If an ecosystem *loses its keystone* species, there will be problems..

For Example:

The loss of a keystone species makes this ecosystem very unstable!

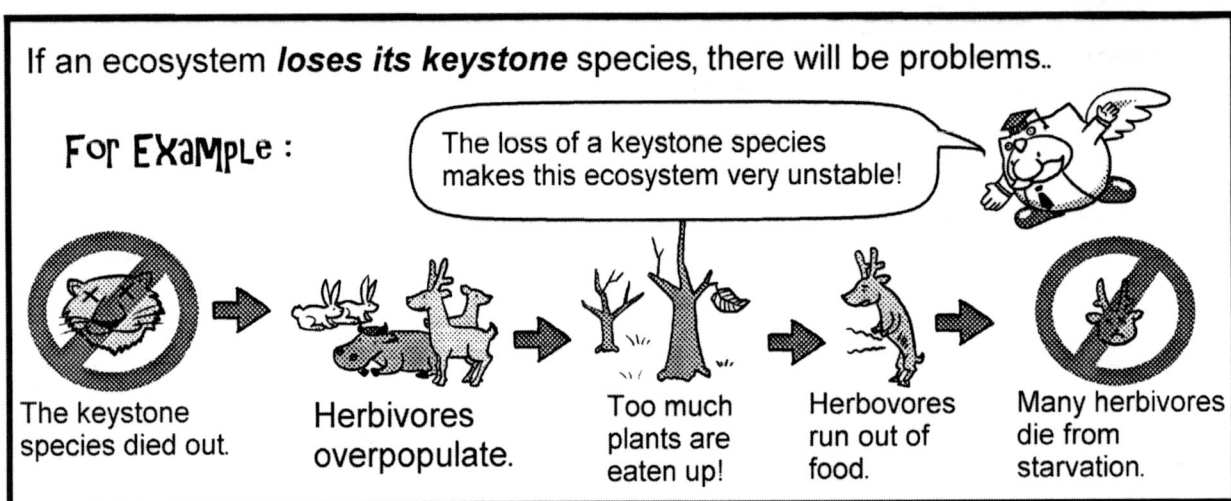

The keystone species died out. → Herbivores overpopulate. → Too much plants are eaten up! → Herbovores run out of food. → Many herbivores die from starvation.

Chapter 5: Changes in An Ecosystem

Science Standard Bio/LS . 6 . b :
Students know how to analyze changes in an ecosystem resulting from changes in climate, human activity, introduction of nonnative species, or changes in population size.

NEW VOCABS

* **Acid Rain:** Rain with a low pH (acidic). It is formed when air pollution is mixed with normal rain, which turns the normal rain into an acidic liquid.

* **Air Pollution:** Pollution in the atmosphere (air).

* **Biomagnification:** This principle states that: In a food chain, the higher level consumers will accumulate more pollutants in its body.

* **Catch-and-Release:** A common way to "sample" a population. It works by capturing and marking some individuals in the population, then release them. When you capture a sample of the population again, you count the number of previously-marked individuals to estimate the full size of the whole population.

* **Direct Survey:** 1 of the field observation techniques, in which a scientist watches the organisms directly to study them.

* **Ecological Footprint:** The amount of land needed to provide an organism with enough resources, shelter, and waste disposal.

* **Field Observation:** A style of studying an ecosystem, in which scientists are physically present in an ecosystem to study it. (as opposed to working indoors).

* **Global Warming:** The increase of earth's global temperature due to an overabundance of Greenhouse Gases.

* **Greenhouse Gases:** Gases that raise the earth's temperature by absorbing energy from sunlight. Common examples are carbon dioxide, methane, and water vapor.

* **Habitat Fragmentation:** Turning an organism's habitat from 1 big land into small fragments of lands.

* **Indicator Species:** Species that are especially sensitive and vulnerable to problems in the ecosystem. Scientists tend to observe these species to get a forewarning of any problem in the ecosystem.

* **Indirect Survey:** 1 of the field observation techniques, in which a scientist searches for signs (footprints, fur…etc.) of the organisms' presence. Indirect surveys are usually used for organisms that are hard to find/keep track of.

* **Model:** A method of studying the ecosystem in which scientists use computer, math, or other methods to simulate how an ecosystem would behave.

* **Nonnative Species:** An introduced, foreign, non-indigenous, species in an ecosystem. Scientists sometimes also call them: "invasive species" or "introduced species".

* **Nonrenewable Resource:** Resources that cannot replenish its supply—once it is used up, it is gone forever. (Examples: Fossil fuel, natural gases…etc.)

* **Particulates:** Microscopic solids that float in the air. Examples are: pollen, lint, dusts…etc.

* **Pollution:** Substances or products that have an adverse effect on the environment. Examples are: toxic chemicals, toxic metals …etc. Even heat and noise count as pollution.

* **Primary Succession:** A succession that starts out from a barren land (with no vegetation and little or no soil).
This process requires pioneer species to begin.

Unit 1: Chapter 5

* **Radio telemetry:** 1 of the field observation techniques, in which scientists capture an animal, plant a radio-signal emitter on the animal, release the animal, then track the radio signal to know the animals' whereabouts at all times.

* **Renewable Resource:** Resources that replenishes itself over time (fish, trees…etc.) or simply does not run out(sunlight, wind…etc.). However, they are not indestructible, so if a resource is destroyed by abuse, it would be gone forever.

* **Sample the population:** 1 of the field observation techniques, in which scientists study a small part of the population to get an idea of what the whole population is like. There are many ways to "sample", but a very popular way to do it is by "catch-and-release."

* **Secondary Succession:** A succession that takes place right after a severe disturbance (such as fire, flood, hurricane…etc.) happens in an ecosystem. In this process, the ecosystem builds itself back up. It is a faster process and does not require pioneer species to begin.

* **Smog:** A type of air pollution. Smog is made when exhaust from combustion (from vehicles, factories…etc.) react with sunlight to form even more harmful substances.

* **Succession:** A process by which an ecosystem changes into different types of ecosystems, stage by stage, over time. There are 2 types of successions: Primary and Secondary.

* **Sustainable Development:** Using the earth's resources in such a way as to not damage the resource or the environment.

* **Water Pollution:** Pollution in water.

There are **4 things** that can **change** an ecosystem:

(1) Human activities

(2) Change in climate

A forest may turn into a grassland if its climate becomes drier.

(3) Introduction of *__nonnative species.__
(*It's also called "invasive species". Species that are from a foreign ecosystem)

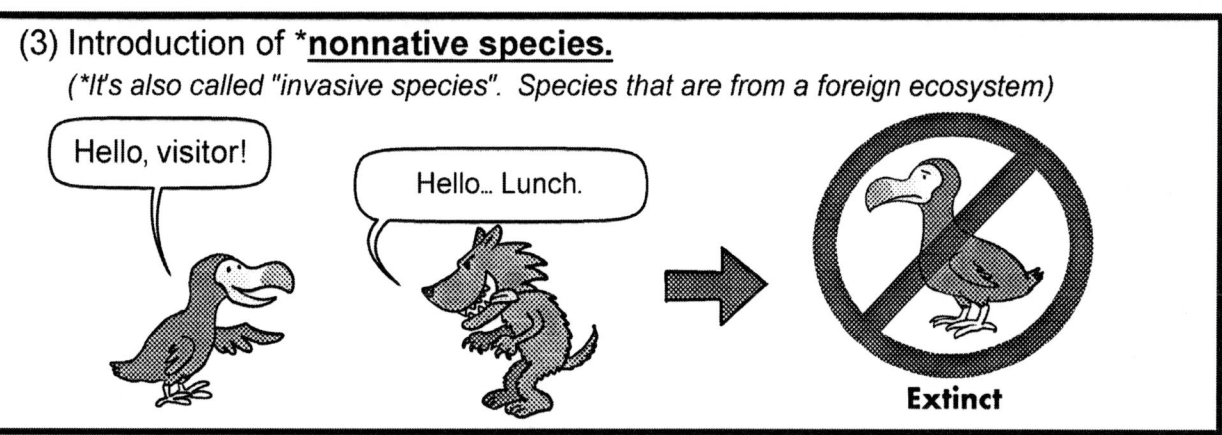

(4) Changes in population size

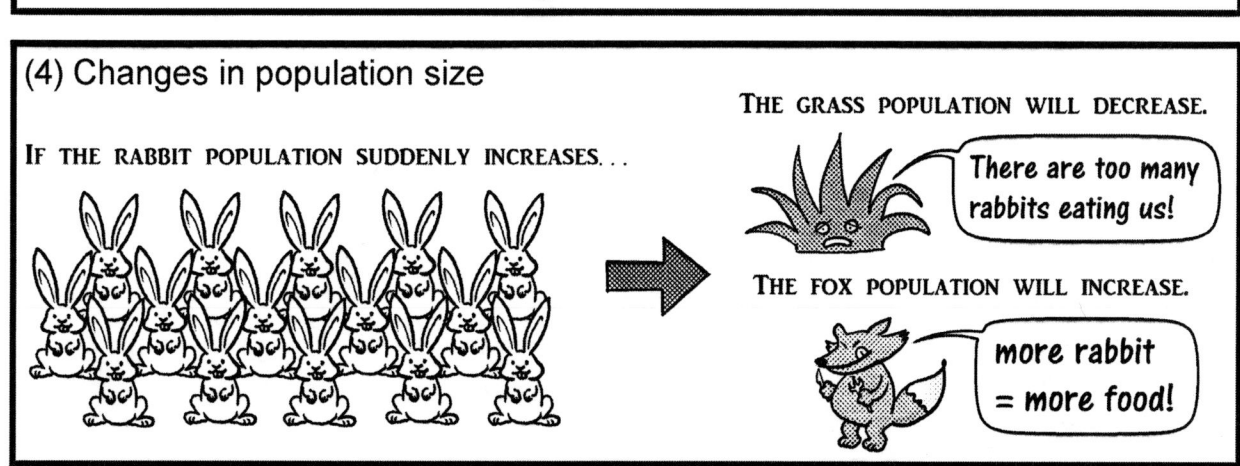

Scientists can study an ecosystem using the following ways:

Field Observation

In field study, the scientist is physically present in the ecosystem in order to observe it.

There are four typical ways for field observation. They are listed below:

(1) Direct Survey: Watch the organisms directly.

(2) Indirect Survey: (For organisms that are hard to keep track of.) Search for signs of the organism's presence.

(3) Radio Telemetry: Plant an object (such as collar) on an animal. The object emits a radio signal, which tells us the animal's whereabouts.

Step 1. Capture the animal

Step 2. Install a radio-signal emitter.

Step 3. By tracking the radio signal, we can track the animal's location.

(4) Sample the population: Study a small part of the population to get an idea of the whole population.

A common sampling method is "catch-and-release".

Step 1: Capture the animal.

Step 2: Put labels on the animals you captured.

Step 3 is on the next page

Put an "X" on this fish!

page 36

Unit 1: Chapter 5

Next Topic: Succession

Over time, an ecosystem can slowly change from 1 type to another type. We call this process **succession**.

Any of these 4 things can cause succession to take place.

Human Activities

Change in Population Size

Climate Change

Introduction of Nonnative Species

THERE ARE 2 TYPES OF SUCCESSIONS :

1st Type: Primary Succession:

1. **Primary Succession** starts on a barren land that has **no soil** or **plant**. (Example: a land completely destroyed by a volcano.)

2. **"Pioneer species"**, which are species that can somehow survive this type of tough environment, move in.

We are tough!

*When we die, our body will turn into **soil**, which allows other plants to survive here.*

Examples of pioneer species: lichen, marram grass...etc.

Lichen

Marram Grass

3. Over time, succession will *continue to develop* the land into a more complex ecosystem... until the land reaches its maximum potential or encounters a disturbance.

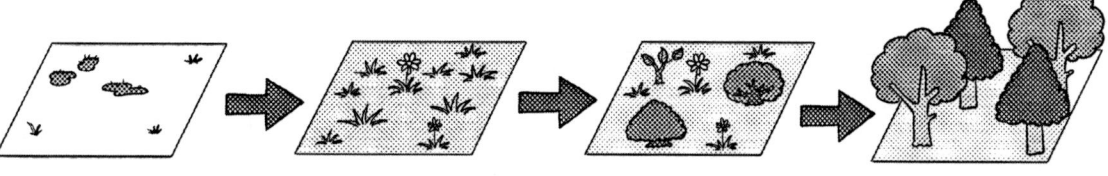

Unit 1: Chapter 5

2nd Type: Secondary Succession:

If **Primary succession** is about building an ecosystem from *scratch*...

Secondary succession is about ***rebuilding*** a damaged ecosystem.

Steps to Secondary Succession:

1. An ecosystem is disturbed or disrupted. (By fire, flood, human activities...etc.)

2. Since this ecosystem *already has* **soil**, grass, shrubs, and seedlings start growing immediately.

In **Secondary Succession**, you do *NOT* need pioneer species.

3. The ecosystem ***rebuilds itself back up.*** This process is similar to that of Primary Succession...
Except that it ***does not require pioneer species*** to start the process because the land ***already has soil.***

New Topic: Protecting the Ecosystems

There are 2 types of **resources**:

Renewable Resources: Resources that can replenish itself.

EXAMPLES: wind, tree, fish, sun

While renewable resources do not run out, they can be **destroyed** by human **abuse.** (Example: Over-fishing a fish to extinction)

Nonrenewable Resources: Resources that cannot replenish itself because once they are used up, they're gone forever.

EXAMPLES: propane, fossil fuel

Once a nonrenewable resource is used up, it is gone forever.

Because most resources require *land* to exist, we measure how much resources an organism needs *by the size of land required for it to survive.*

For EXAMPLE:

I am just a small fox, but I need about 3 acres of land to provide me with enough food, water...etc.

A human takes up at least 4.4 acres of land to have enough food, water electricity...etc.

The *amount of land* needed to provide an organism with enough resources, shelter, and waste disposal, is called the organism's: **Ecological Footprint**

Humans have a very big Ecological footprint!

page 41

Unit 1: Chapter 5

One of the reasons why humans have such a big ecological footprint is because we produce a lot of **pollution**.

Pollution's defnition:
Undesirable factor(s) added to the environment

EXAMPLES:

Animal wastes   Metals

Radioactive Waste

Carbon Dioxide

Each 1 of these things is called a **"Pollutant"**.

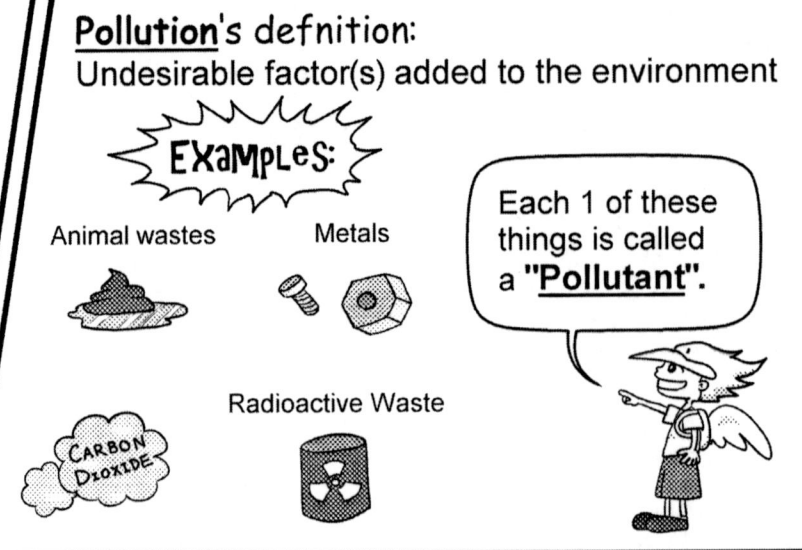

When pollution exists in the air, we call it **Air Pollution**.

Smog is a type of air pollution. It is formed by the exhaust from vehicles (or from industries) reacting with sunlight to form even more harmful substances.

How to make smog:

Exhaust is produced → Sunlight react with the exhaust → Exhaust turns into smog.

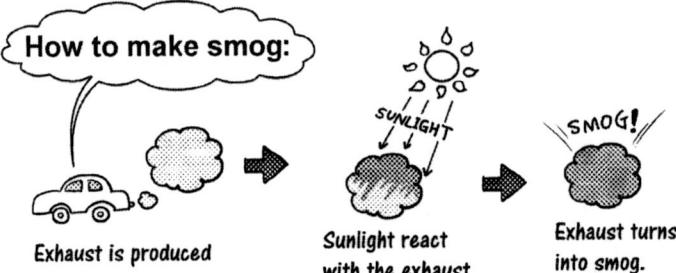

Particulates are microscopic solids floating in the air. Certain particulates are *harmful* to our health.

Pollen is an example of particulates!

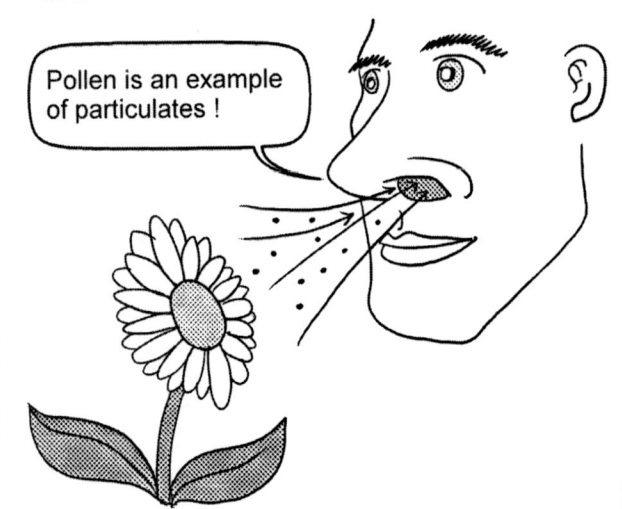

Smog contains many harmful *particulates.*

page 42

Unit 1: Chapter 5

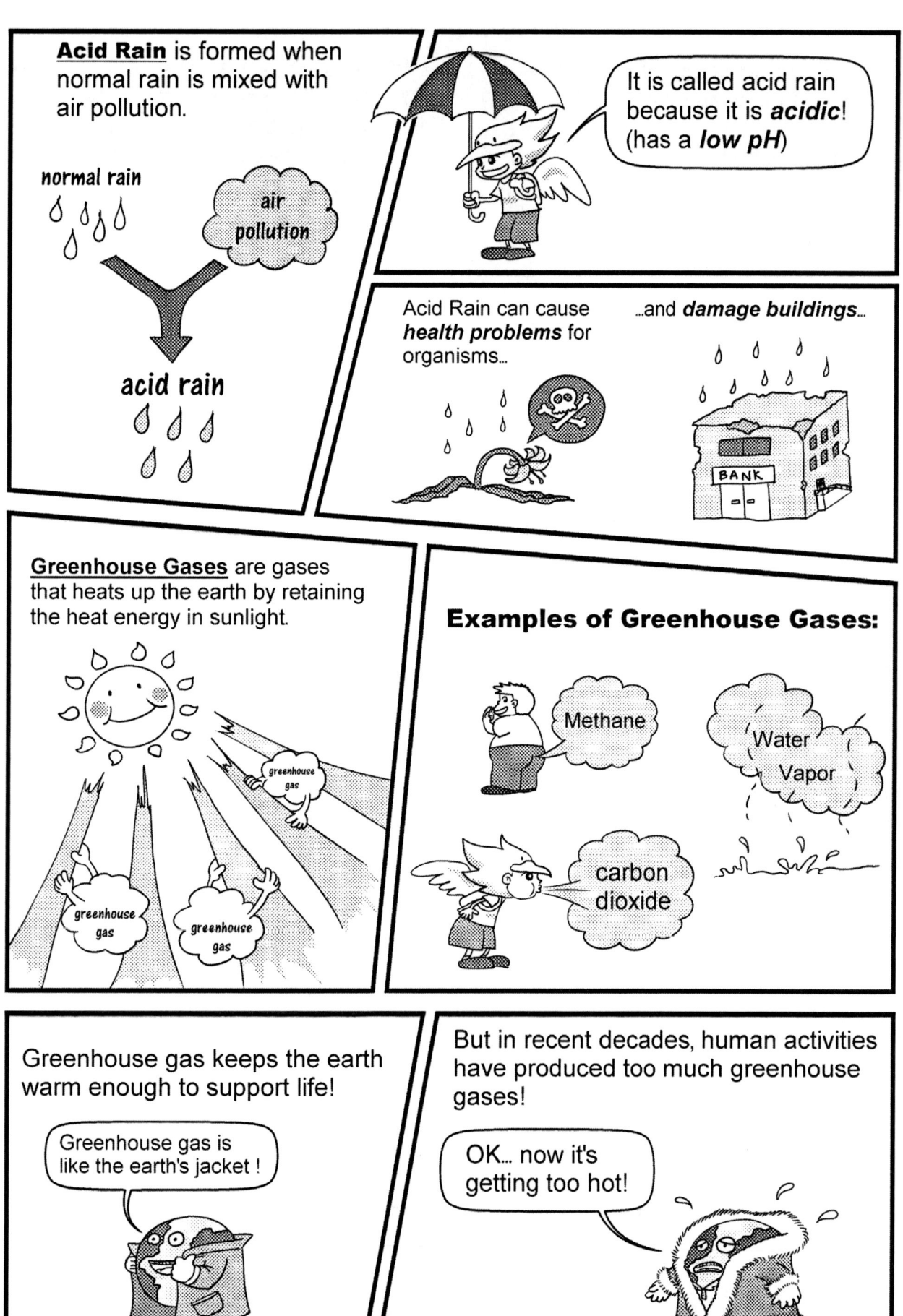

Human activities that increase greenhouse gases:

raising cattles

burning coal

burning fossil fuel

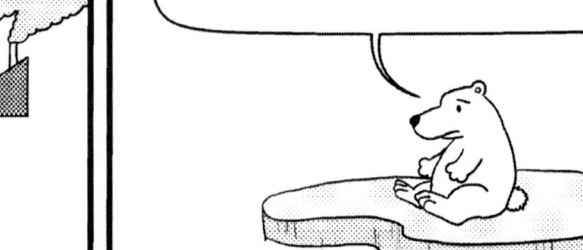

The increase of temperature by "too much greenhouse gases" is called **Global Warming**. And it damages our ecosystems!

Global warming → Polar ice melt → Polar bear has less ice to hunt seal on.

New Concept: Water Pollution

Water pollution means pollution in water.

Water pollution harms organisms.

Frogs are especially vulnerable to water pollution due to their *semi-permeable* skin.

(*Some substances can easily bypass it.)

Organisms that are vulnerable to ecological problems (such as water pollution) are called **Indicator Species**.

Uh oh~ this sick frog indicates that there is pollution here.

The idea of indicator species is like how miners used to keep a canary in the mines. Canaries are very vulnerable to toxic gases. Therefore, they are used to detect gas leaks.

The canary is dead!

Run for your lives!

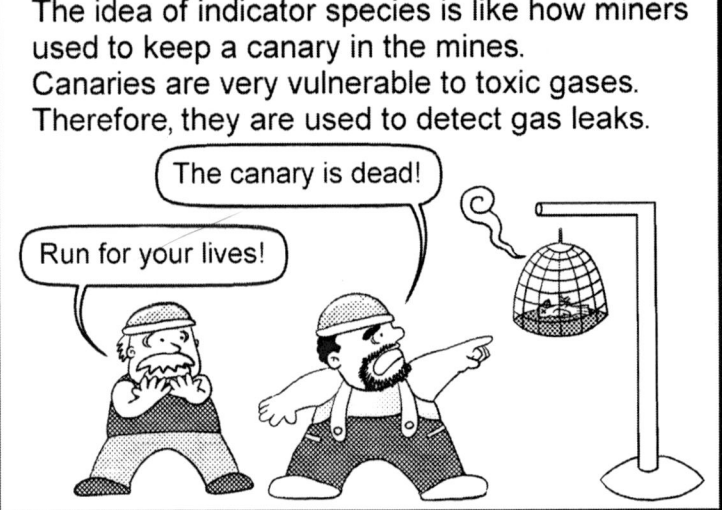

3 New Vocabularies

1. Biomagnification: The tendency for pollutants to accumulate in larger concentrations in the higher-level consumers in a food chain.

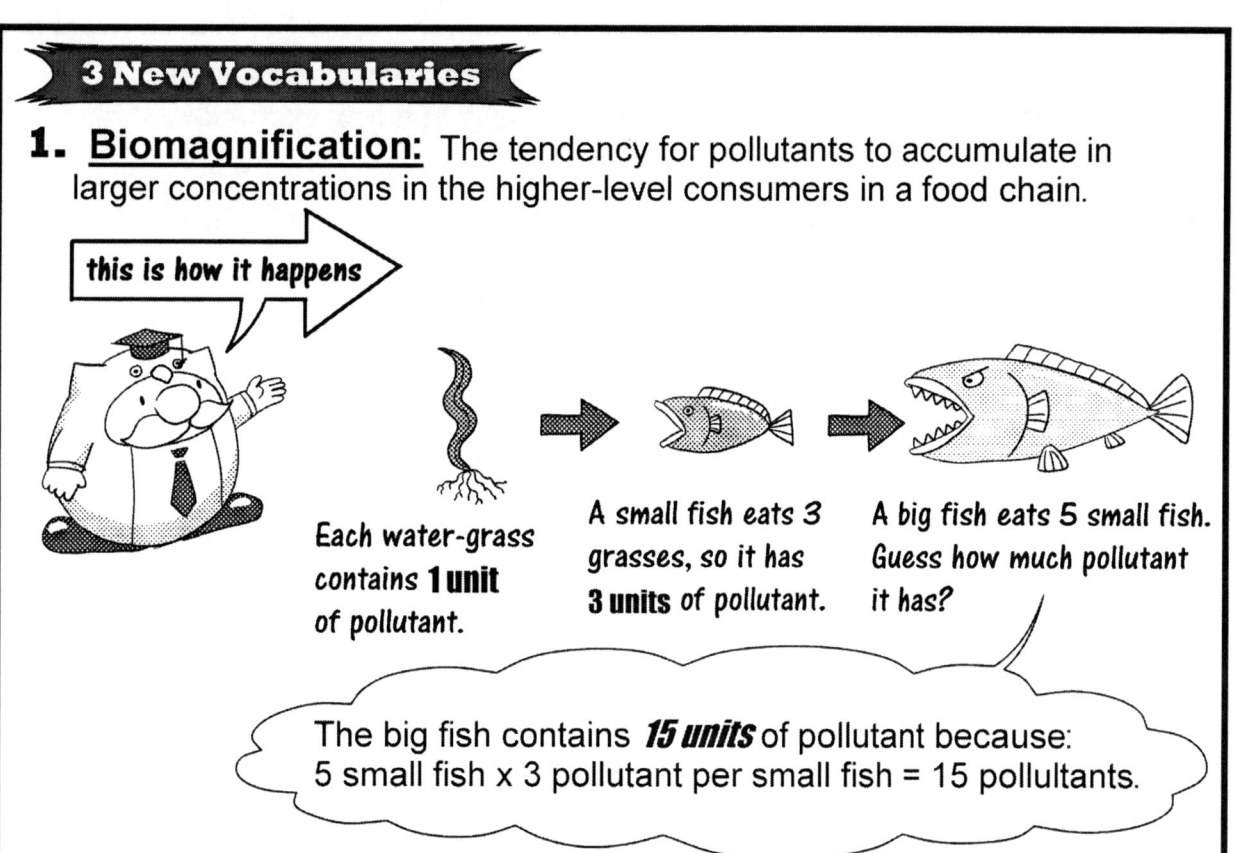

Each water-grass contains **1 unit** of pollutant.

A small fish eats 3 grasses, so it has **3 units** of pollutant.

A big fish eats 5 small fish. Guess how much pollutant it has?

The big fish contains **15 units** of pollutant because:
5 small fish x 3 pollutant per small fish = 15 pollutants.

2. Habitat Fragmentation: Turning an organism's habitat from 1 big land into many small fragments of lands (usually due to human activity). Habitat fragmentation is harmful to the organisms who live there.

EXAMPLE

Many deers live in a forest

Then, humans build roads through the forest.

The roads prevented some males from mating with the females.

Hey ~ Come over!

I am too scared!

And some of deers that try to cross the road got hit by cars.

Oh, no!

page 45

Unit 1: Chapter 5

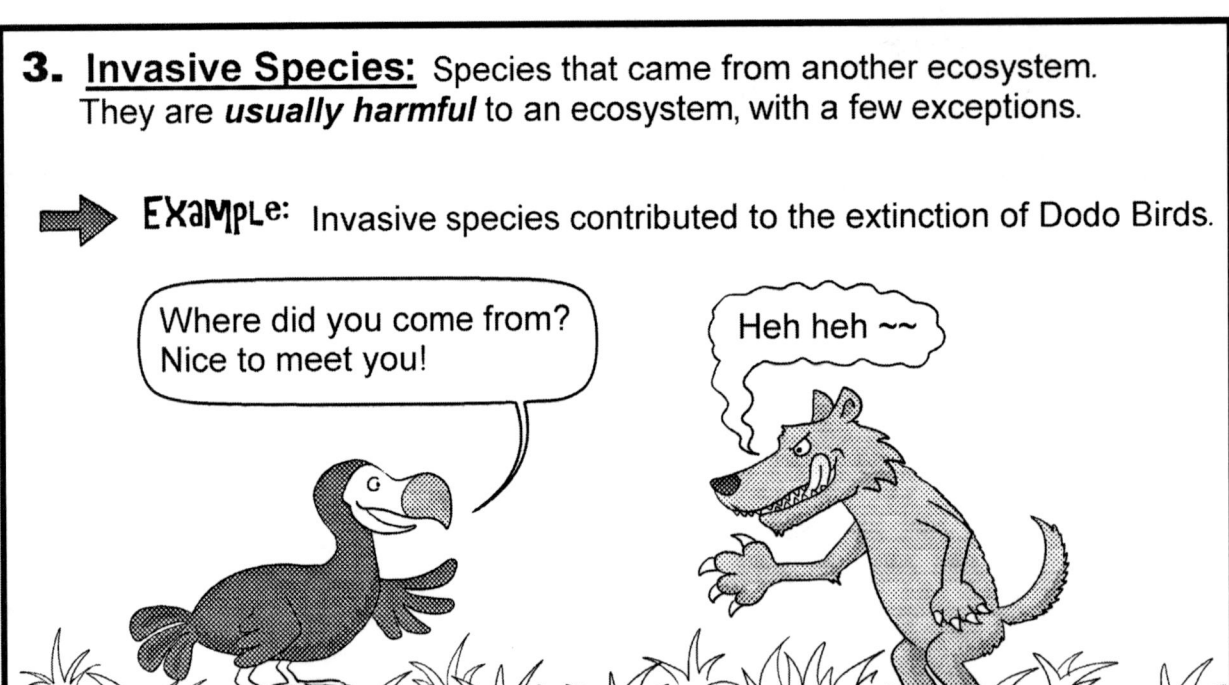

Chapter 6: Population Size

Science Standard Bio/LS . 6 . c :
Students know how fluctuations in population size in an ecosystem are determined by the relative rates of birth, immigration, emigration, and death.

NEW VOCABS

* **Carrying capacity:** The maximum number of individuals of a population an area can support.

* **Density-dependent limiting factor:** These are limiting factors that have a stronger influence on limiting a population's size if the population density is high (crowded).

* **Density-independent limiting factor:** These are limiting factors that can affect a population regardless of its population density.

* **Dispersion:** The manner at which individuals in a population are distributed in an area. There are 3 types of dispersions: clumped, uniform, or random.

* **Emigration:** A situation in which individuals move OUT of an area, which decreases the population.

* **Exponential growth (graph):** A type of population growth in which the population increases faster and faster over time. Its graph features a curve going upward.

* **Immigration:** A situation in which individuals move INTO an area, adding to the population.

* **Limiting factors:** Environmental factors that limit a population's growth and maximum size.

* **Logistical growth (graph):** A type of population growth in which it started out with exponential growth. Then the growth start to slow down as the population size approaches the area's carrying capacity (limit). Eventually, the growth comes to a stop at the area's carrying capacity.

* **Population Density:** The number of individuals per specific amount of living space.

Review:

Population:
A group of the same type of organism living in an area.

A population could be big.

× 10,000

...Or it could be small.

× 20

It is OK to have a *big* popluation, as long as you have a *big* area for it to live on.

But problems can occur if the area is **not big enough** to support a population.

It's too crowded!!

The "crowdedness" of a population is calculated by its **Population Density**.

Population Density: The number of individuals per specific amount of living space.

Population Density is *calculated* like this:

$$\frac{\text{Total Population Size}}{\text{Total Area Size}} = \text{Population Density}$$

A high Population Density means it is very *crowded*.

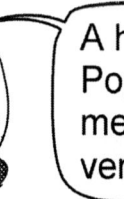

page 49

Unit 1: Chapter 6

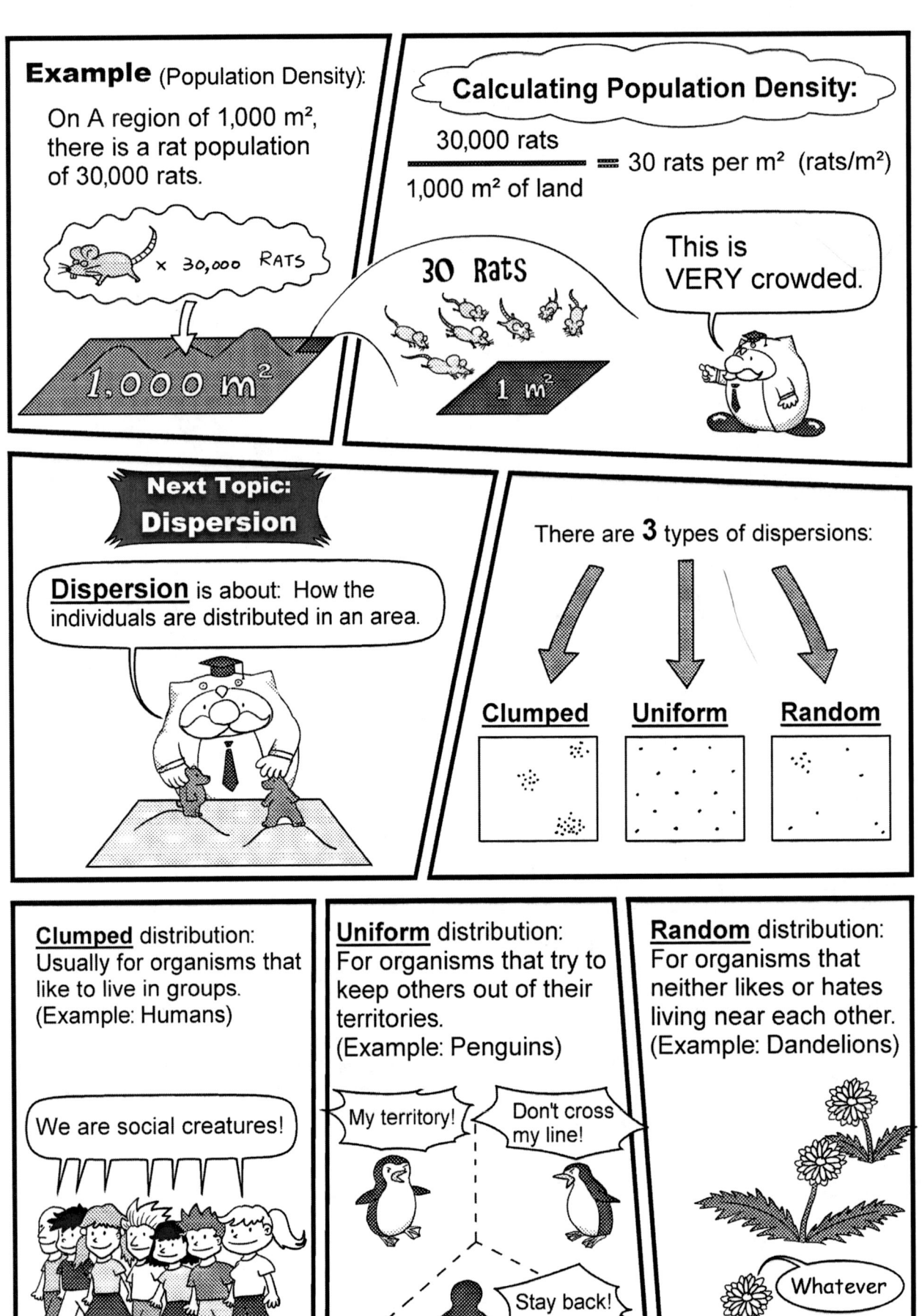

Next Topic: Change in Population Size

There are 4 things that can change a population's size.

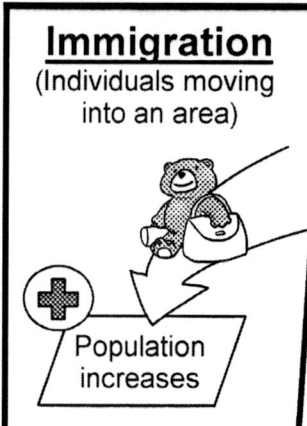

Immigration
(Individuals moving into an area)
➕ Population increases

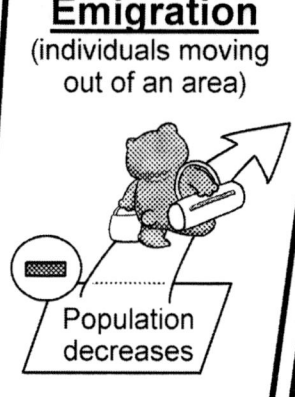

Emigration
(individuals moving out of an area)
➖ Population decreases

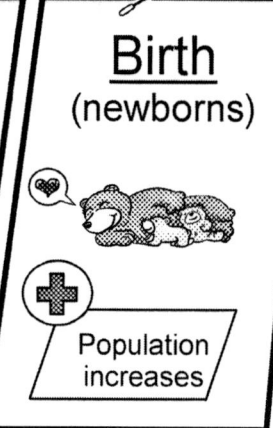

Birth
(newborns)
➕ Population increases

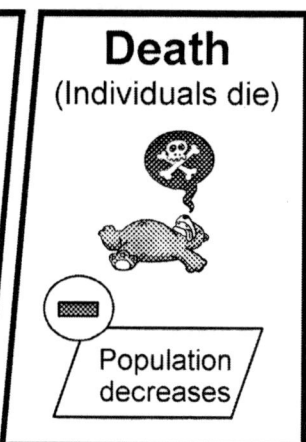

Death
(Individuals die)
➖ Population decreases

These factors cause the population size to change over time.

In 1990, the population decreased.

In 2010, the population increased.

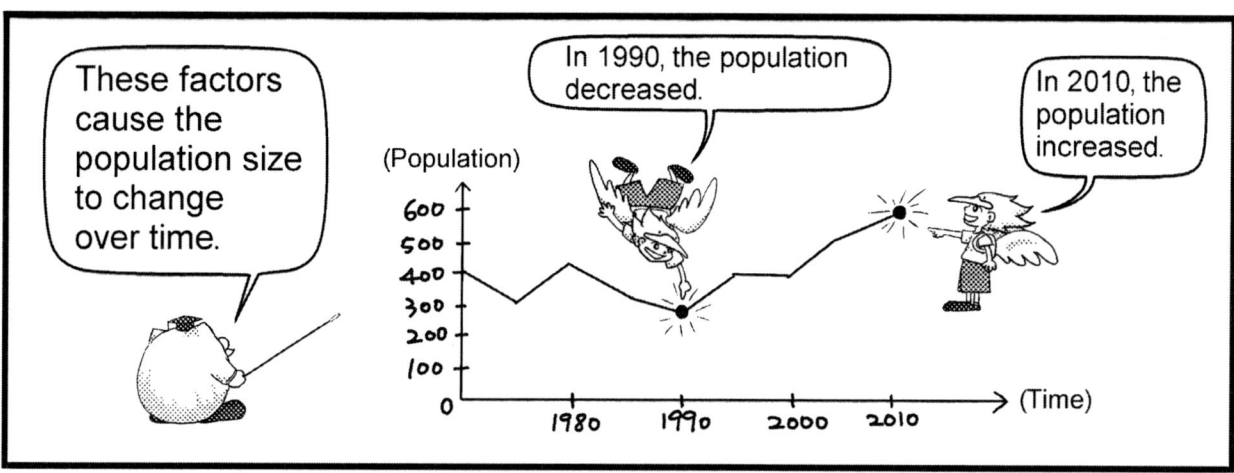

When a graph has this shape, we call it a **"Exponential graph."**

In exponential graph, the population grows *faster and faster* every year!

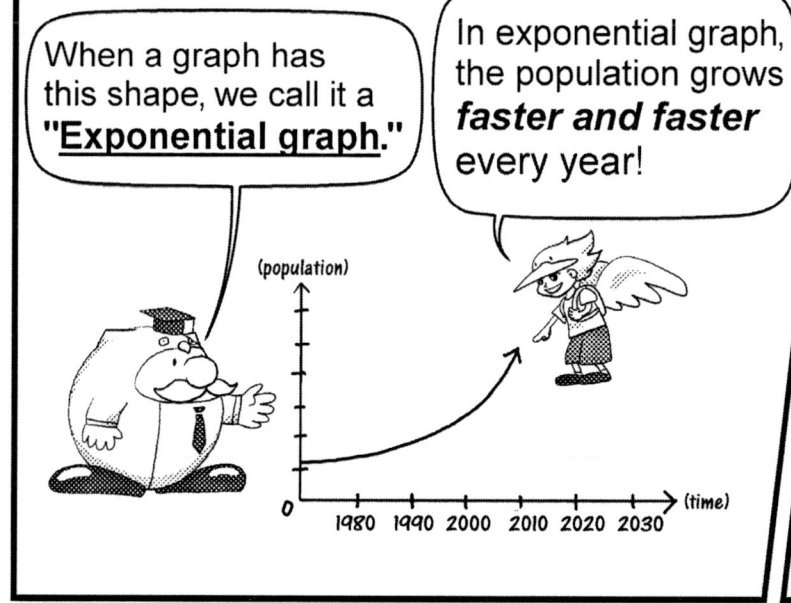

Note:

A straight line will mean that the population changes at a *constant rate.*

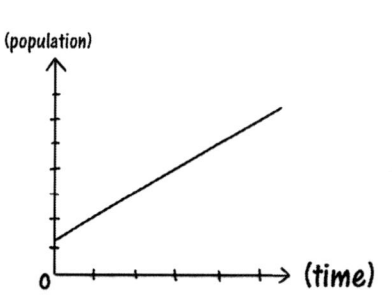

Unit 1: Chapter 6

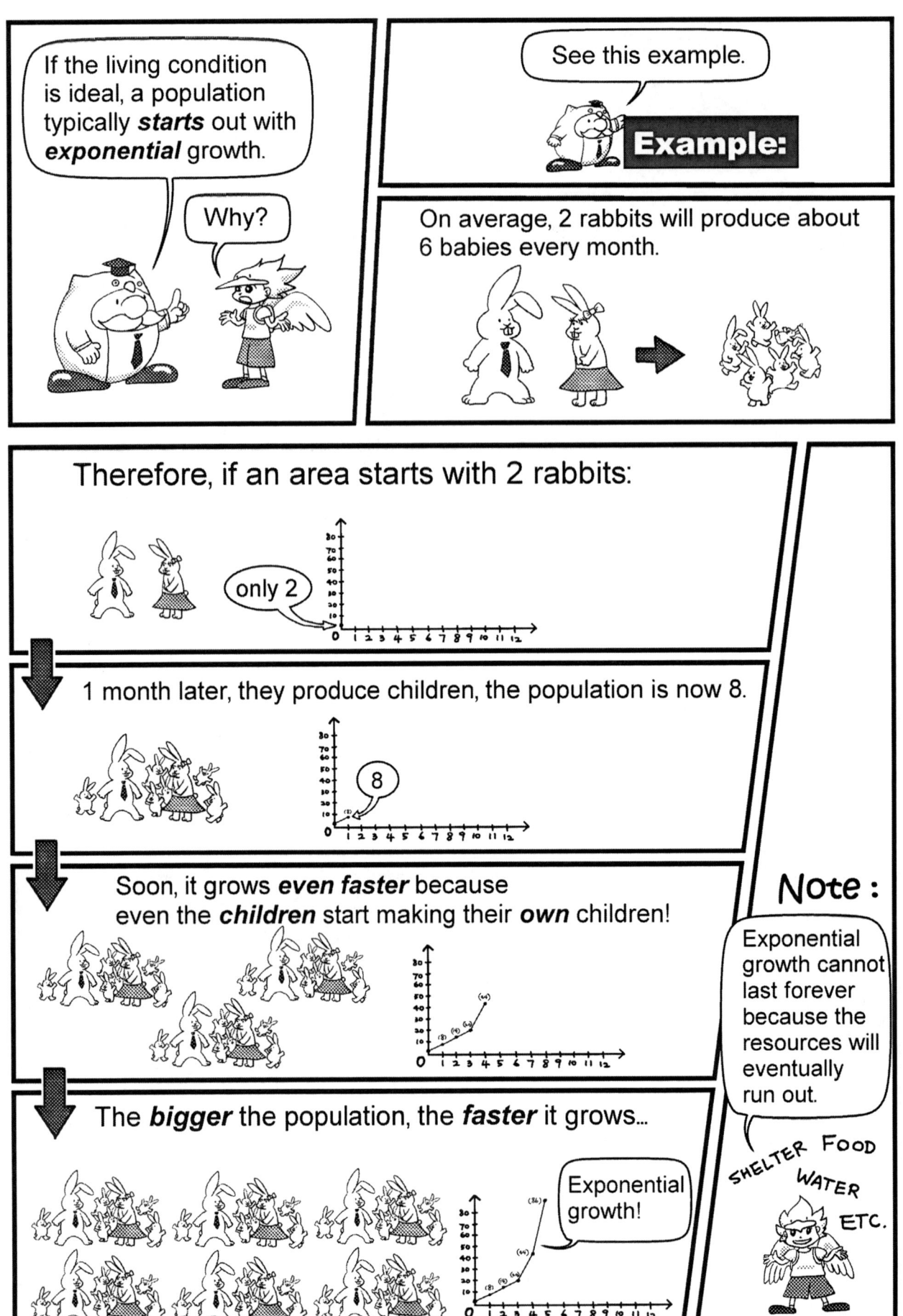

Population Size: Increase, Decrease, or Staying the Same

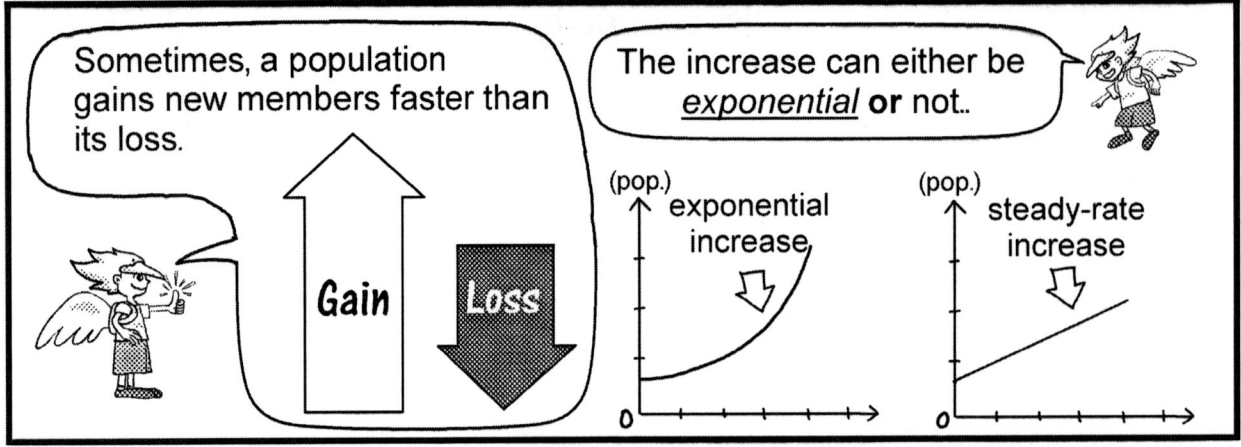

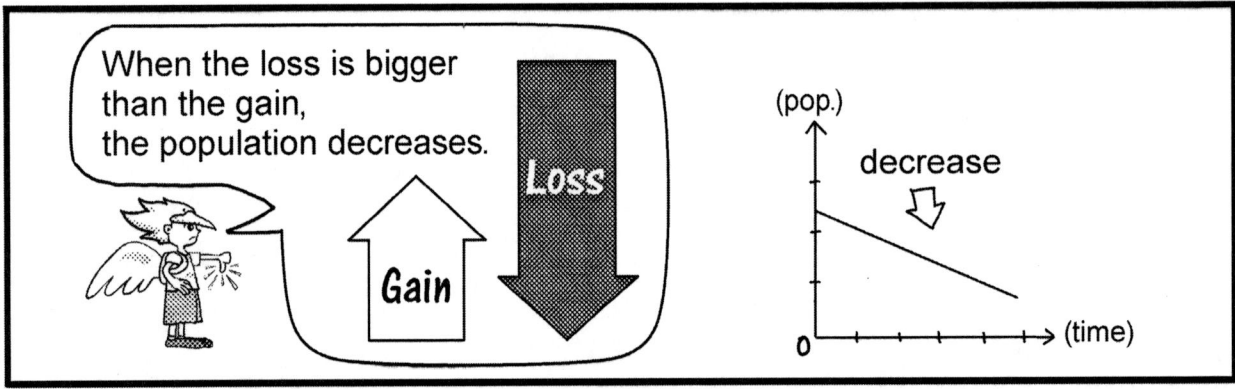

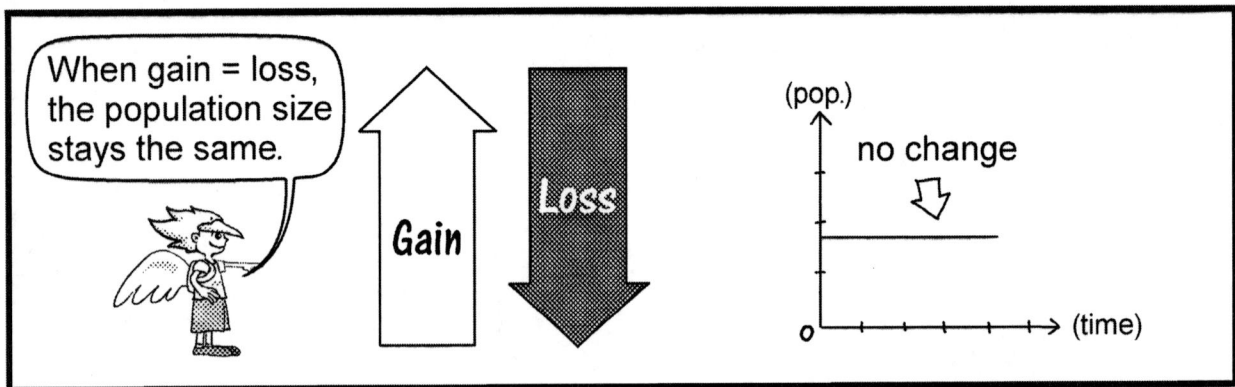

2 Types of Limiting Factors:

Density-Dependent Limiting Factors

These factors are more *effective* when there is *high* population *density*.

EXAMPLE: COMPETITION FOR FOOD

The higher the population density, the more intense the competition.

"This grass is mine!"

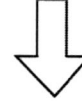

At high population density, starvation will take a bigger toll.

Winner / Starved to death

Density-Independent Limiting Factors

Their effectiveness does *not* depend on population *density*.

EXAMPLE: FOREST FIRE

Forest fire kills and destroys regardless of the population density.

"Run for your lives!"

"I will kill them... whether they are crowded or not!"

page 55

Unit 1: Chapter 6

More Examples of **Density-Dependent** Limiting Factors:

Disease

Diseases spread faster when there is a high population density.

(because everyone is closer)

Predators

Predators catch and kills more effectively when the population density is high.

"The more crowded they are, the easier it is to catch them!"

More Examples of **Density-Independent** Limiting Factors:

Avalanche

Natural Disasters (such as flood)

Chapter 7: The Biosphere (and Other Spheres)

Science Standard Bio/LS . 6 . d :
Students know how water, carbon, and nitrogen cycle between abiotic resources and organic matter in the ecosystem and how oxygen cycles through photosynthesis and respiration.

(Note: For additional background knowledge on this topic, see Chapter 1 (of unit 1) and chapter 11 (unit 2).)

NEW VOCABS

* **Atmosphere:** All the various types of gases that cover and surround the earth.

* **Biosphere:** All the earth's ecosystems and their organisms. In other words, all the areas of the earth where life exists as well as the living organisms that live in these areas.

* **Biota:** The sum of all the living organisms. In a global context, the earth's biota is the sum of all the organisms on earth. In the context of a single ecosystem, its biota is the sum of all the organisms that live in that ecosystem.

* **Geosphere:** The combined total of the solid parts of the earth, from the earth's surface to the earth's core.
(including molten magma, rocks, sand/dirt...etc.)

* **Hydrosphere:** The combined total of the earth's water (including solid ice, liquid, or water vapor).

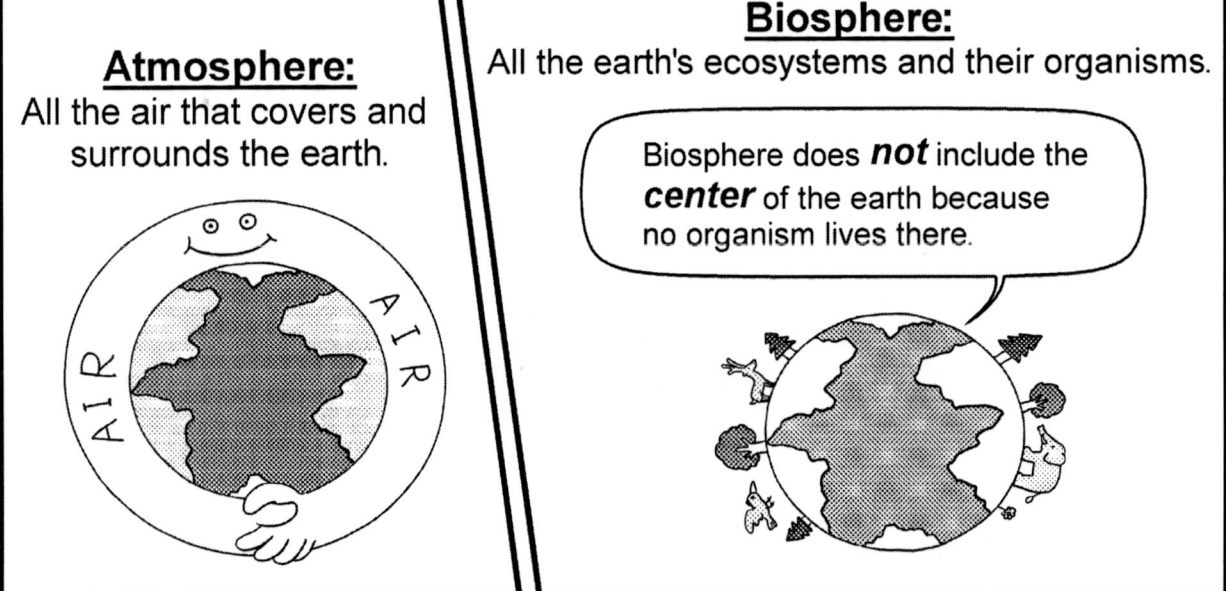

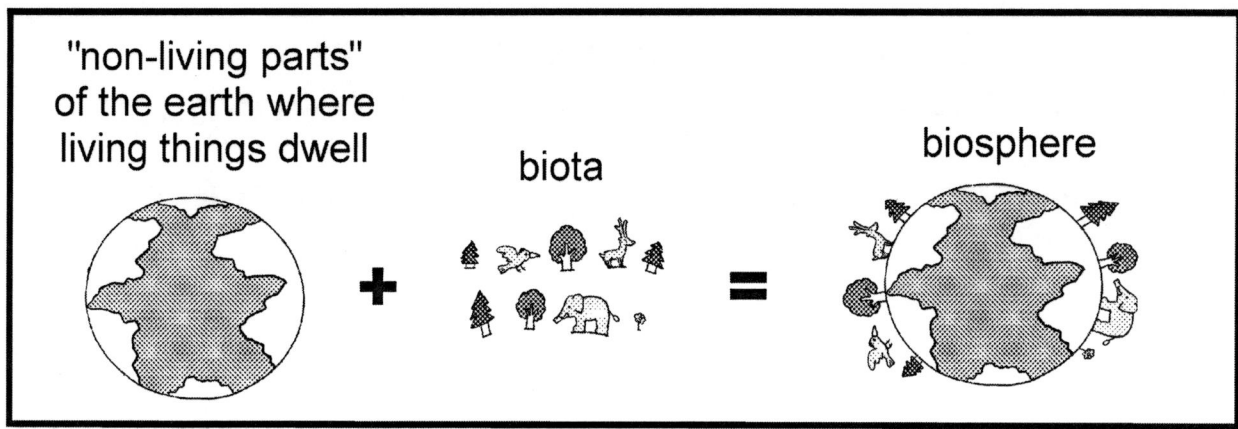

Review from Chapter 1 (Material Cycles):
Ecosystems cycles all the materials
(Such as: **Water**, **Carbon**, **Nitrogen**, and **Oxygen**)
between the **environment** and the **organisms**.

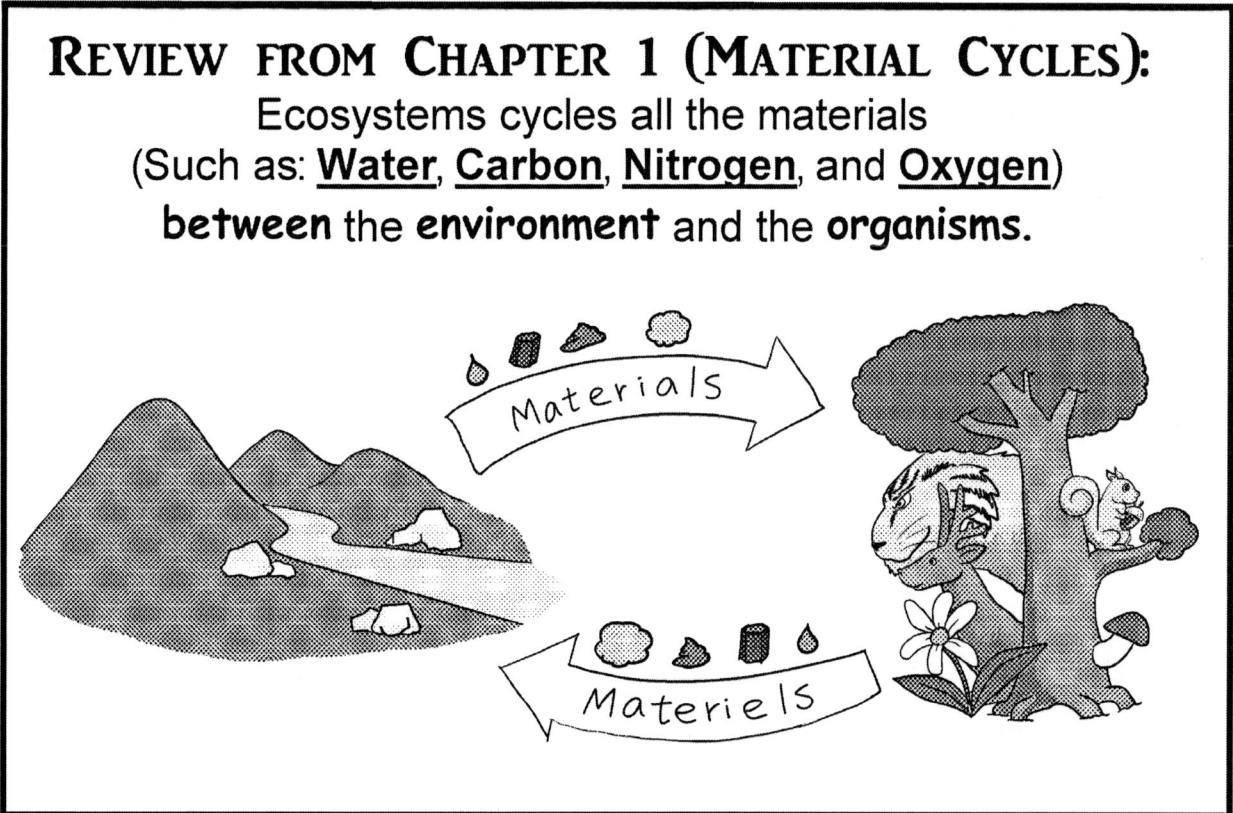

Oxygen cycles between plants and animals by **Photosynthesis** and **(Cellular) Respiration**

(Note: For in-depth knowledge on these processes, see Chapter 11 of Unit 2.)

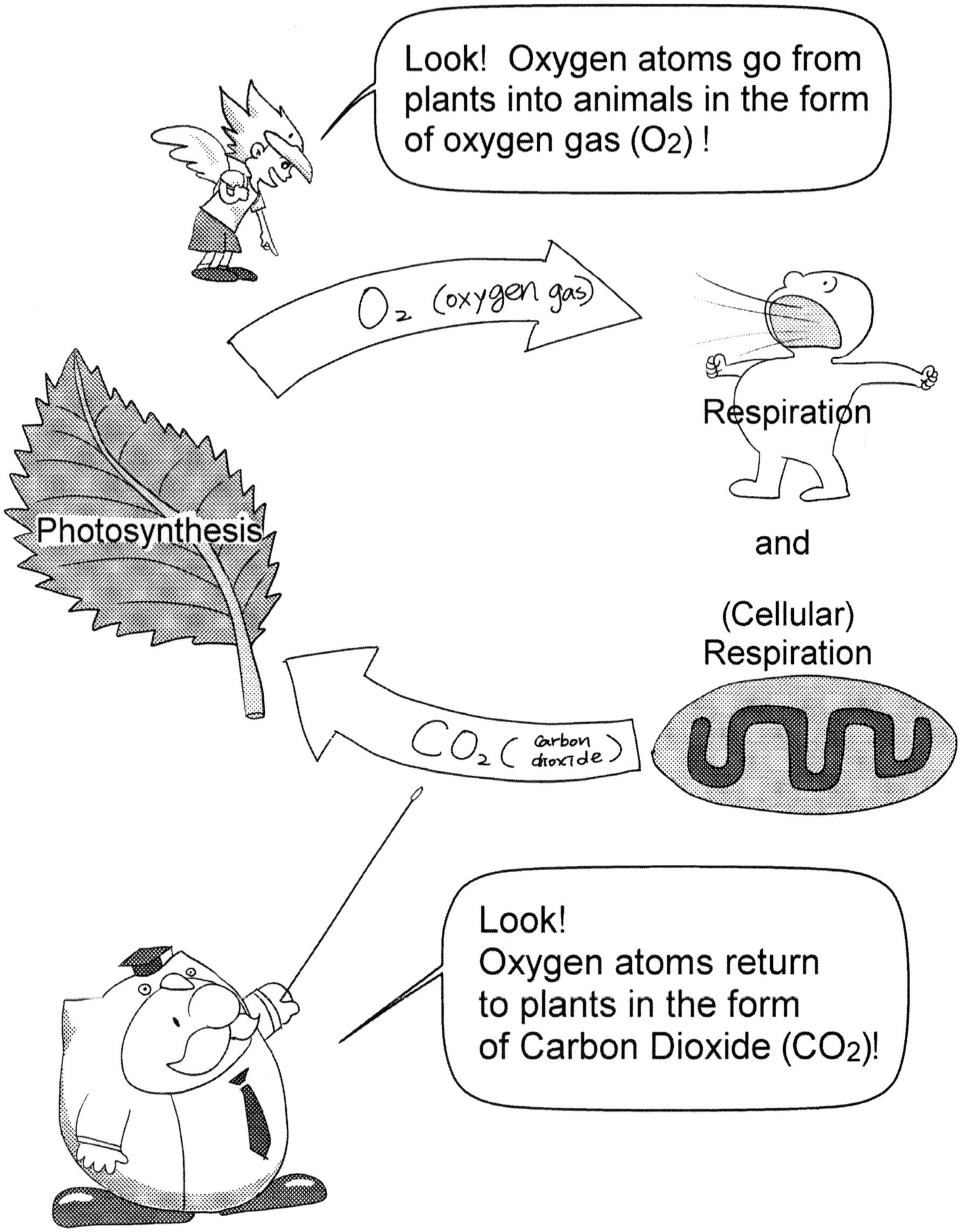

Look! Oxygen atoms go from plants into animals in the form of oxygen gas (O_2)!

O_2 (oxygen gas)

Photosynthesis

Respiration

and

(Cellular) Respiration

CO_2 (carbon dioxide)

Look! Oxygen atoms return to plants in the form of Carbon Dioxide (CO_2)!

Chapter 8: Producers and Decomposers

Science Standard Bio/LS . 6 . e :
Students know a vital part of an ecosystem is the stability of its producers and decomposers.

NEW VOCABS

* **Consumers:** Also known as heterotrophs. Organisms that must eat other organisms to survive.

* **Decomposers:** Organisms that chemically break down substances for food. Examples of decomposers are bacteria, fungus, and protists such as amoebas.

* **Producers:** Also known as autotrophs. Organisms that can make their own food by photosynthesis (or chemosynthesis).

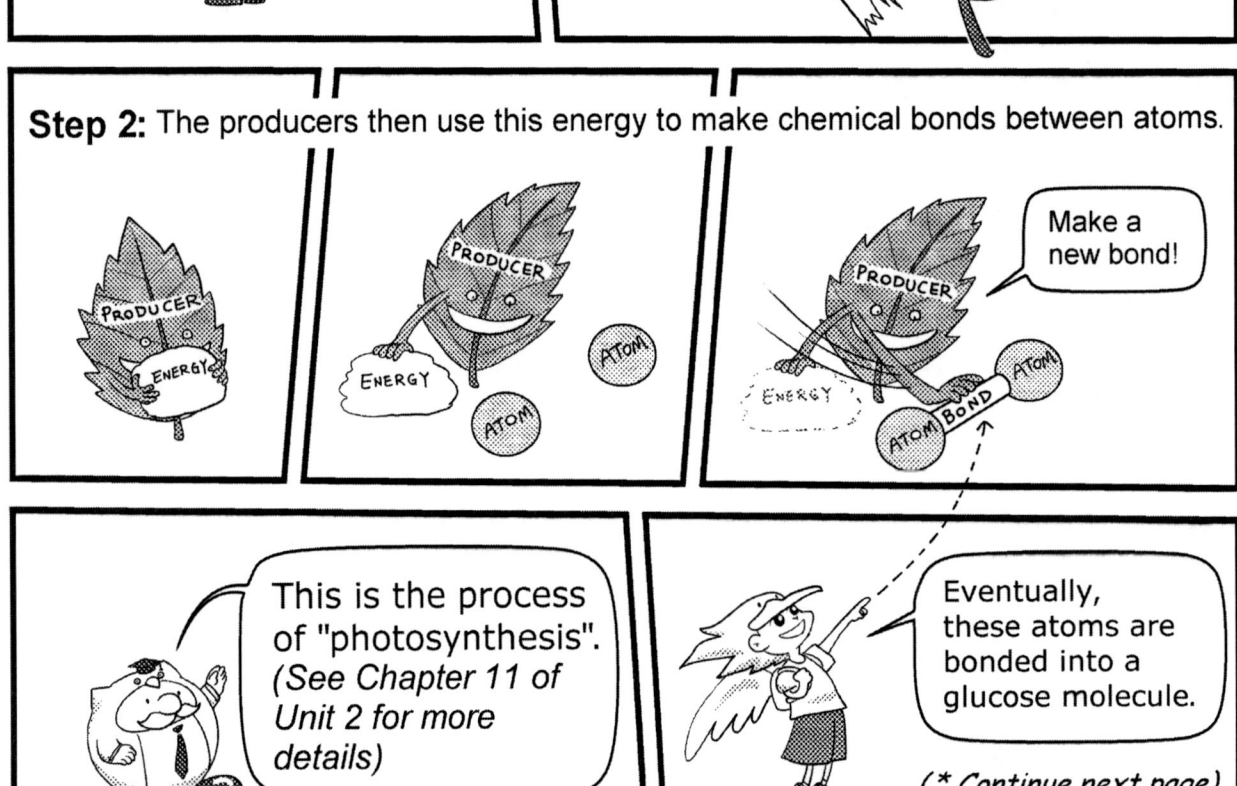

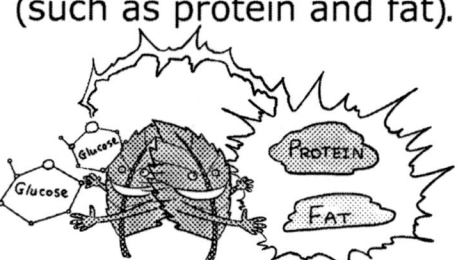
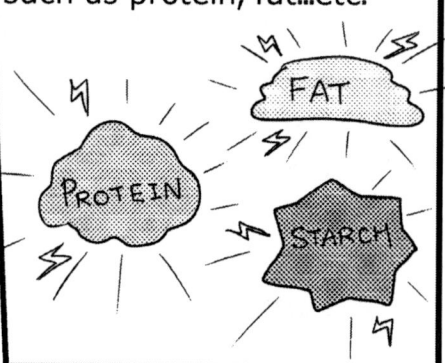

Therefore, plants are very nutritious!

Consumers cannot capture the energy from sunlight

"All animals are consumers!"

So they must get their energy by eating plants.

When we eat other organisms...

We are really eating a bunch of bonded atoms

We break down many of the bonds to get energy out of them.

Then...poop out the remaining atoms and bonds as feces!

"Feces *still* contains a lot of energy-rich bonds!"

*continue next page

Unit 1: Chapter 8

Decomposers are important because they help break molecules down, so **producers** can easily *absorb* these molecules.

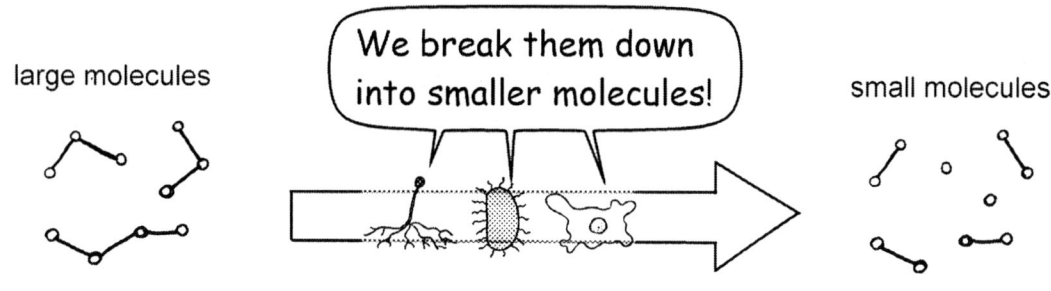

In addition to breaking down feces, decomposers *also* help break down:

Urine

Dead Organisms

Miscellaneous Wastes

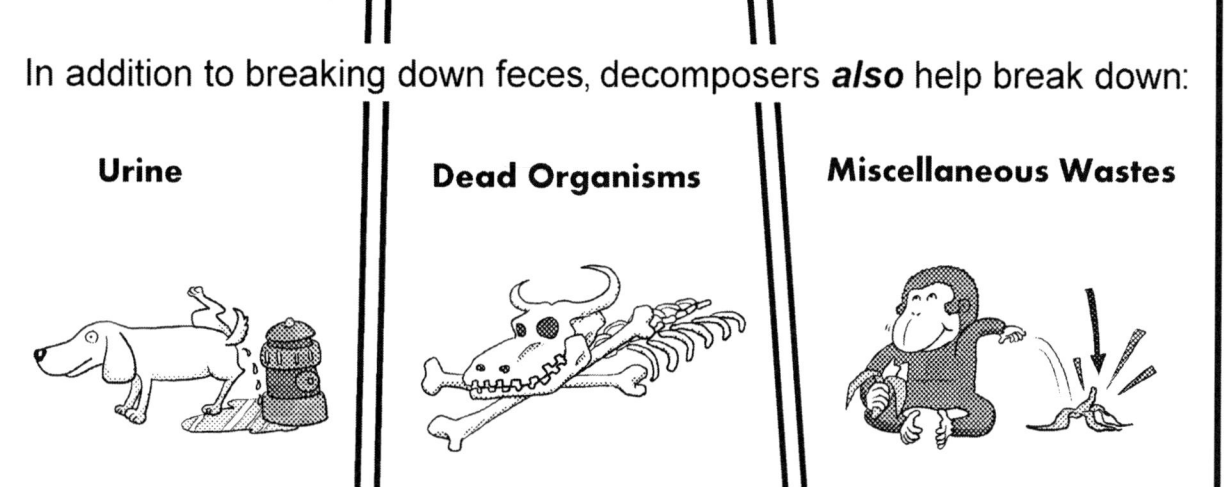

Producers and *decomposers* work together to keep an ecosystem **stable**, because *decomposers help* prepare the molecules to be *recycled* back into the producers to use in *photosynthesis*!

On the other hand, the **consumers** do *not* play as crucial of a role in stabilizing an ecosystem.

page 65

Unit 1: Chapter 8

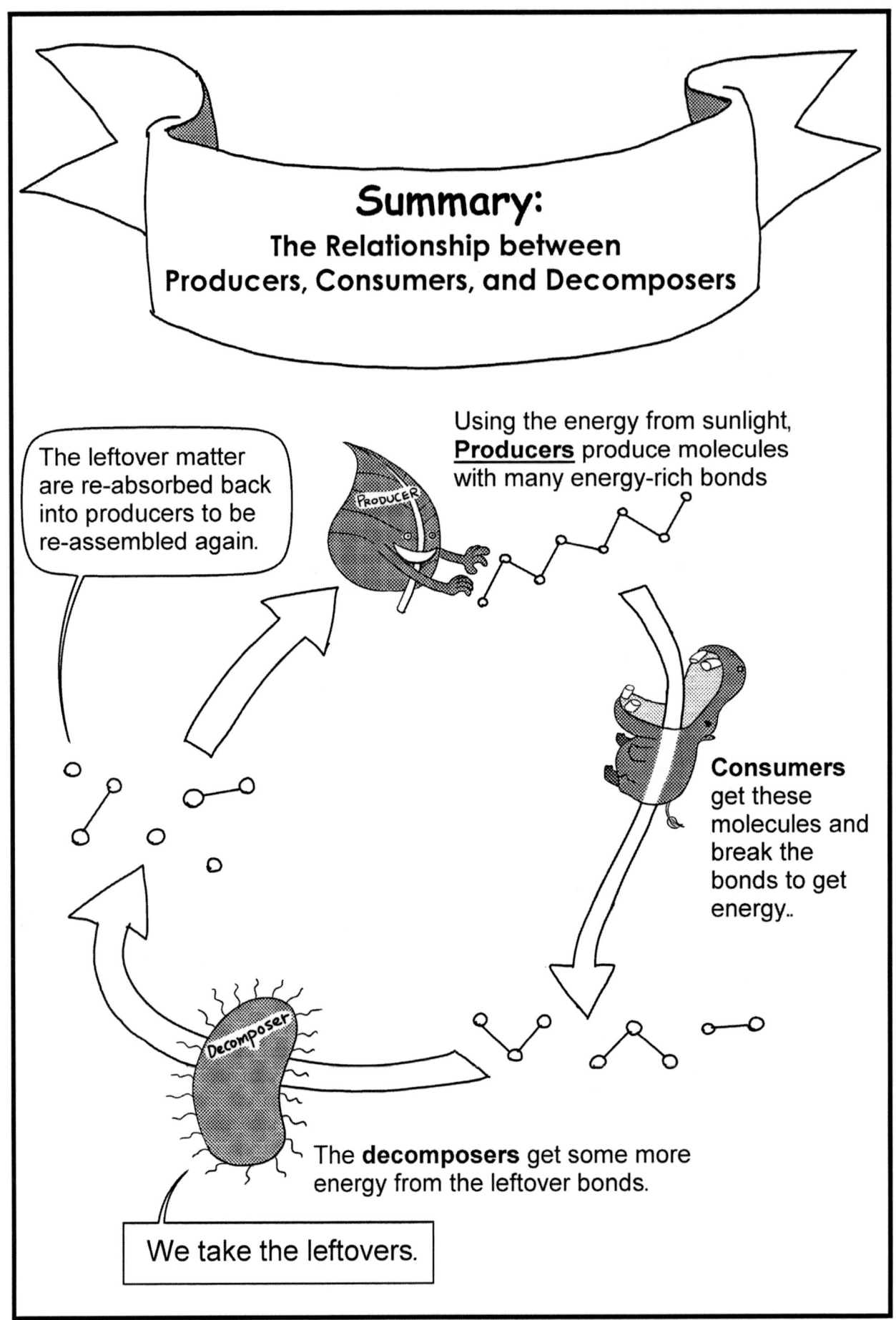

Chapter 9: Energy-Flow in an Ecosystem

Science Standard Bio/LS . 6 . f :
Students know at each link in a food web some energy is stored in newly made structures but much energy is dissipated into the environment as heat. This dissipation may be represented in an energy pyramid.

NEW VOCABS

* **Biomass:** The mass of all the living organisms (by weight), be it in an ecosystem, biome, or globally speaking.

* **Biomass Pyramid:** A pyramid-shaped diagram that shows the biomass of each trophic level in an ecosystem.

* **Energy Pyramid:** A pyramid-shaped diagram that shows the energy (specifically, chemical energy) in each trophic level in an ecosystem. (Note: The ratio of energy at each upper level is usually 10% of the lower level. But the number 10% is not exact and the exact value differs in each ecosystem)

* **Primary Consumers:** Consumers who eat the producers. In other words, the herbivores.

* **Pyramid of Numbers:** A pyramid-shaped diagram that shows the number of individual organisms in each trophic level in an ecosystem.

* **Secondary Consumers:** Consumers (carnivores) who eat the primary consumers.

* **Tertiary Consumers:** Consumers (carnivores) who eat the secondary consumers.

Unit 1: Chapter 9

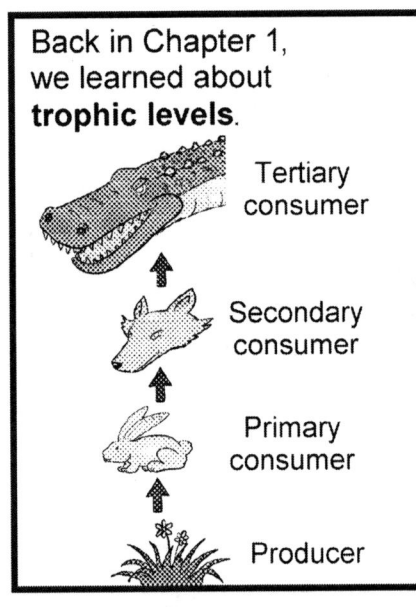
Back in Chapter 1, we learned about **trophic levels**.

Energy always flows from lower levels to higher levels

Let's take a close look!

Primary Consumers (Herbivores) get energy by eating the producers...

However, *most* of the energy from the grass was *lost/wasted* in the transfer..

When secondary consumers (carnivores) eat herbivores, the energy is transferred again.

Again, most of the energy was lost/wasted.

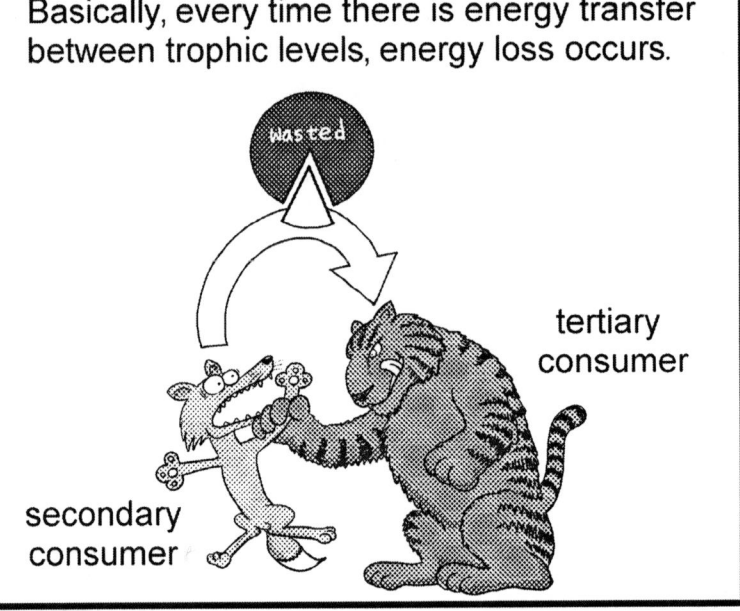
Basically, every time there is energy transfer between trophic levels, energy loss occurs.

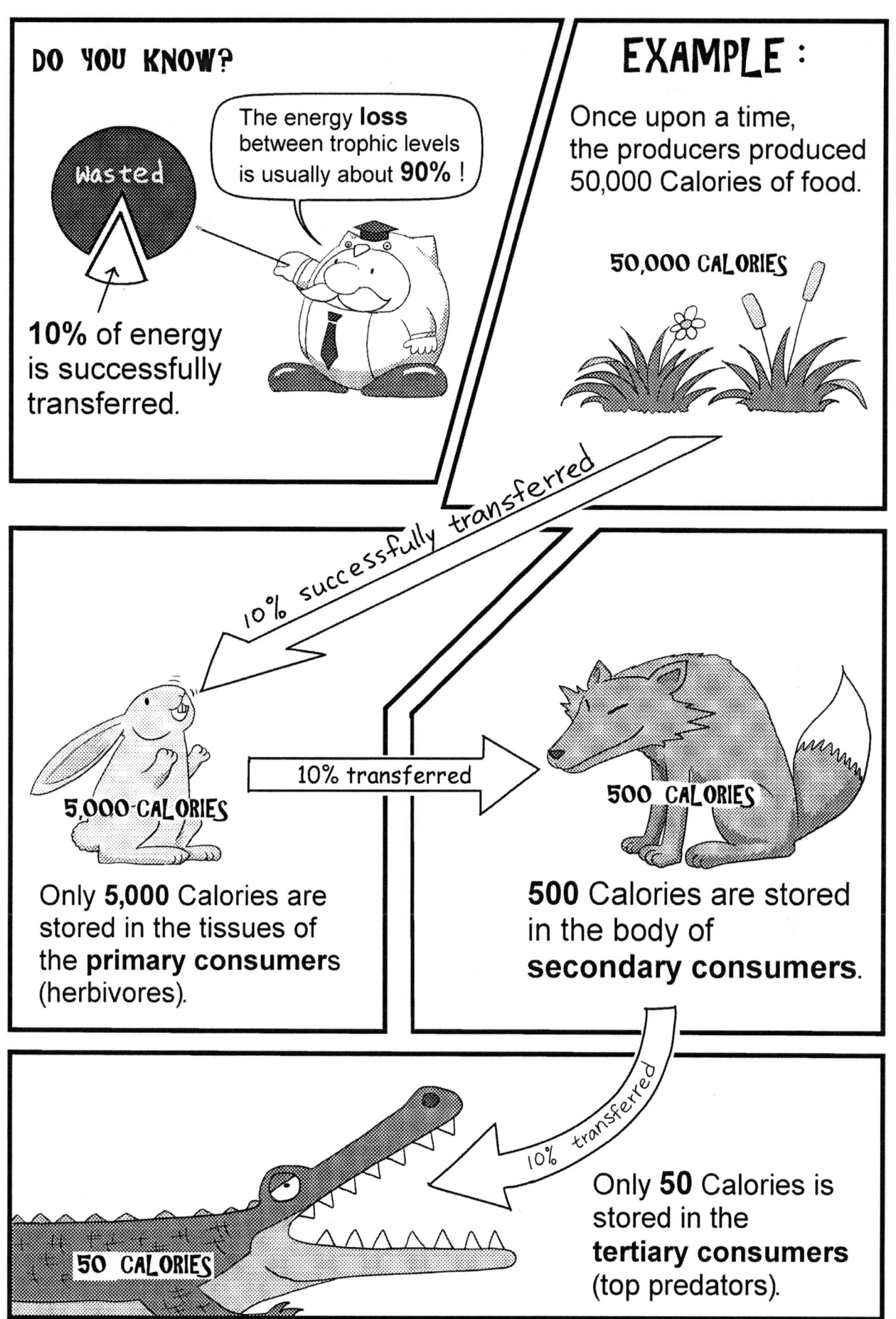

To illustrate that there is less energy in the higher trophic levels, ecologists draw an **"Energy Pyramid"**.

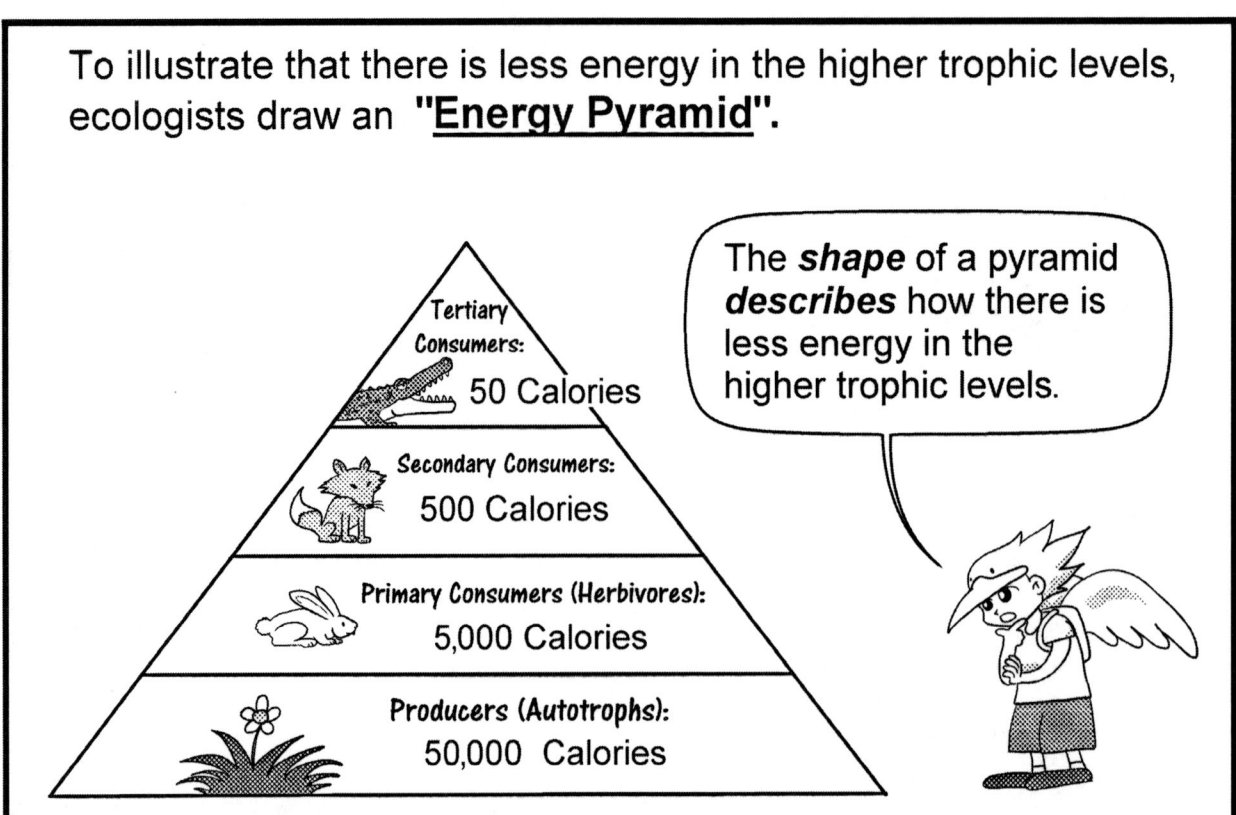

The *shape* of a pyramid *describes* how there is less energy in the higher trophic levels.

Tertiary Consumers: 50 Calories
Secondary Consumers: 500 Calories
Primary Consumers (Herbivores): 5,000 Calories
Producers (Autotrophs): 50,000 Calories

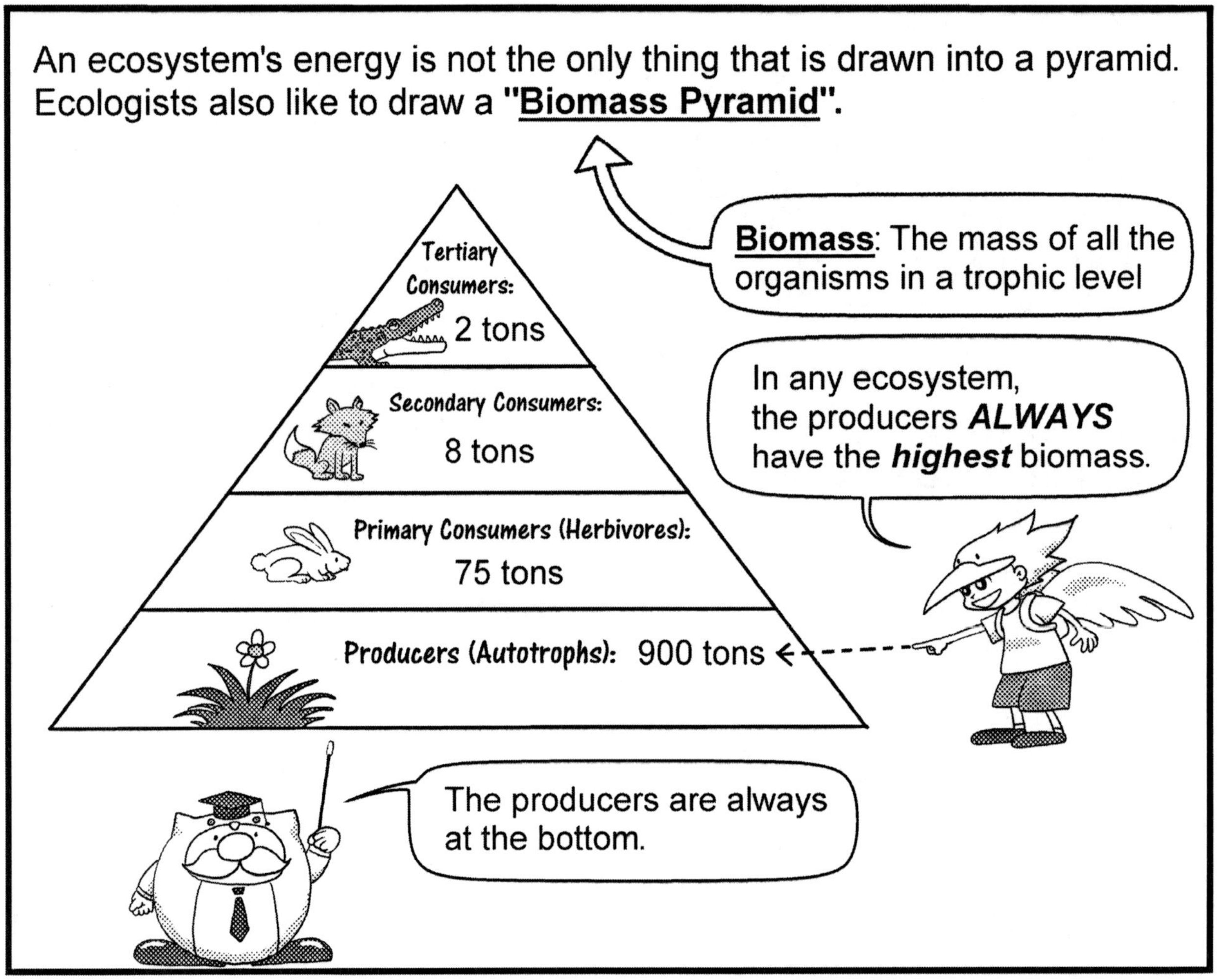

An ecosystem's energy is not the only thing that is drawn into a pyramid. Ecologists also like to draw a **"Biomass Pyramid"**.

Tertiary Consumers: 2 tons
Secondary Consumers: 8 tons
Primary Consumers (Herbivores): 75 tons
Producers (Autotrophs): 900 tons

Biomass: The mass of all the organisms in a trophic level

In any ecosystem, the producers **ALWAYS** have the *highest* biomass.

The producers are always at the bottom.

Unit 1: Chapter 9

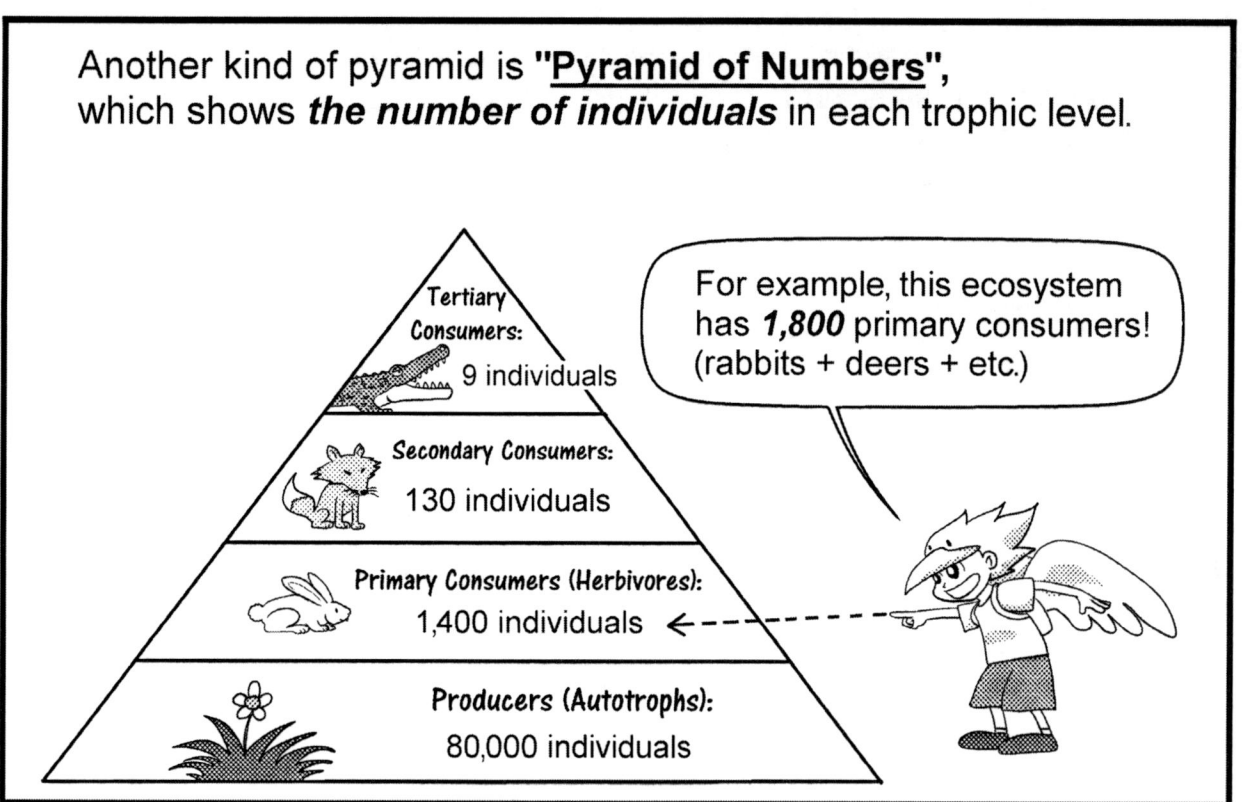

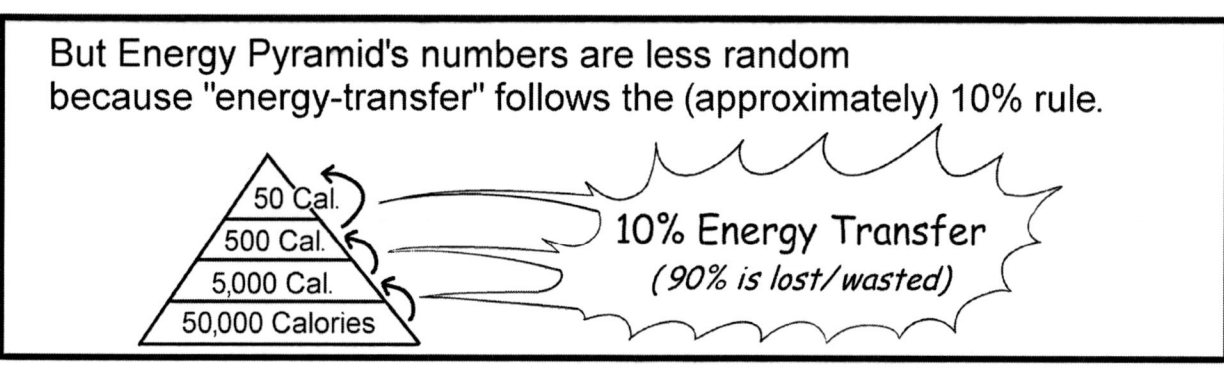

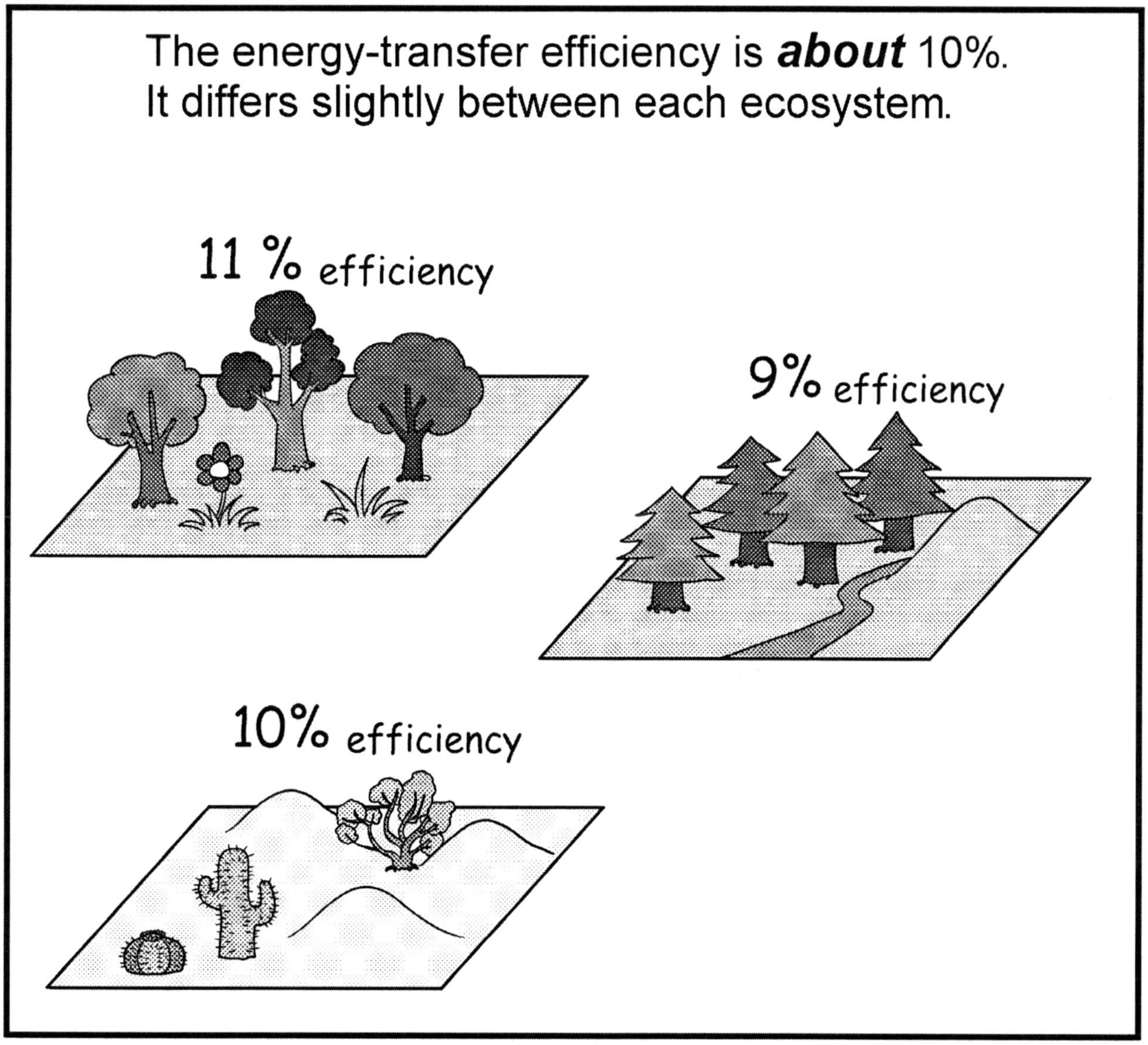

Unit 2:
Cell Biology

Chapter 10: The Nucleus Stores Genetic Information

Science Standard: 7.1.c
Students know the nucleus is the repository for genetic information in plant and animal cells.

NEW VOCABS

* **Cell:** The basic unit that makes up all living things.

* **D.N.A.:** Abbreviation for "Deoxyribonucleic Acid." It is the molecule that makes up an cell's genetic information.

* **Nuclear membrane:** Also known as the "nuclear envelope." It is the membrane that forms the surface of the nucleus.

* **Nucleus:** A cellular structure where the cell's genetic information is stored.

* **Organism:** A living thing.

* **Trait:** An inheritable (abled to be passed down from parent to child) characteristic of an organism.

Before we start, know this vocabulary:

The word **Organism** means **"living thing."**

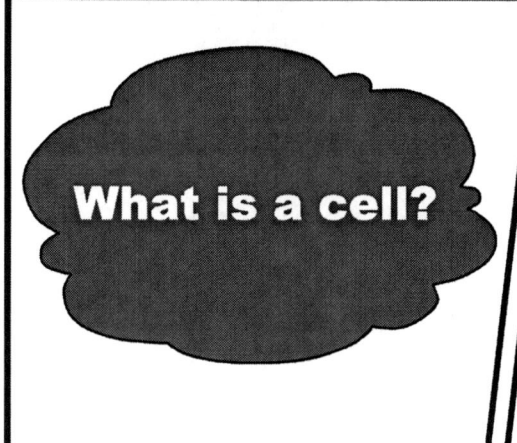

What is a cell?

Cells are the basic unit of life.

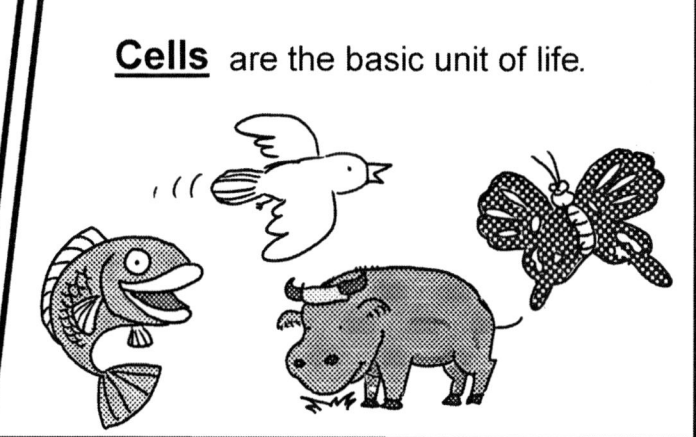

All living things ...
(such as humans)

... are made of cells!

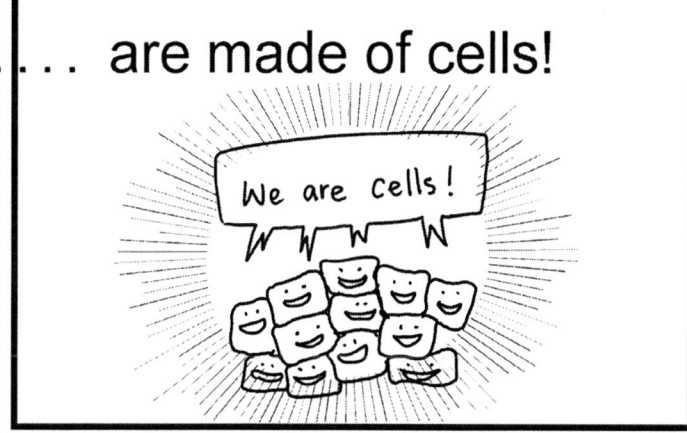

We are cells!

A cell is also the smallest living thing on earth.

LIFE ... IS ... GOOD ...

Anything smaller (such as "half a cell") does not count as living.

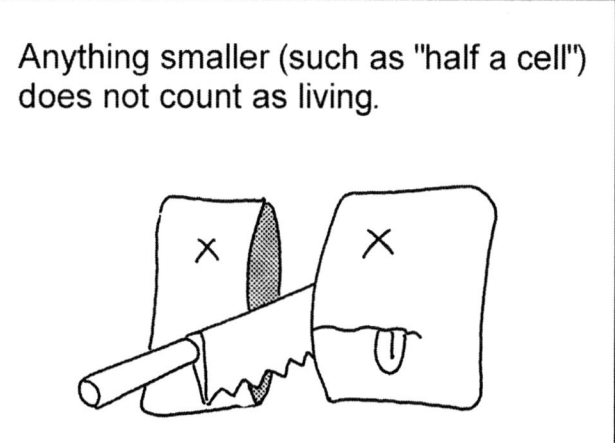

page 77

Unit 2: Chapter 10

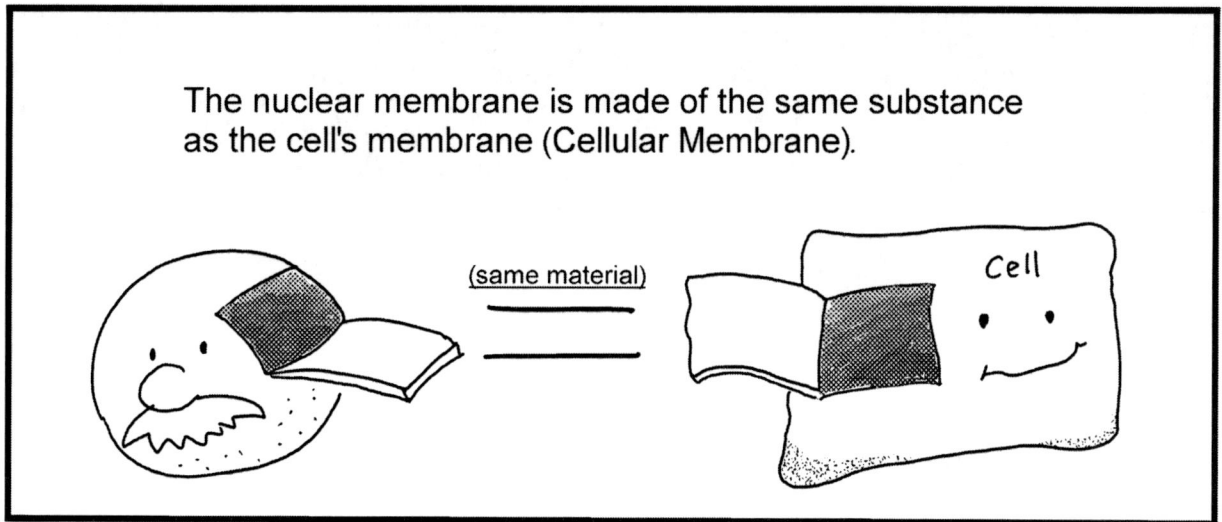

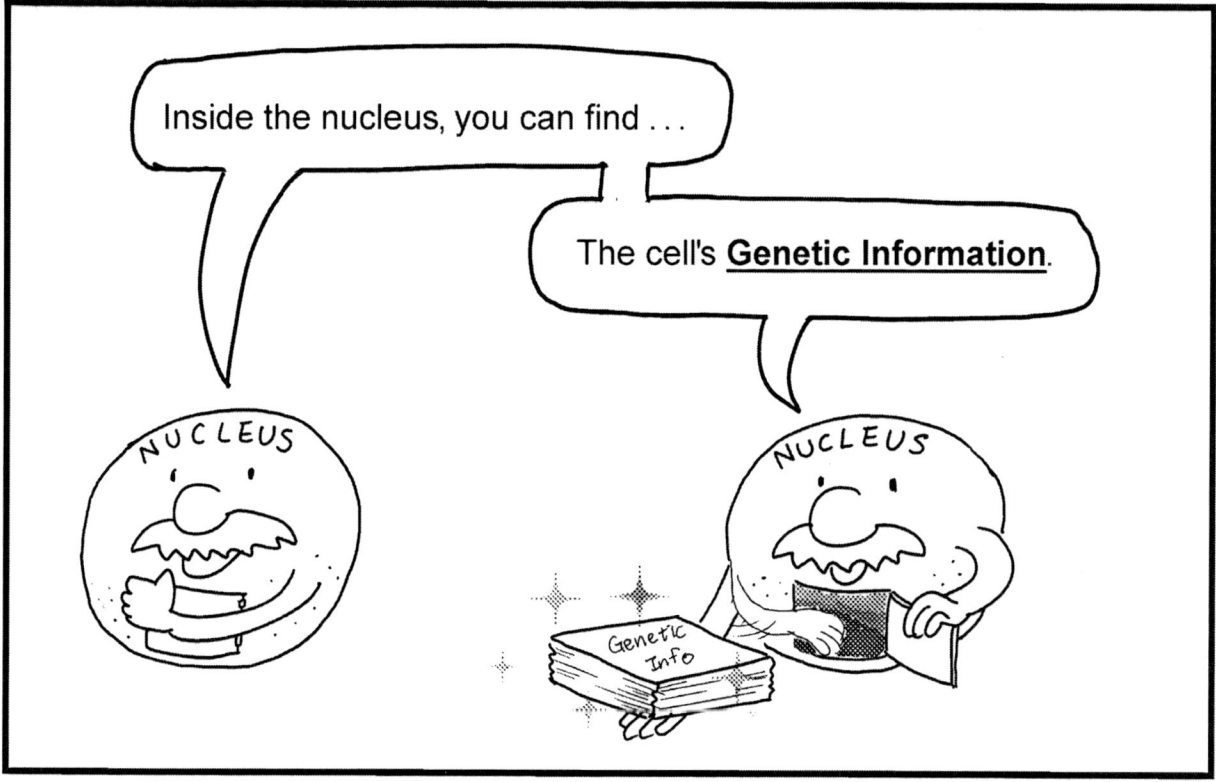

What is "Genetic Information?"

It is made of special molecules called "**D.N.A.**"

Genetic information controls ALL of your **"Traits."**
(Traits = characteristics.)

IMPORTANT

Because Genetic Info is made of D.N.A., you can use these 2 terms interchangeably.

"D.N.A." = "Genetic Info" = "Genes"

Different genes control different "Traits."
A "Trait" is an outwardly-shown quality of an organism.

For example, the traits of my neighbor's pet Golden Retriever are:

* Floppy Ears
* Long Fur.
* Friendly
* Likes to Swim.
* Prone to Arthritis.
* Golden Fur.
* Big Paws.

Notice that even personality and natural-born gifts count as traits!

Genetic Info not only causes one to have "Traits," it also tells the cell what to do in every situation.

For example, I can look up Genetic Info for instructions on "How to store fat."

I can also find info on "How to mend broken bones."

Genetic Info is passed down from 1 generation to the next.

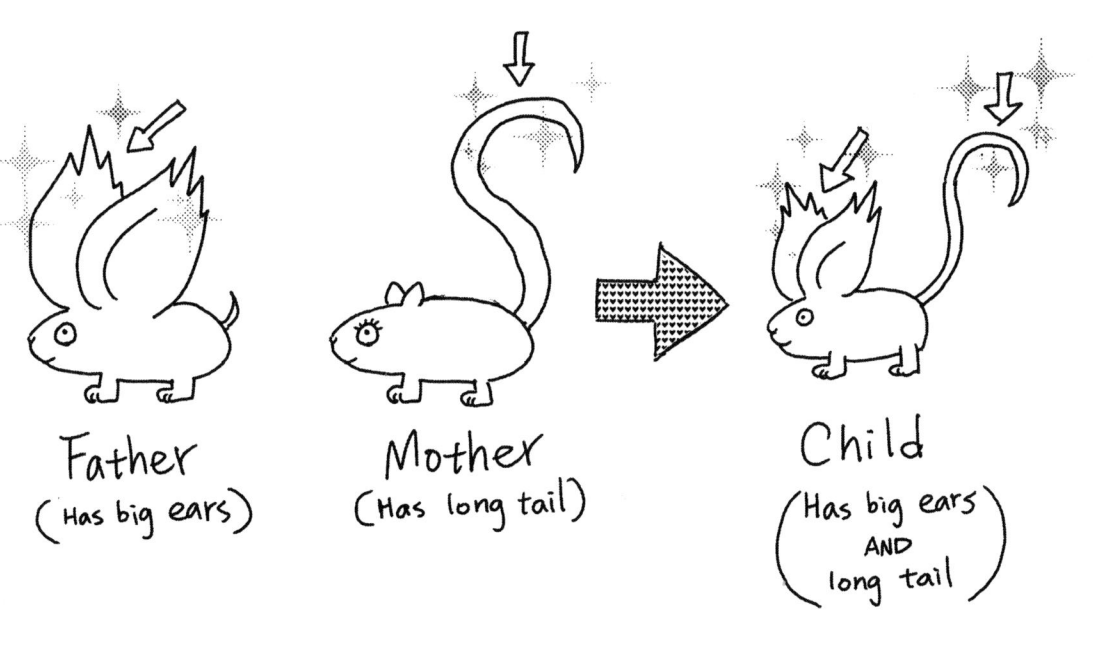

Father (Has big ears)

Mother (Has long tail)

Child (Has big ears AND long tail)

Unit 2: Chapter 10

Review:

Nucleus is like a cell's command center.

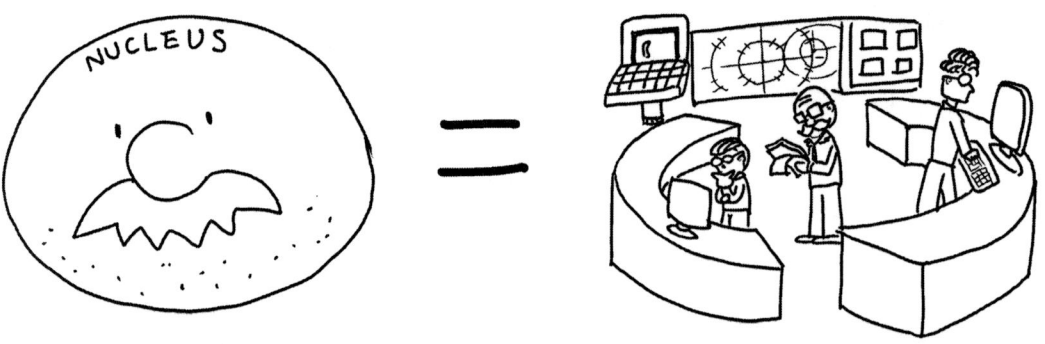

Genetic info is like a cell's instruction manual.

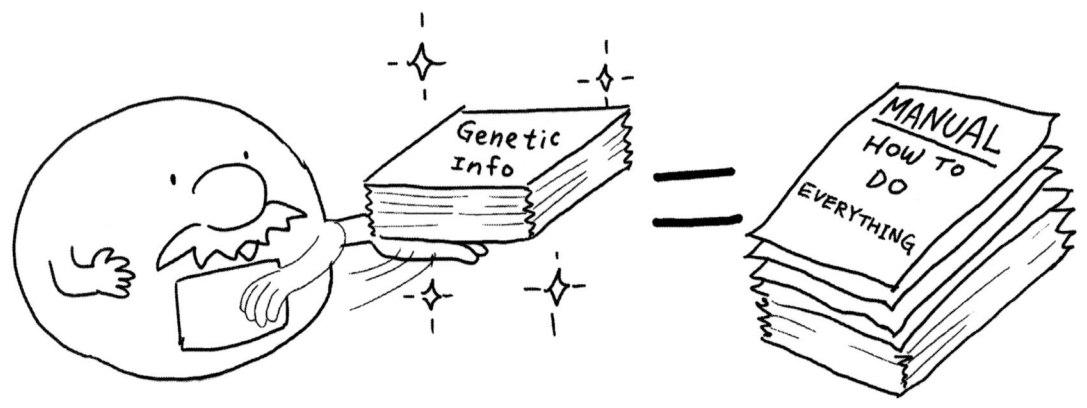

Doesn't it make sense to store the manual inside the command center?

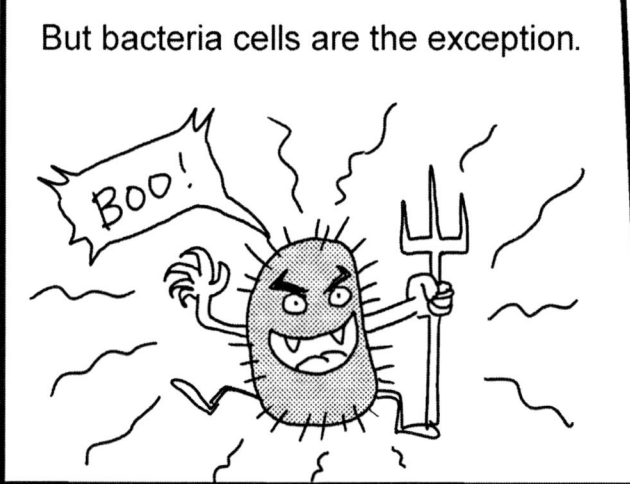

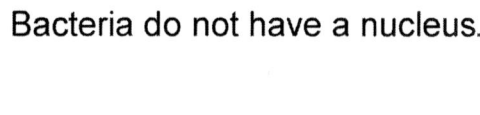

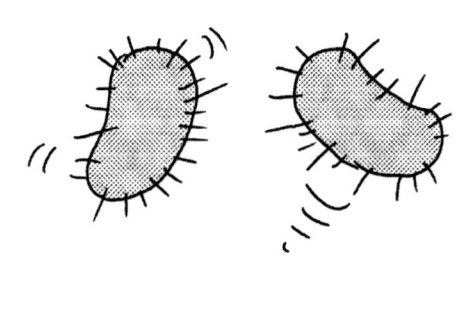

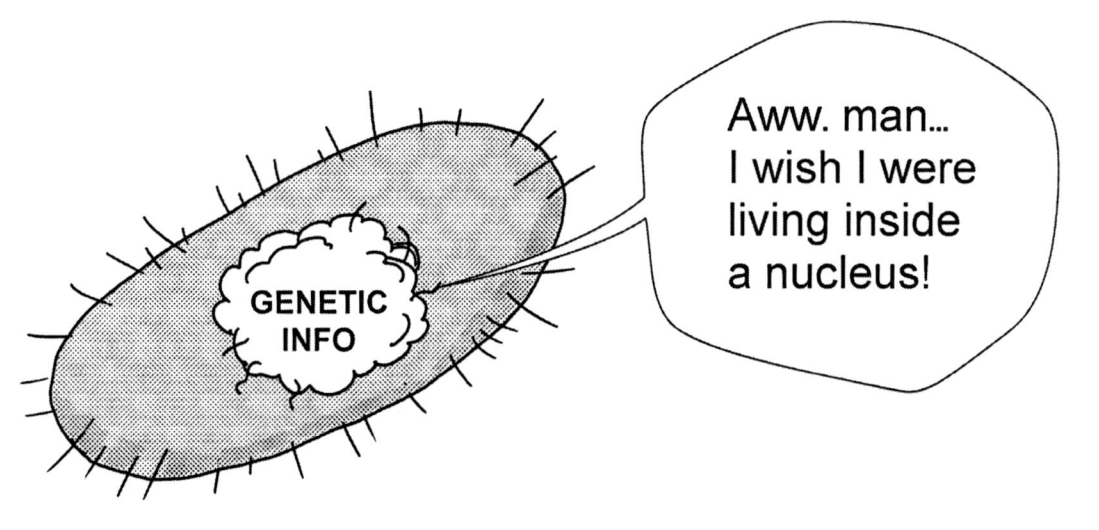

Chapter 11: Cellular Respiration and Photosynthesis

Science Standard: 7.1.d
Students know that mitochondria liberate energy for the work that cells do and that chloroplasts capture sunlight energy for photosynthesis.

NEW VOCABS

* **ADP:** Abbreviation for "adenosine diphosphate." It contains almost no energy but can be transformed into ATP (which is rich in energy) during cellular respiration.

* **ATP:** Abbreviation for "adenosine triphosphate." It is rich in energy. But when its energy is used up, it downgrades into ADP.

* **Cellular respiration:** A process in which glucose and oxygen are consumed to produce energy for a cell.

* **Cellulose:** Typically known as "plant fiber."
  Plants can make cellulose out of many glucose molecules.

* **Chlorophyll:** It is a pigment molecule that can absorb light energy. There are a lot of chlorophylls inside a chloroplast.

* **Chloroplast:** A cell structure found in the plant cells.
  It performs photosynthesis in order to produce glucose.

Unit 2: Chapter 11

* **Fermentation:** It is a process that produces only 2 ATP. There are 2 types of fermentations—one type produces alcohol; another type produces lactic acid.

* **Glucose:** The simplest form of sugar, made of 6 carbons + 12 hydrogens + 6 oxygens. It is rich in energy.

* **Mitochondria:** (singular: mitochondrion) A cell structure that performs cellular respiration to produce energy for the cell.

* **Photosynthesis:** A process that uses light energy to convert carbon dioxide and water into glucose (and oxygen gas).

* **Pigment:** A molecule that reflects a certain type of light.

* **Respiration:** The act of "breathing."

And every city needs a Powerplant.

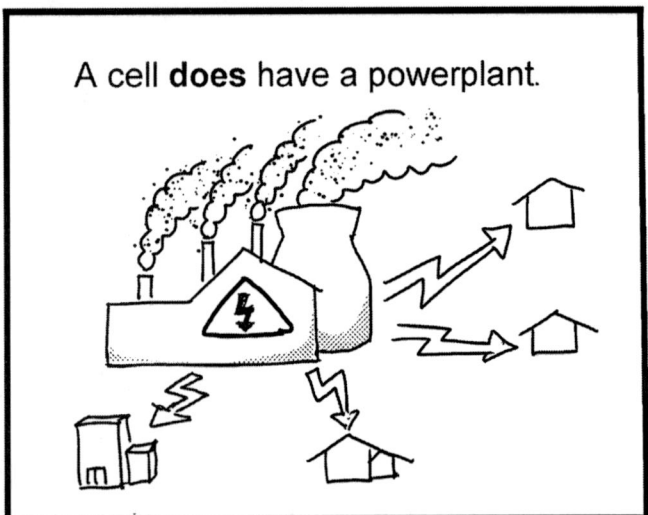
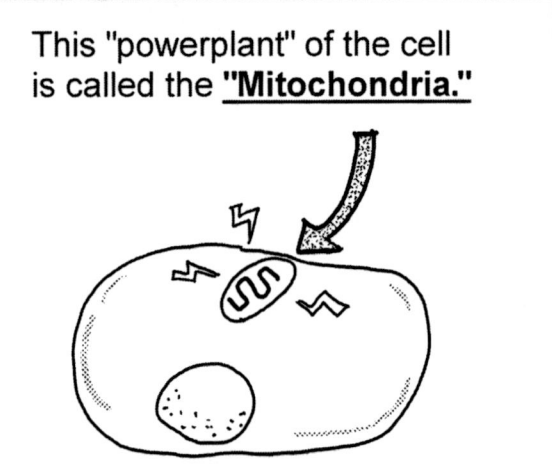
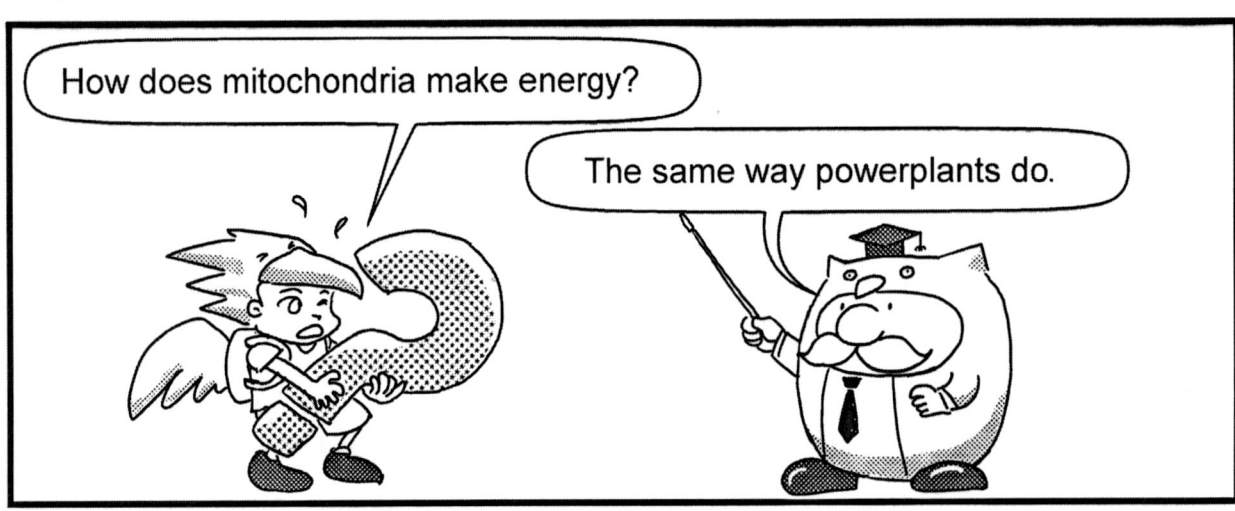

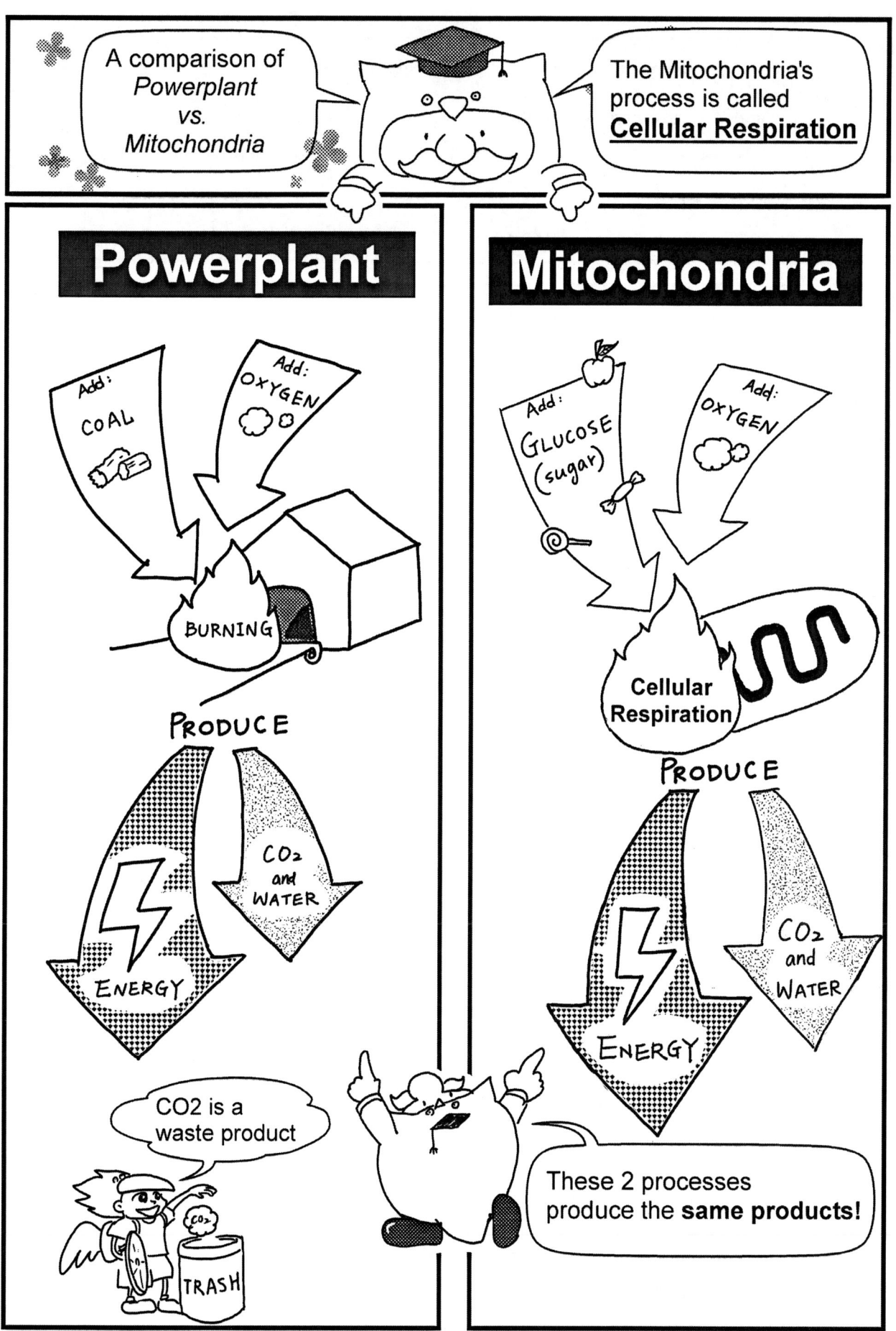

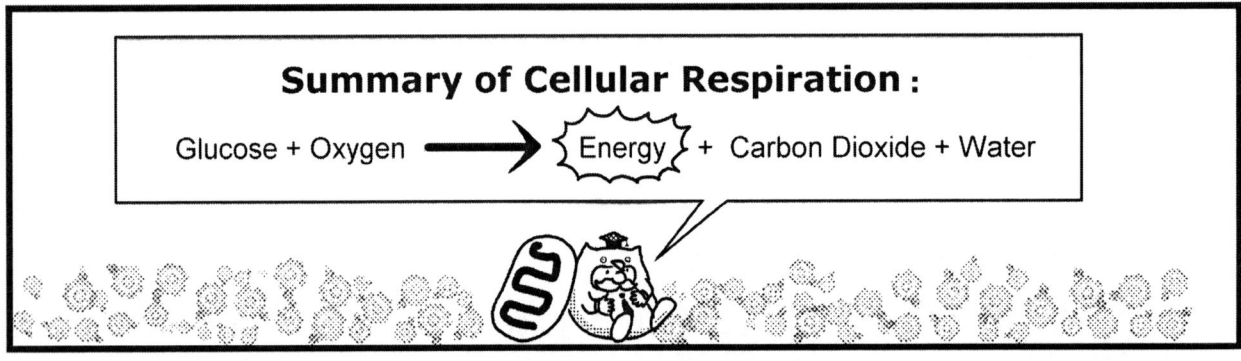

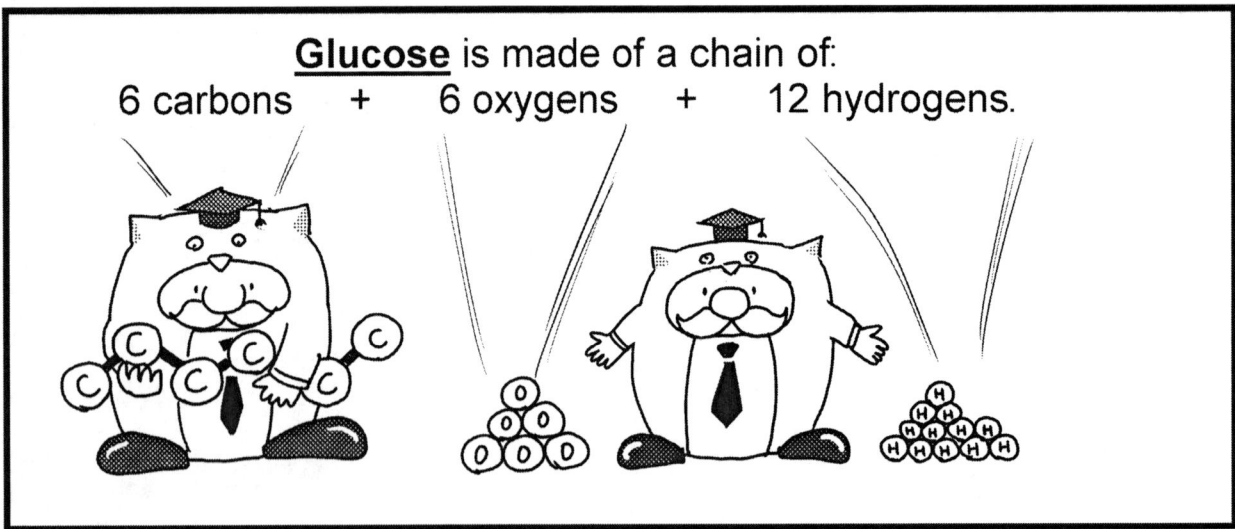

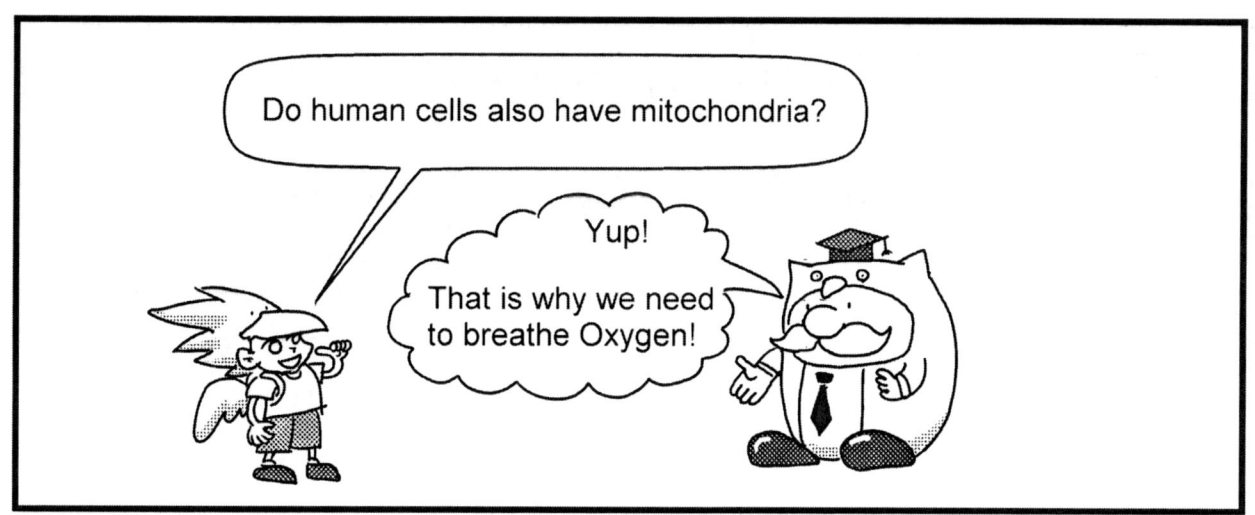

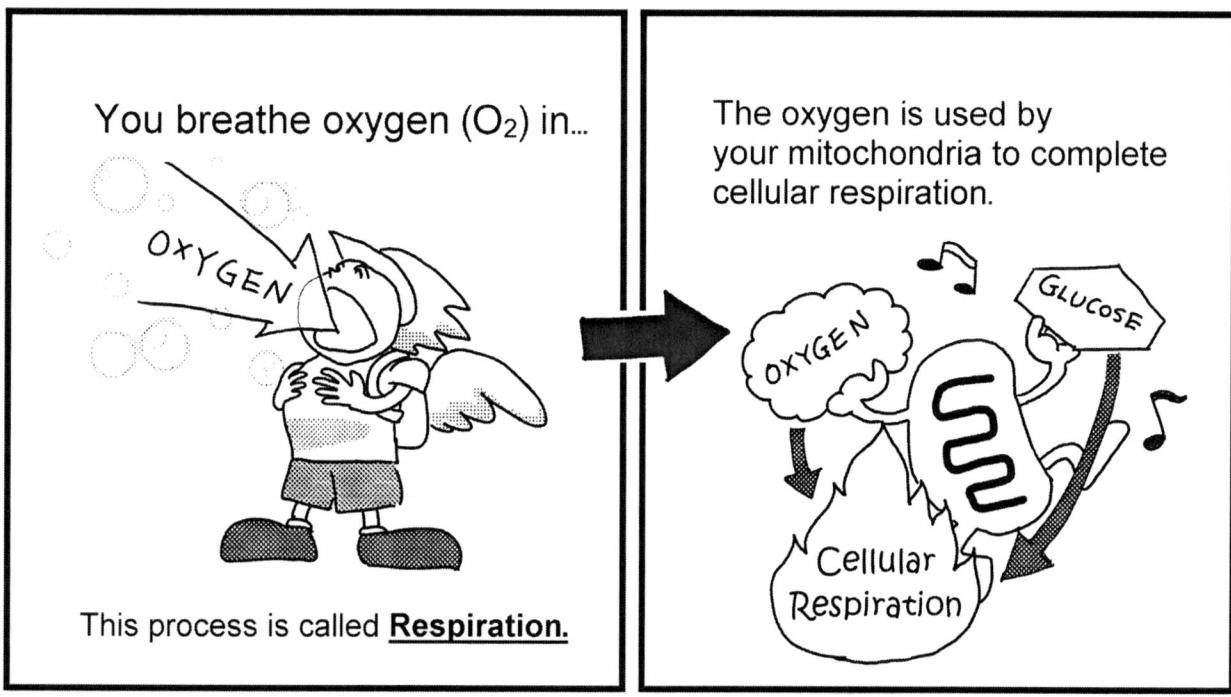

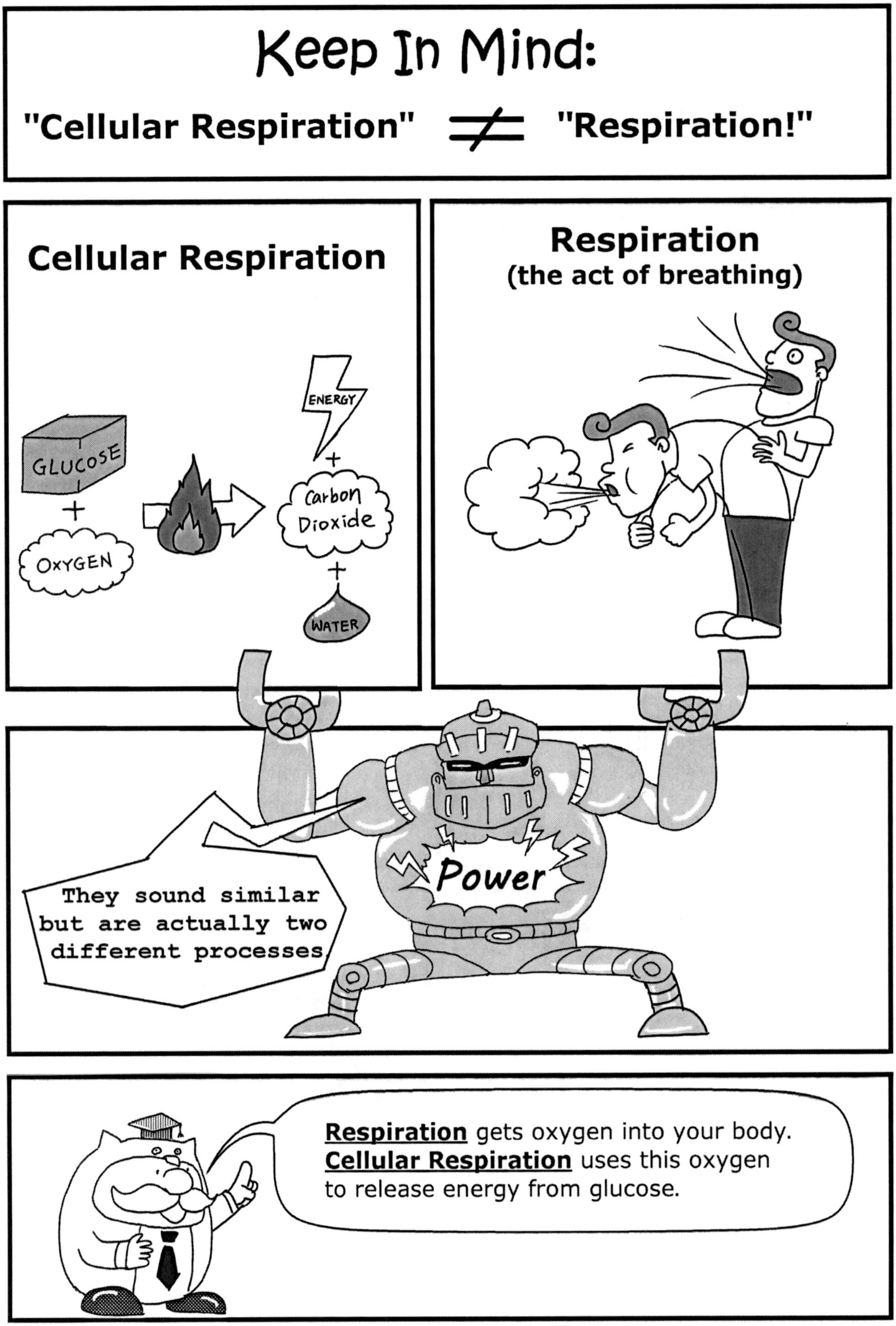

Keep In Mind:

"Cellular Respiration" ≠ "Respiration!"

Cellular Respiration

GLUCOSE + OXYGEN → ENERGY + Carbon Dioxide + WATER

Respiration (the act of breathing)

They sound similar but are actually two different processes

Power

Respiration gets oxygen into your body. Cellular Respiration uses this oxygen to release energy from glucose.

The Relationship Between Respiration and Cellular Respiration

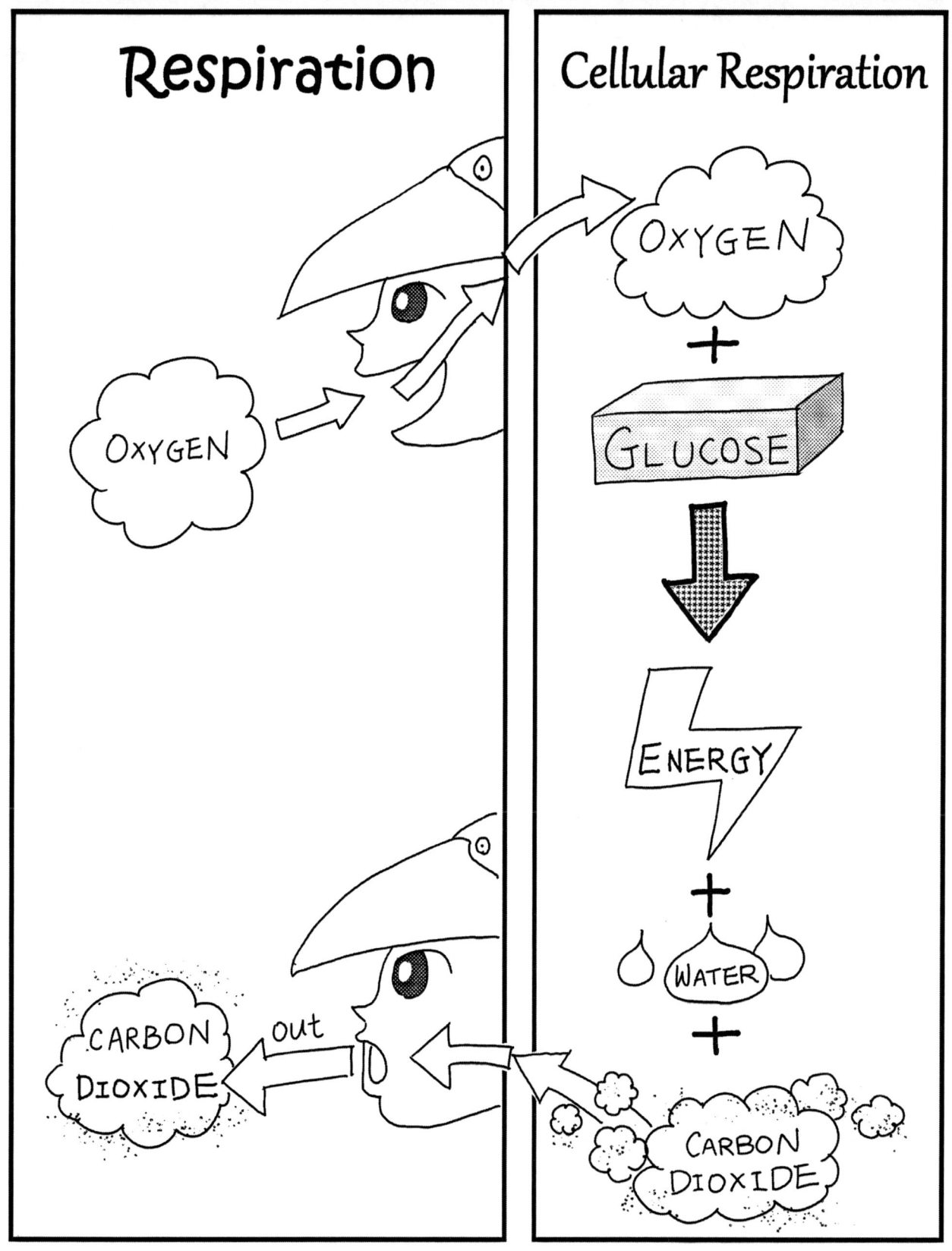

page 91
Unit 2: Chapter 11

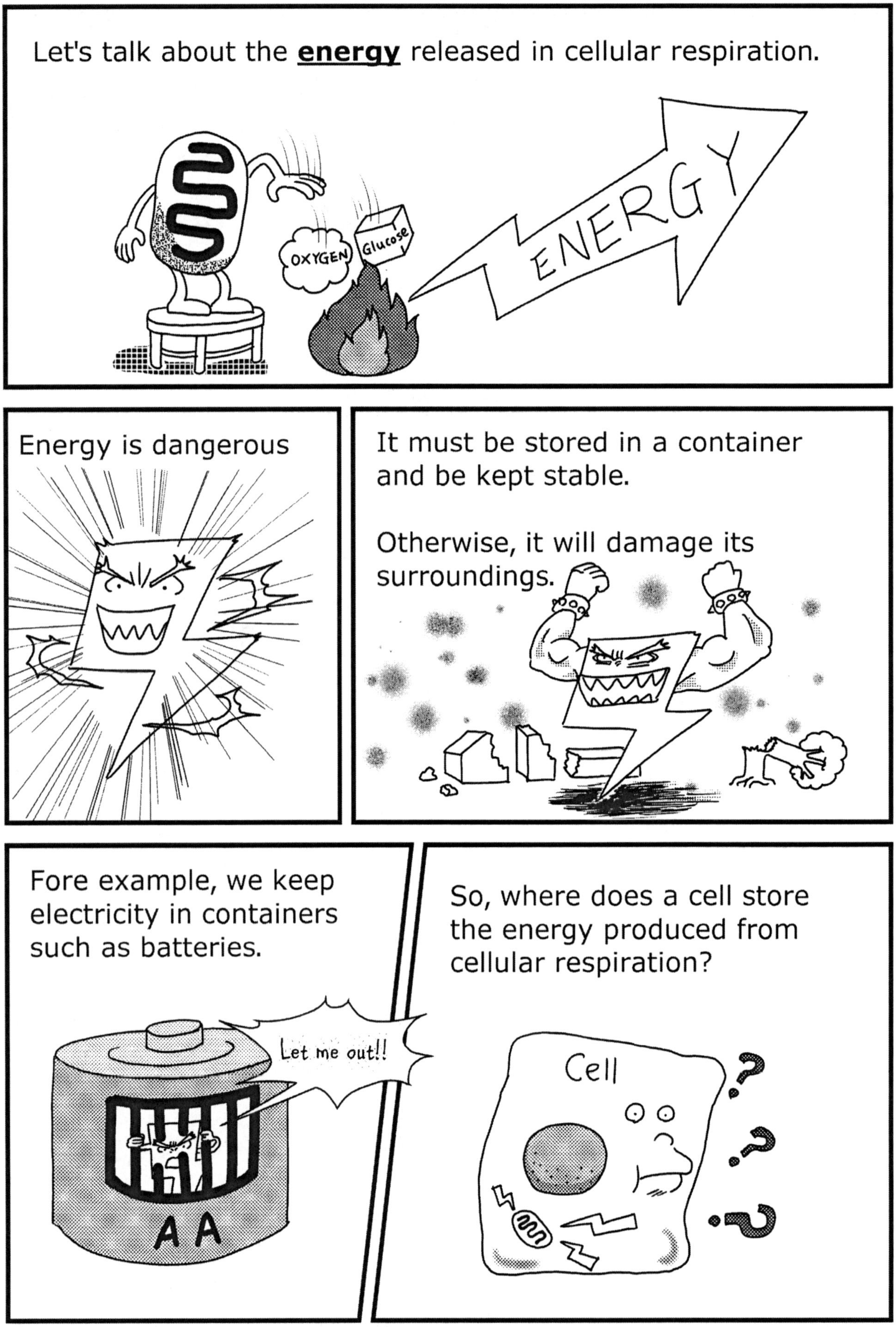

Let's talk about the **energy** released in cellular respiration.

Energy is dangerous

It must be stored in a container and be kept stable.

Otherwise, it will damage its surroundings.

Fore example, we keep electricity in containers such as batteries.

Let me out!!

So, where does a cell store the energy produced from cellular respiration?

page 92

Unit 2: Chapter 11

The energy released by mitochondria is *stored* in molecules called **ADP**, which, when charged with energy, turns into **ATP**.

Here is how ADP/ATP works

You start with ADP
(ADP = **A**denosine **D**i-**P**hosphate)

"Di" means "2"

You can view ADP as the "empty battery."

Energy is added

When you add energy to ADP, it will turn into ATP
(ATP = **A**denosine **T**ri-**P**hosphate)

ATP is like a fully charged battery.

Unit 2: Chapter 11

ADP has only **2** phosphates.
ATP has **3** phosphates and is rich in energy.

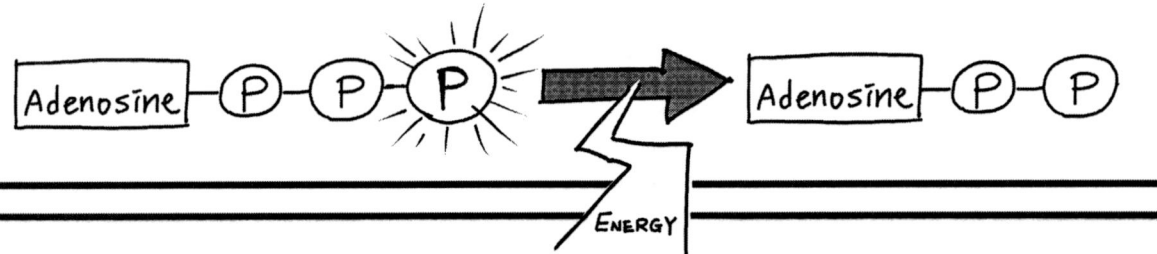

Of course, when ATP's energy is used up, it will down-grade itself into ADP.

The released energy is used to do useful work.

By the way, 1 round of cellular respiration can charge 36 ADPs into 36 ATPs!

36 ATP

I am good at this!

A Different Topic: FERMENTATION

Cellular Respiration is not the only process that makes ATPs.

Fermentation can also make ATPs. But it only makes 2 ATPs.

Compare Fermentation with Cellular Respiration...

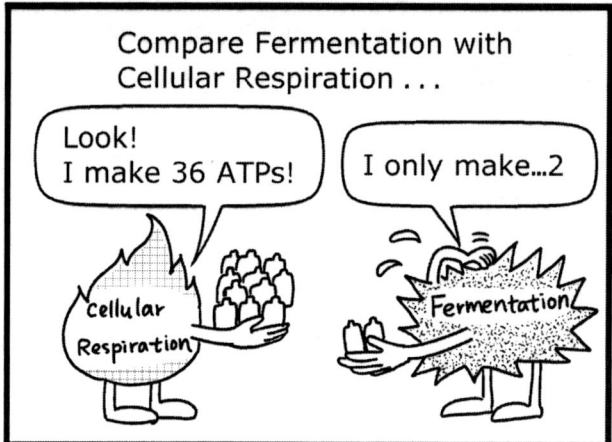

"Cheer up! Fermentation has its own unique advantage!"

Unlike cellular respiration, Fermentation needs **no oxygen**.

Fermentation also happens more quickly than cellular respiration.

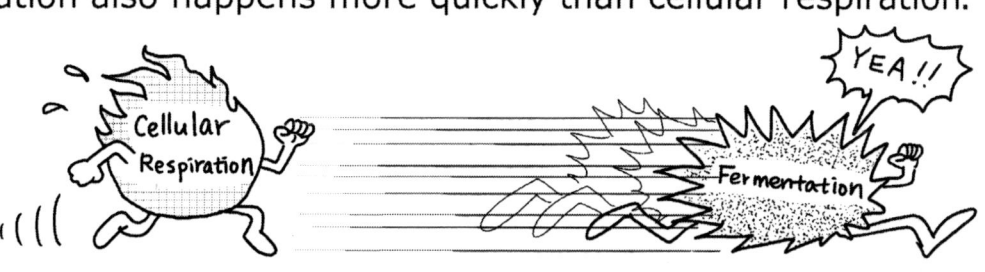

BOTH fermentation and cellular respiration produce **Carbon Dioxide** as waste.

page 95

Unit 2: Chapter 11

For that reason, breadmakers use fermentation to create carbon dioxide air bubbles in the bread to make it fluffy.

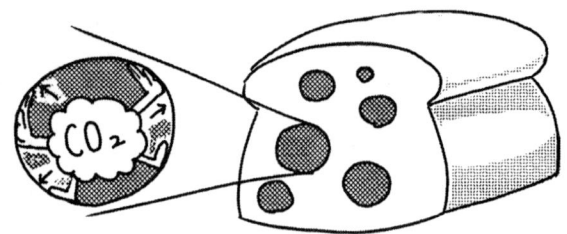

Fermentation also produces acid, which humans use to make the following sour-tasting food:

Certain fermentation do not make acid, but they make alcohol.

Hmm ~ exquisite wine.

Animals normally get their energy from cellular respiration. However, during emergencies, animals will resort to fermentation.

Relaxed: cellular respiration

Intense/**Emergency:** Fermentation

Fermentation can happen very *quickly.* This allows the organism to deal with sudden emergencies.

But the acid produced by fermentation gives us an uncomfortable burning sensation in our muscles.

Photosynthesis

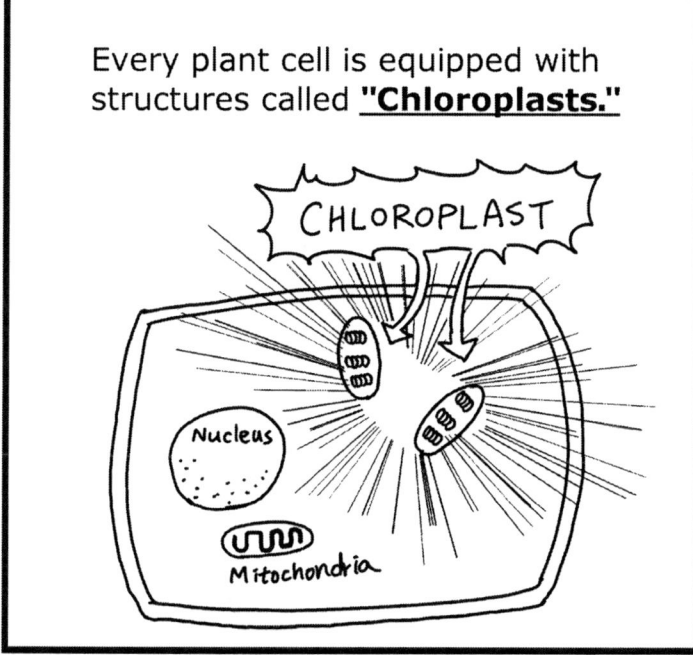

This is how the light energy is used:

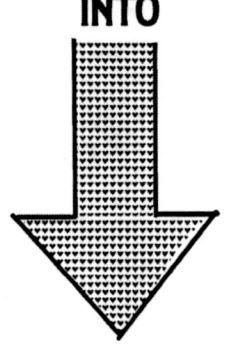

This process is called **"Photosynthesis"**.

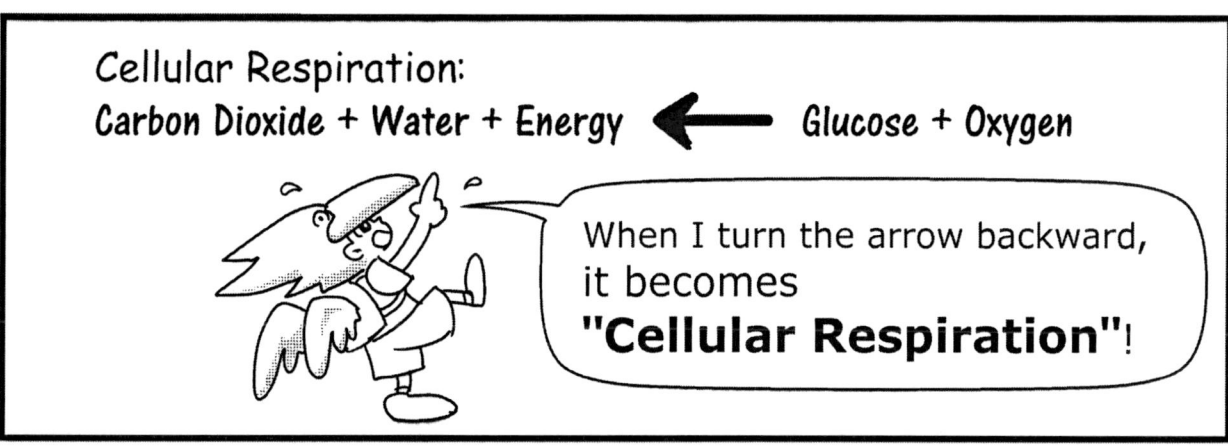

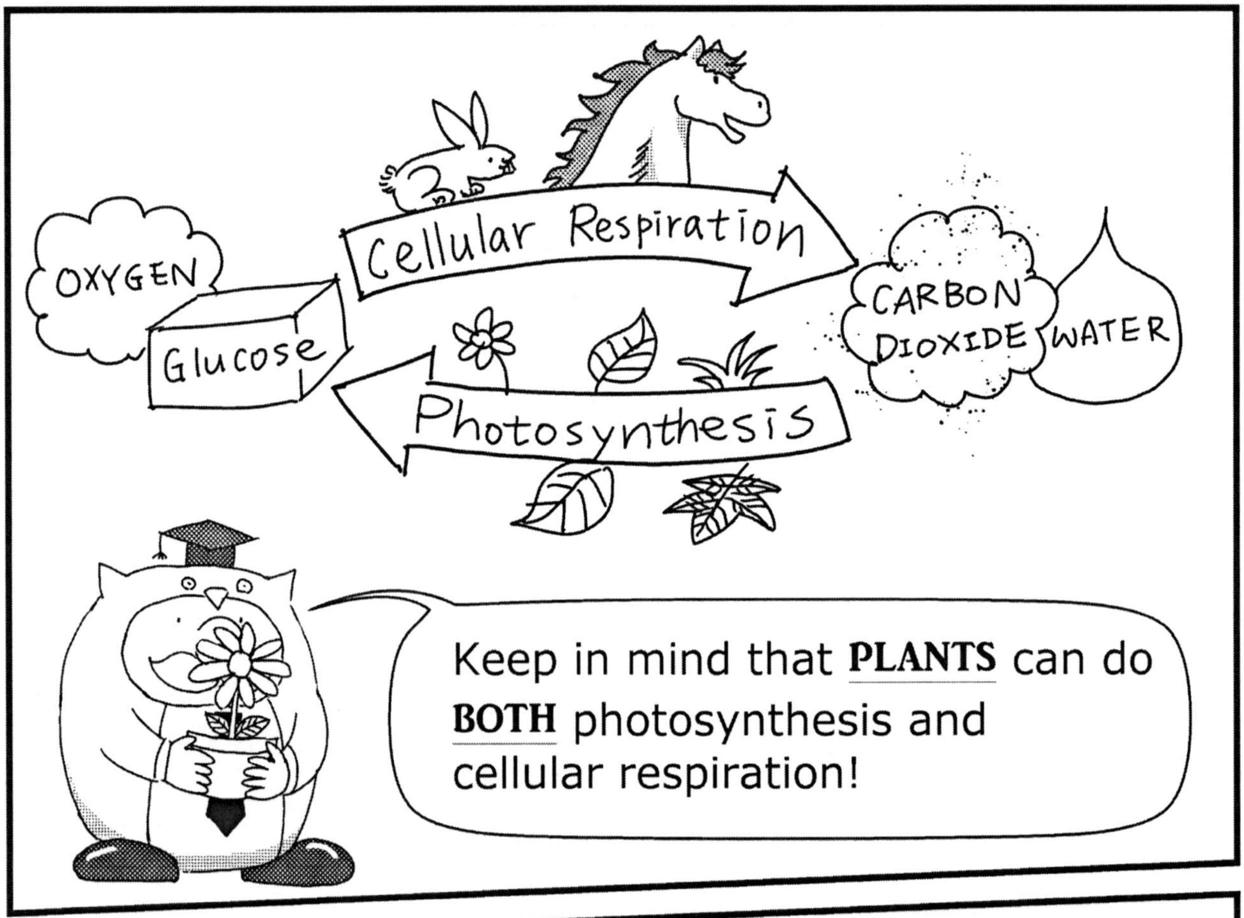

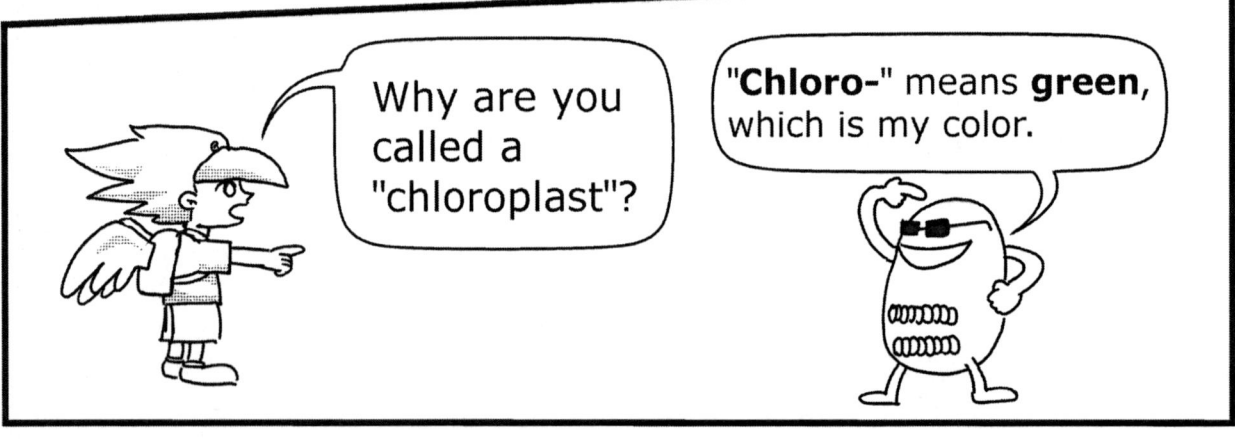

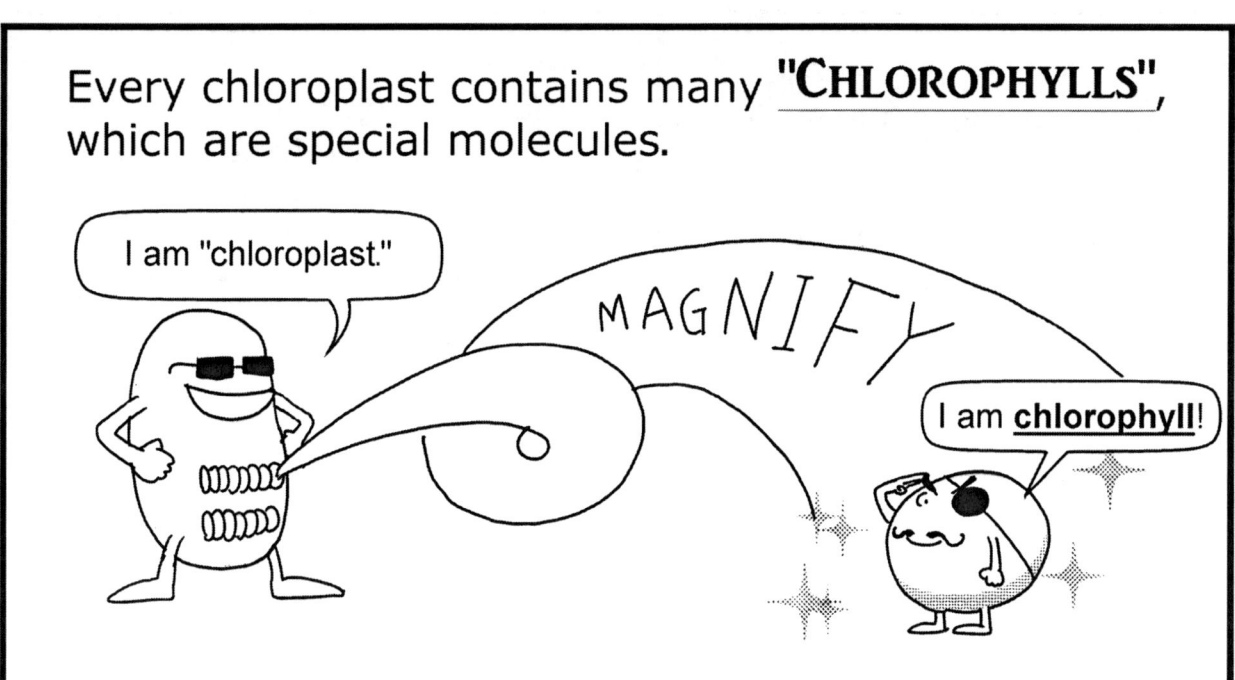

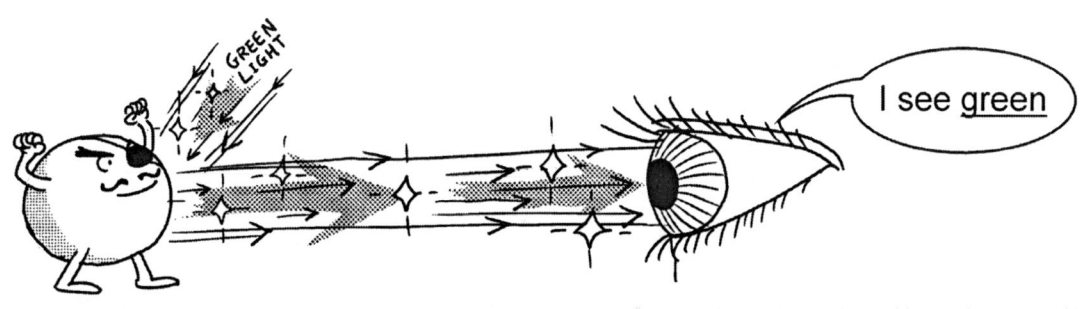

Pigments are the reason why most things have color.

Pigments are everywhere

So, the color of light chlorophylls reflect is ...

Yep!

Green light!

Each chloroplast is filled with many chlorophylls.

A chloroplast looks green because of its chlorophylls.

Most plant cells contain chloroplasts.
(especially leaf cells)

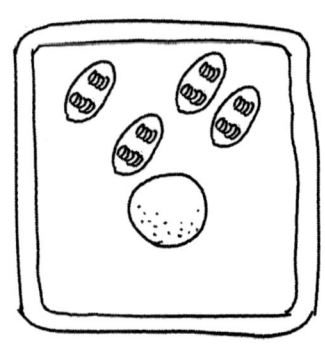

This also causes plant cells to look green.

Each leaf contains many plant cells. 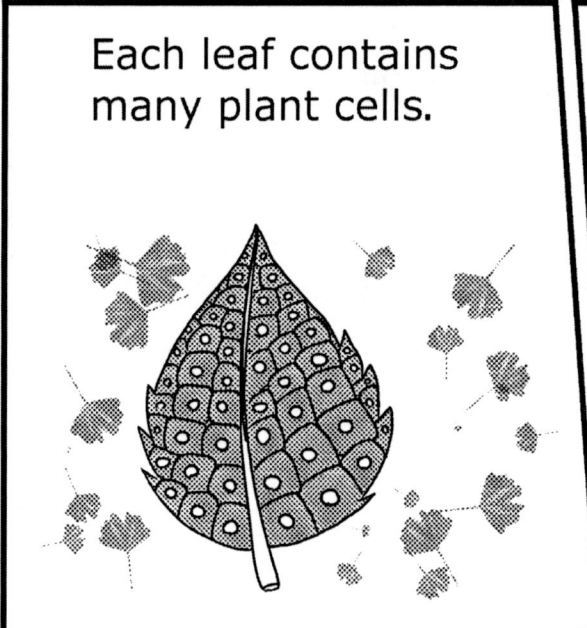	Therefore, leaves look green! 

Fun Fact:

Some organisms, such as algae, have loosely scattered chlorophylls and have no chloroplasts.

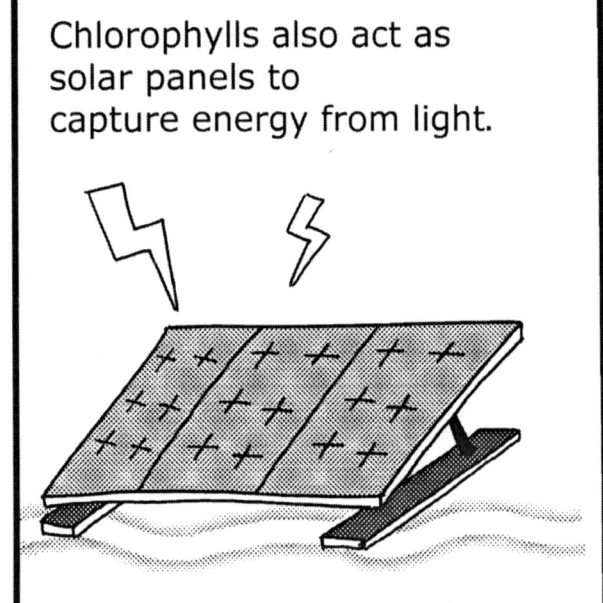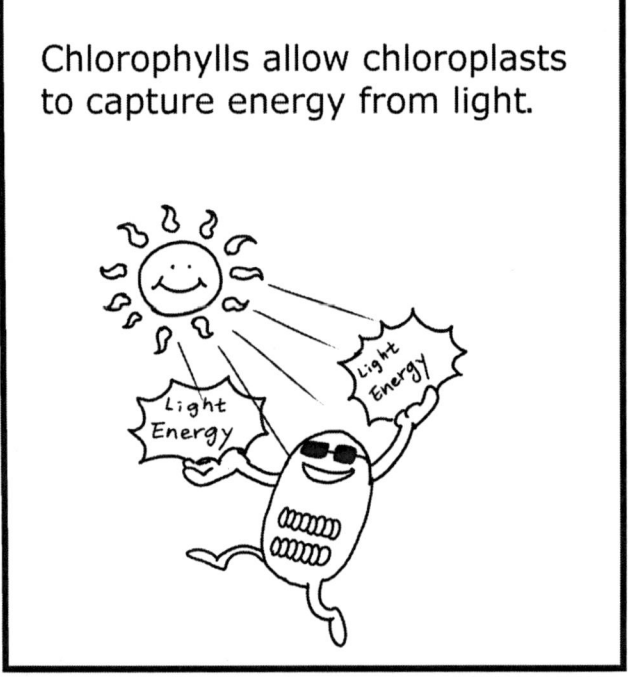

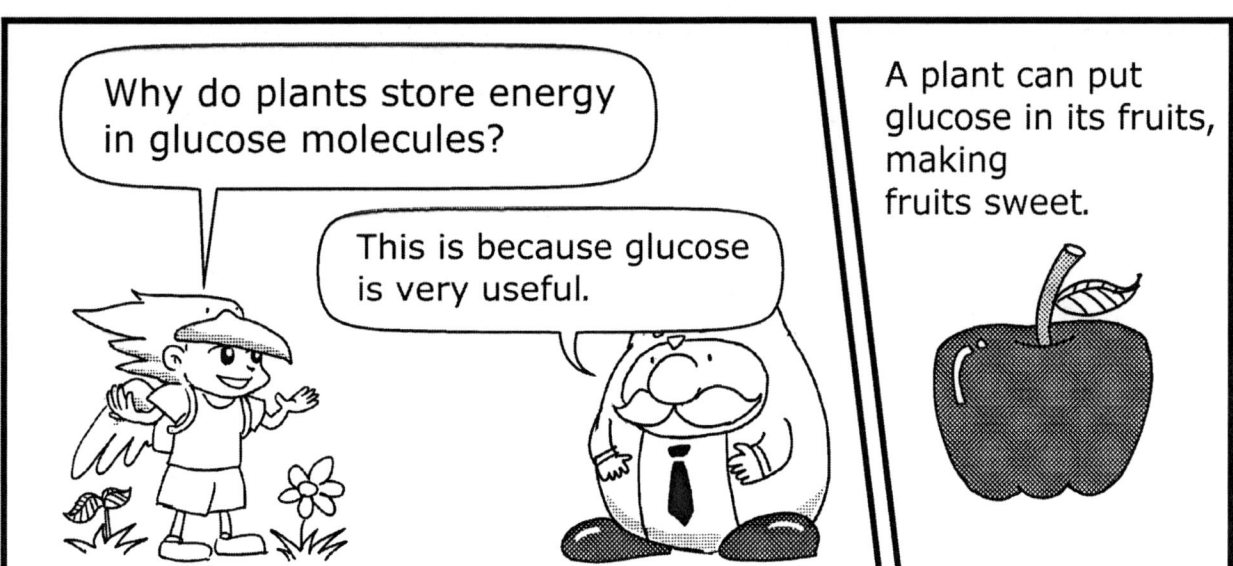

"Why do plants store energy in glucose molecules?"

"This is because glucose is very useful."

A plant can put glucose in its fruits, making fruits sweet.

So animals would want to eat it.

... then help disperse the seeds through feces (poop).

Plants can also store glucose as a backup food source for future use.

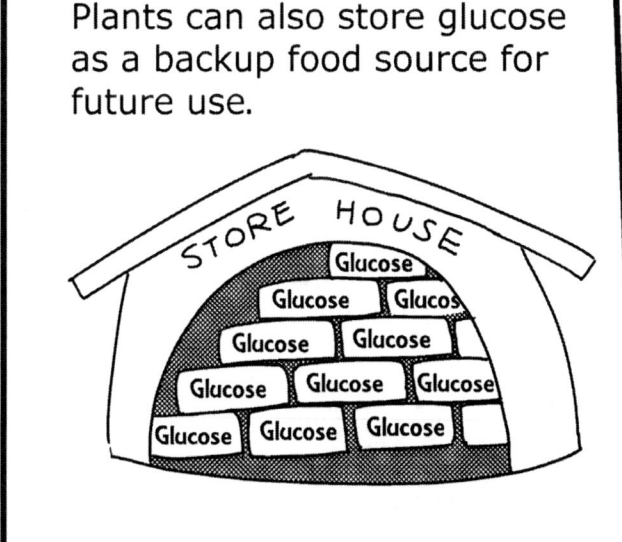

Or use glucose to make cellulose (fiber) and grow bigger.

"And many other uses!"

Chapter 12: Chloroplasts and Photosynthesis

Science Standard Bio/LS . 1 . f :
Students know usable energy is captured from sunlight by chloroplasts and is stored through the synthesis of sugar from carbon dioxide.

NEW VOCABS

* **Calvin Cycle:** See "Dark reaction."

* **Dark reaction:** Also called Calvin Cycle. It is the 2nd part of photosynthesis. It uses the energy captured from the light reaction to assemble carbon dioxide and water into glucose. It takes place in the stroma of a chloroplast.

* **Granum:** (plural: grana) A column of neatly stacked thylakoid is called a granum. Light reaction takes place here.

* **Guard cell:** Guard cells control the opening and closing of a stoma. 2 guard cells are assigned to each stoma.

* **Light Reaction:** The 1st part of photosynthesis. It got its name because it captures light energy in the process.

* **Stoma:** (plural: stomata) Stoma is the pore located on plants. It allows a plant to take in useful gasses such as carbon dioxide.

* **Stroma:** The fluid-filled space in a chloroplast. It contains no chlorophyll. Dark reaction (Calvin Cycle) happens in this space. (Hint: Do not confuse this word with stoma or stomata. They are completely unrelated)

* **Thylakoid:** The small disks located inside of a chloroplast. Each thylakoid contains a lot of chlorophyll.

* **Vascular System:** A network of pipes / canals made of cellulose. It allows water (and other substances) to move throughout a plant.

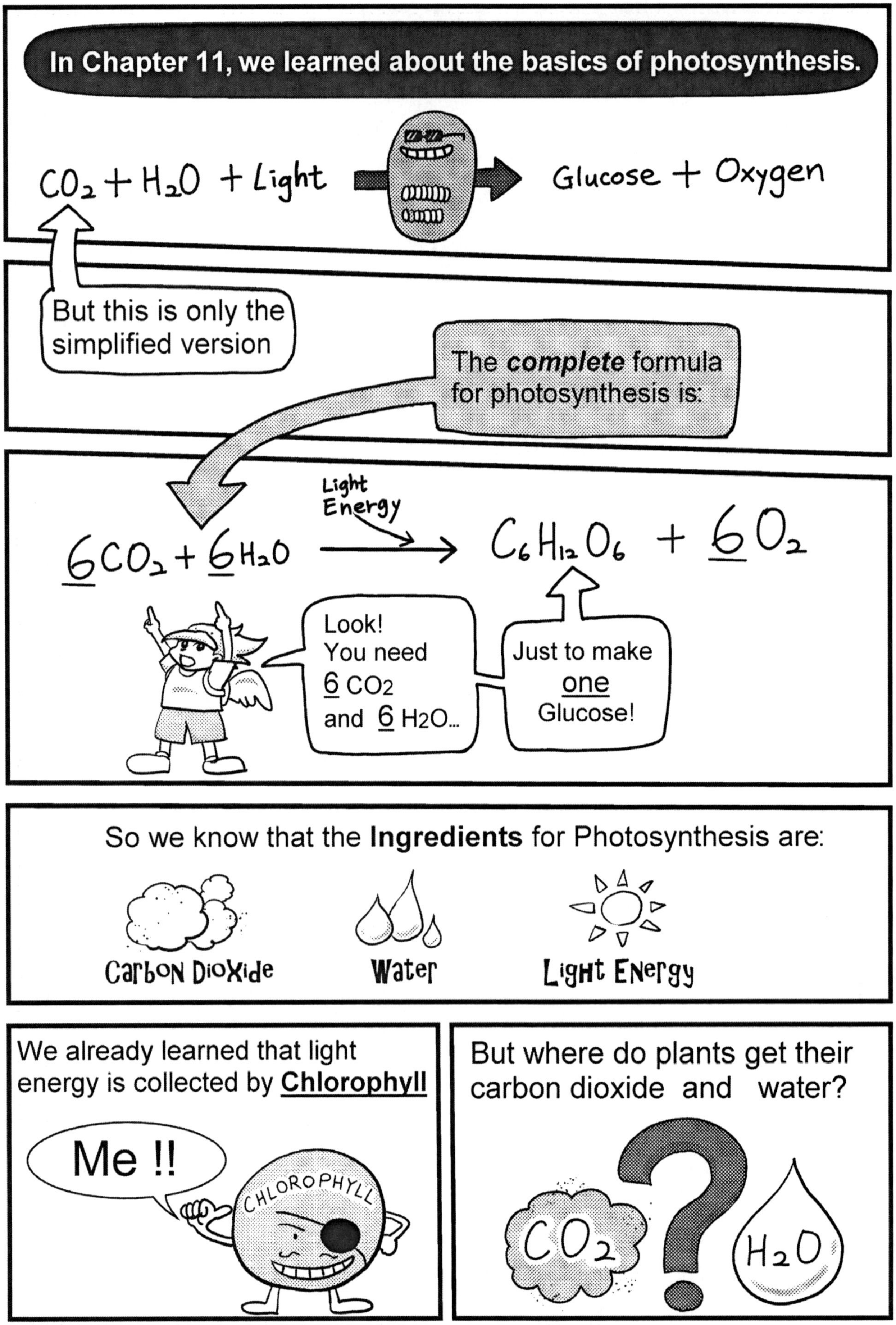

In Chapter 11, we learned about the basics of photosynthesis.

$$CO_2 + H_2O + \text{Light} \rightarrow \text{Glucose} + \text{Oxygen}$$

But this is only the simplified version

The **complete** formula for photosynthesis is:

$$6CO_2 + 6H_2O \xrightarrow{\text{Light Energy}} C_6H_{12}O_6 + 6O_2$$

Look! You need <u>6</u> CO_2 and <u>6</u> H_2O... Just to make <u>one</u> Glucose!

So we know that the **Ingredients** for Photosynthesis are:

Carbon Dioxide — Water — Light Energy

We already learned that light energy is collected by **Chlorophyll**

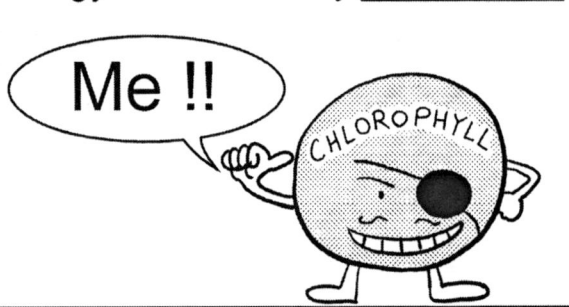

Me!!

But where do plants get their carbon dioxide and water?

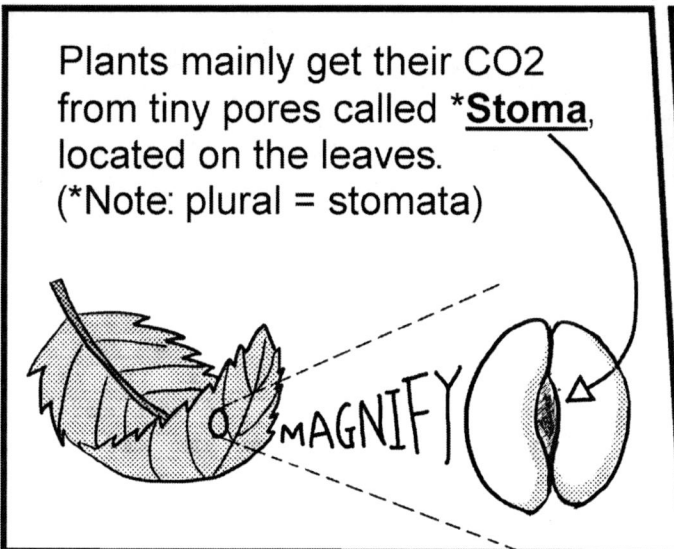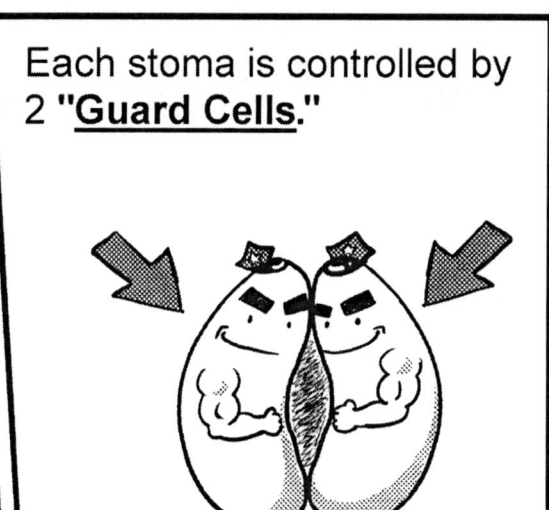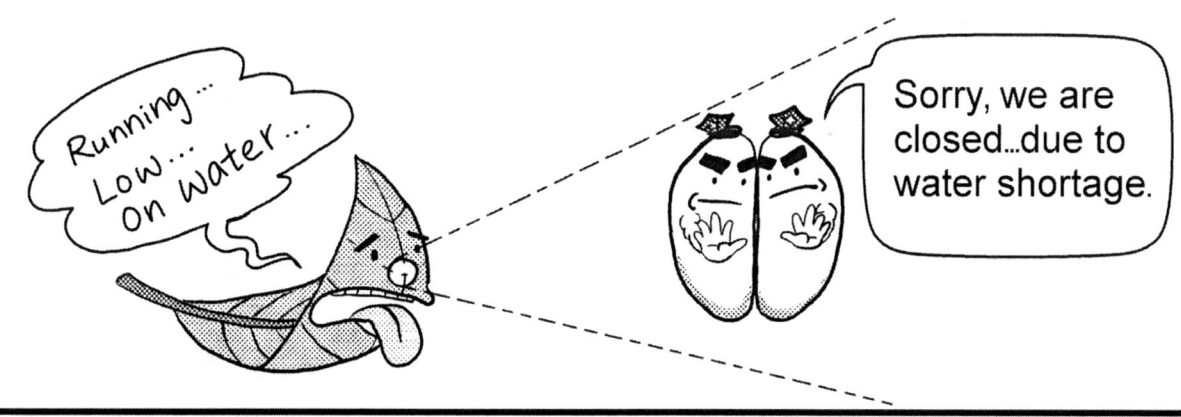

Speaking of water...
Plants get the water for photosynthesis from their roots!

Step 1: The root absorbs water from the soil.

Step 2: Water is transported up and throughout the plant.

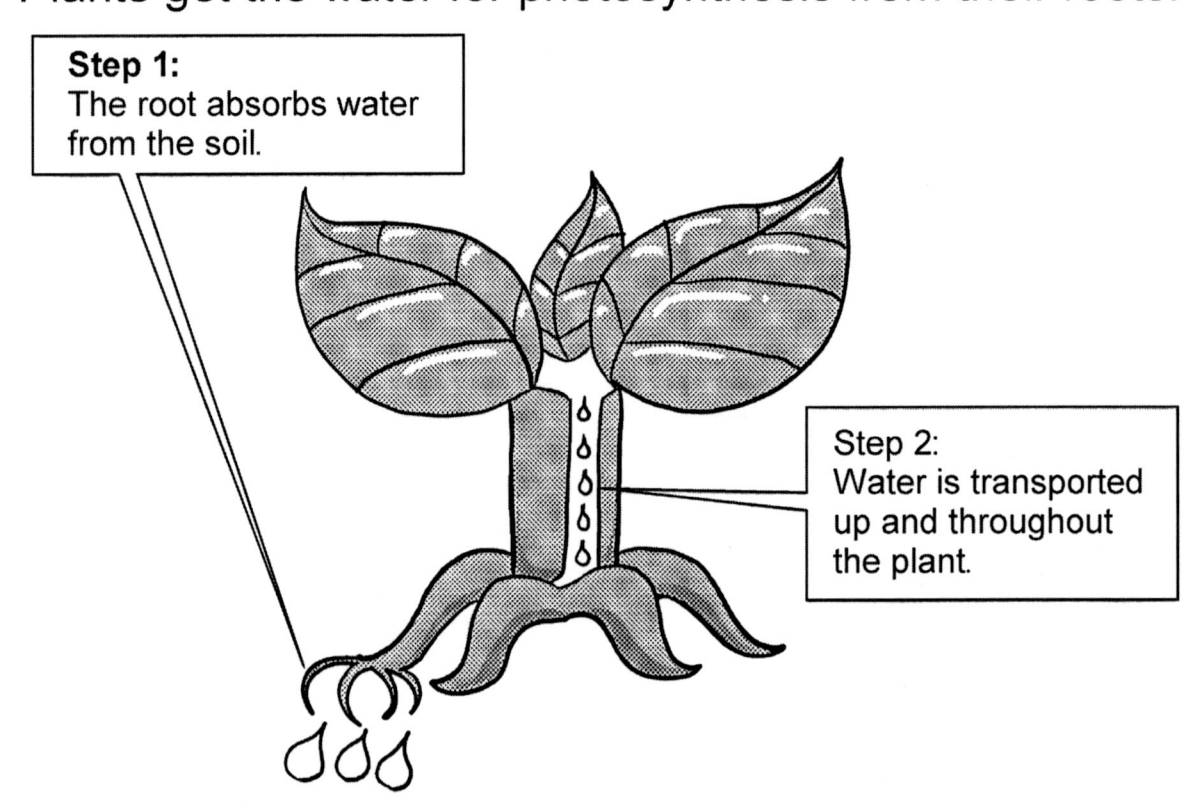

 **How is water transported throughout a plant?**

It is done by the plant's **"Vascular System"**, which is a network of pipes and canals made of *cellulose*.

(* Cellulose is a type of complex carbohydrate. There is more information on cellulose in Chapter 15)

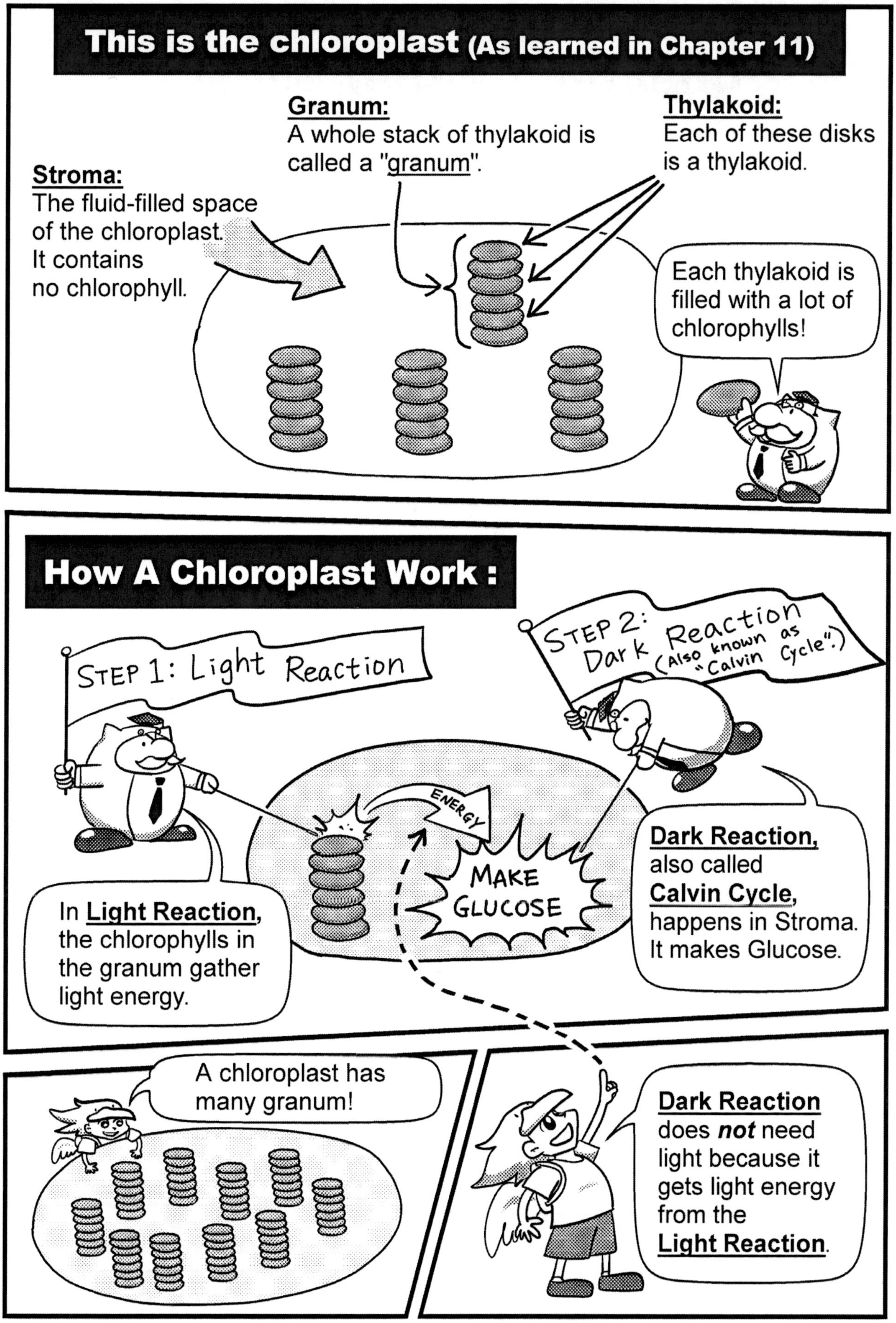

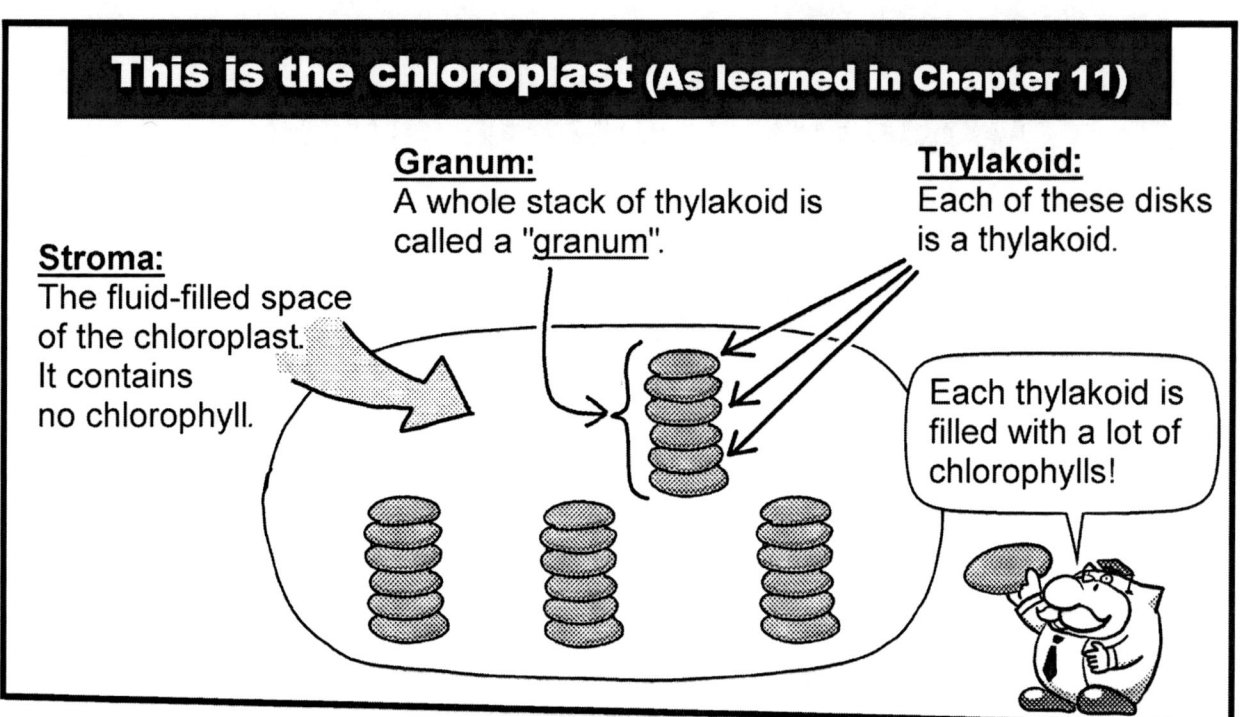

Chapter 13: Cell Division

Science Standard: 7.1.e
Students know cells divide to increase their numbers through a process of mitosis, which results in two daughter cells with identical sets of chromosomes.

NEW VOCABS

- **Anaphase:** The 3rd mitotic phase. Sister chromatids are torn away from each other during this phase.
- **Cell Division:** Mitosis + Cytokinesis
- **Centrioles:** A strucure in a cell that controls spindle fibers.
- **Chromosomes:** DNAs are wrapped around histones (a type of protein) and tightly packaged into chromosomes.
- **Cytokinesis:** The 3rd part of a cell's life when it completely cuts itself into 2 new cells. This takes place right after Mitosis.
- **Daughter Cell:** The new cells produced by cell division.
- **Homologus chromosomes:** 2 of the same type of chromosomes.
- **Interphase:** The 1st part of a cell's life when a cell grows larger and eventually doubles its DNA.
- **Metaphase:** The 2nd mitotic phase. Chromosomes line up during this phase.
- **Mitosis:** The 2nd part of a cell's life when it tries to make 2 of itself.
- **Parent Cell:** The original cell that is trying to make 2 of itself by cell division.
- **Prophase:** The 1st mitotic (adjective of mitosis) phase. The nuclear envelope dissolves during this phase.
- **Sister Chromatids:** 2 duplicated copies of a single chromosome.
- **Spindle Fibers:** A special fiber that is used to move chromosomes around.
- **Telophase:** The 4th mitotic phase. Two new nuclear envelopes reappear during this phase.

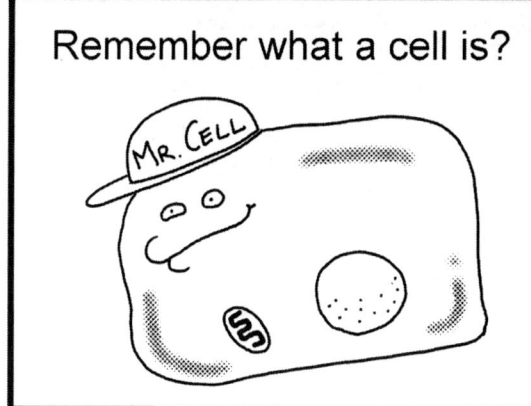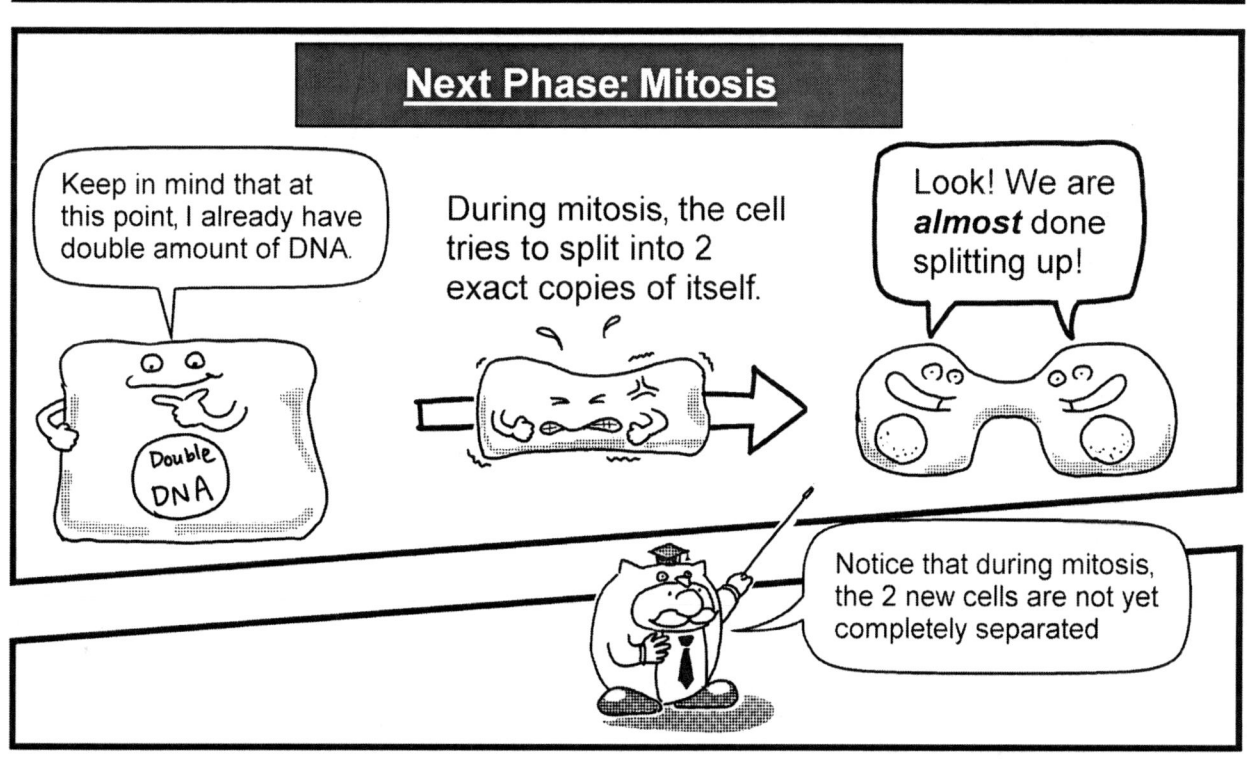

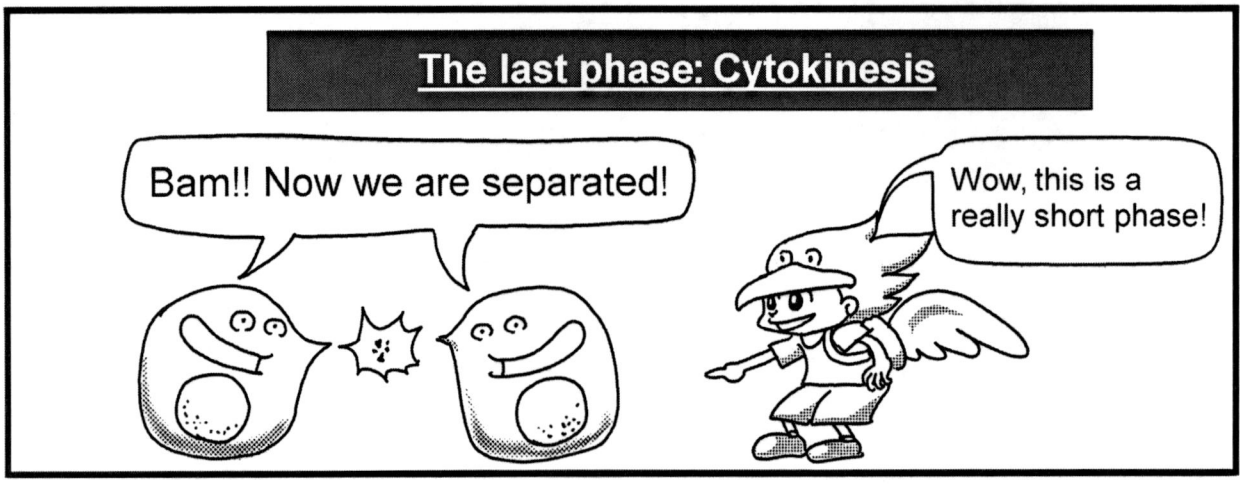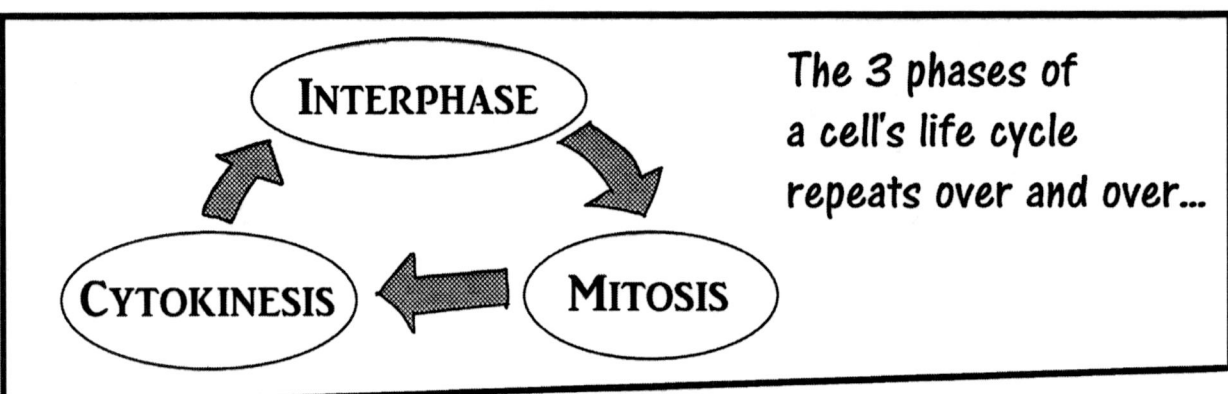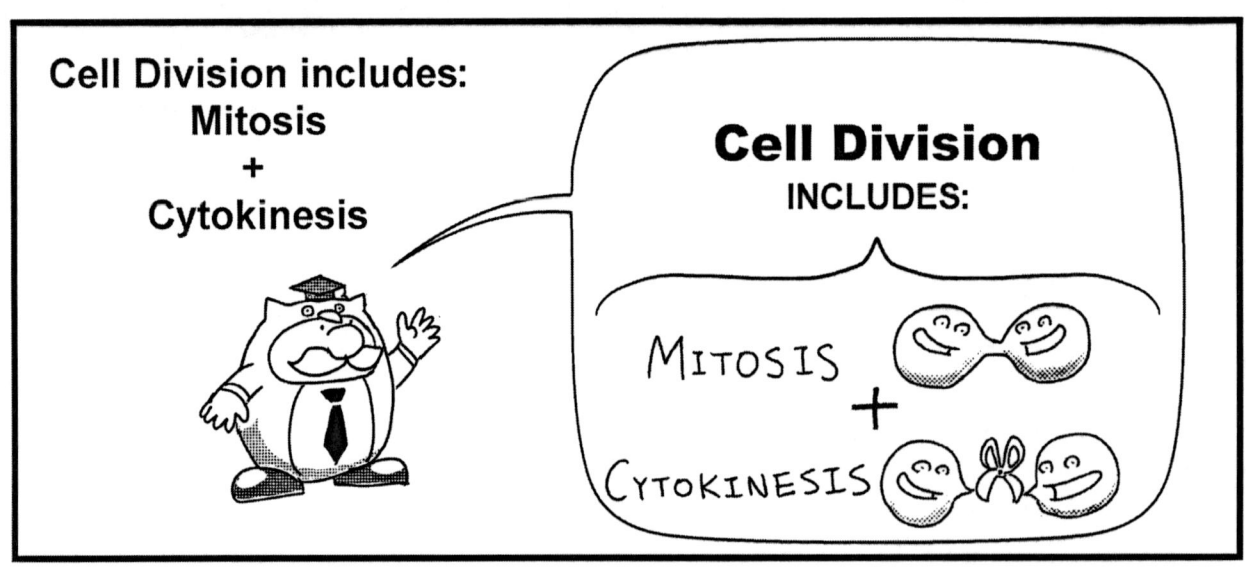

Mitosis is like re-structuring 1 house into 2 houses

But such big projects have a problem.

Problem:
How does a cell protect its fragile DNA during mitosis?

Solution:
At the start of Mitosis, DNA is packed into little packages for **protection** and ease of transport

We call these DNA-packages "Chromosomes"

DNA is wrapped around special proteins called **"Histones."**

...And condensed into **Chromosomes**.

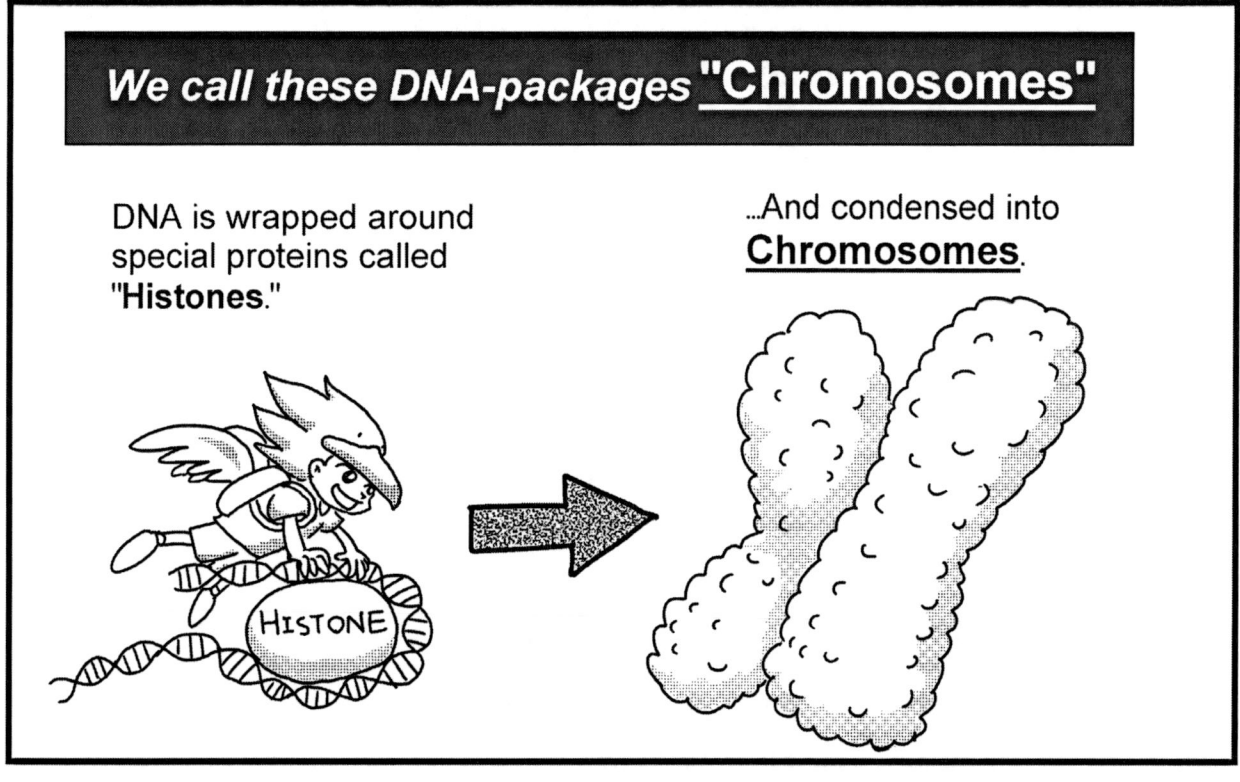

page 115

Unit 2: Chapter 13

MORE ON CHROMOSOMES

You (human) have 23 different types of chromosomes. Each is different in size and functions.

Since each of the 23 chromosome types comes in **pairs**...  → You actually have a total of **46** pieces of chromosomes.

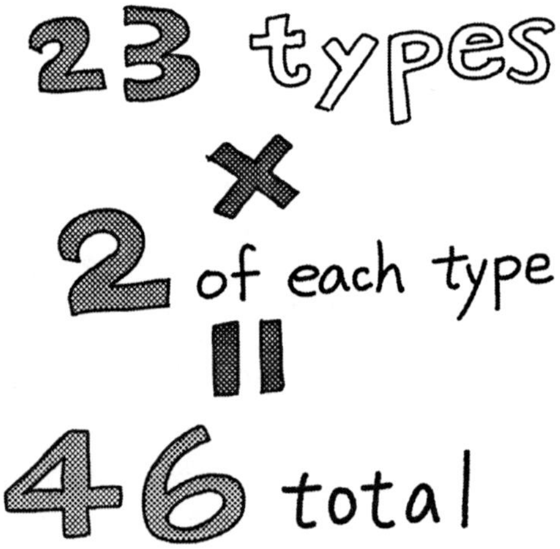

"2 is better than 1!"

23 types × 2 of each type = 46 total

FUN FACT:

Dogs have **78** chromosomes

Unit 2: Chapter 13

DNA during a Cell's Life:

Interphase → Mitosis → Cytokinesis

At the end of Interphase:
a cell's DNA is doubled.

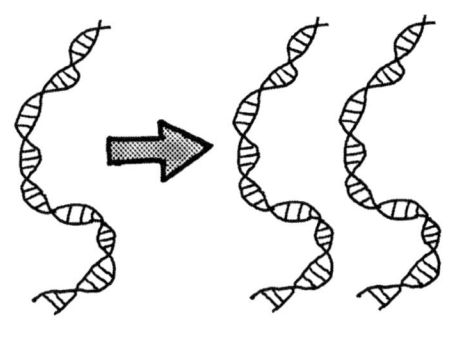

Therefore, when it is time for Mitosis, every chromosome has an *exact copy* of itself.

"I'm seeing doubles!" "Me, too!"

A Close Look at Interphase:
The "**Parent Cell**" (the original cell) doubles its DNA.

Mitosis and Cytokinesis split 1 parent cell into 2 **daughter cells.**

Parent Cell: The original cell.
Daughter Cells: The new cells from cell division.

Unit 2: Chapter 13

Vocabulary Time!

The 2 chromosomes of the same type are called: "**Homologus Chromosomes**."

Example: Every healthy human has a pair of chromosome #19.

"We are both chromosome #19, which controls hairs."

Although Homologus chromosomes are of the same type, they are **NOT** *exact copies* of each other

"Blond!" "Brunette!"

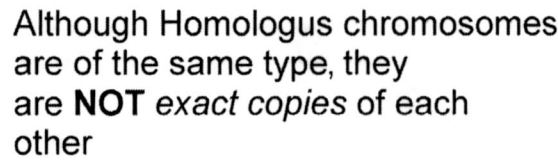

At the end of interphase, each chromosome duplicates into 2 "SISTER CHROMATIDS."

Sister Chromatids are (pretty much) identical copies.

Sister Chromatids → ← Homologus Chromosomes → Sister Chromatids

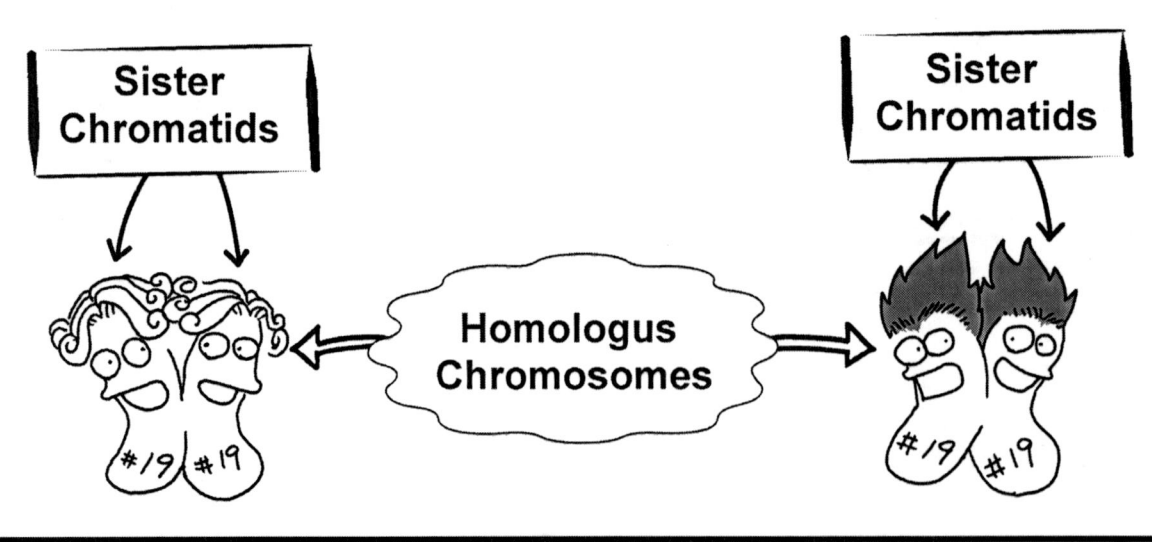

page 118

Unit 2: Chapter 13

Another Example: Human's chromosome #15 controls: eye color and skin tone . . . etc.

Chromosome 15 controls eye color.

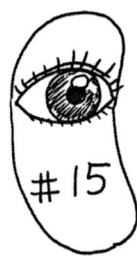

As always, there is a pair of homologus #15 chromosomes.

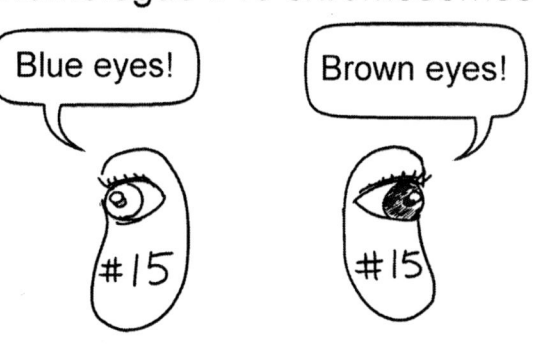

At the end of interphase, the cell's DNA is doubled, Each chromosome becomes 2 sister chromatids.

"Not just me. But every chromosome is doubled!"

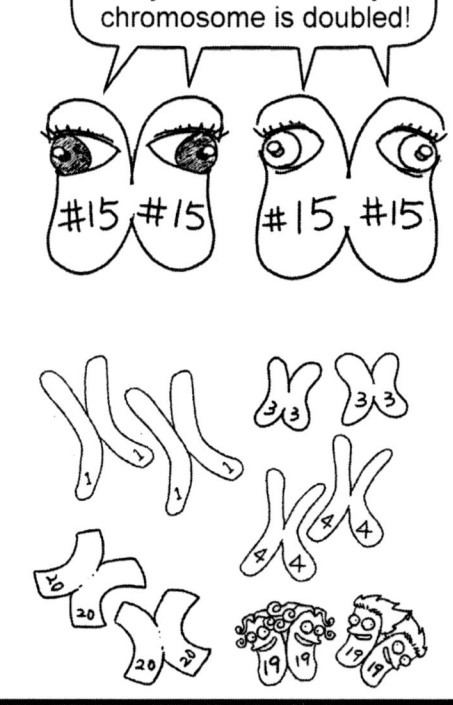

During mitosis, the sister chromatids separte and turn into new cells' chromosomes.

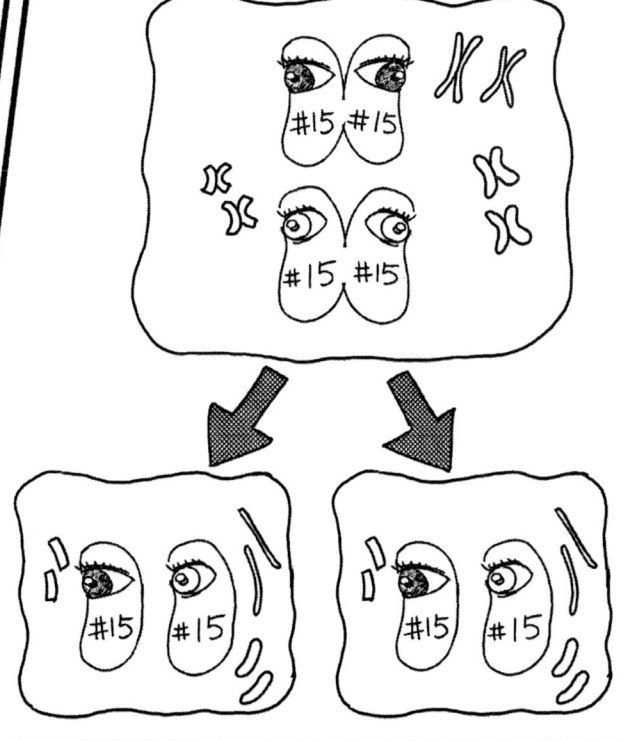

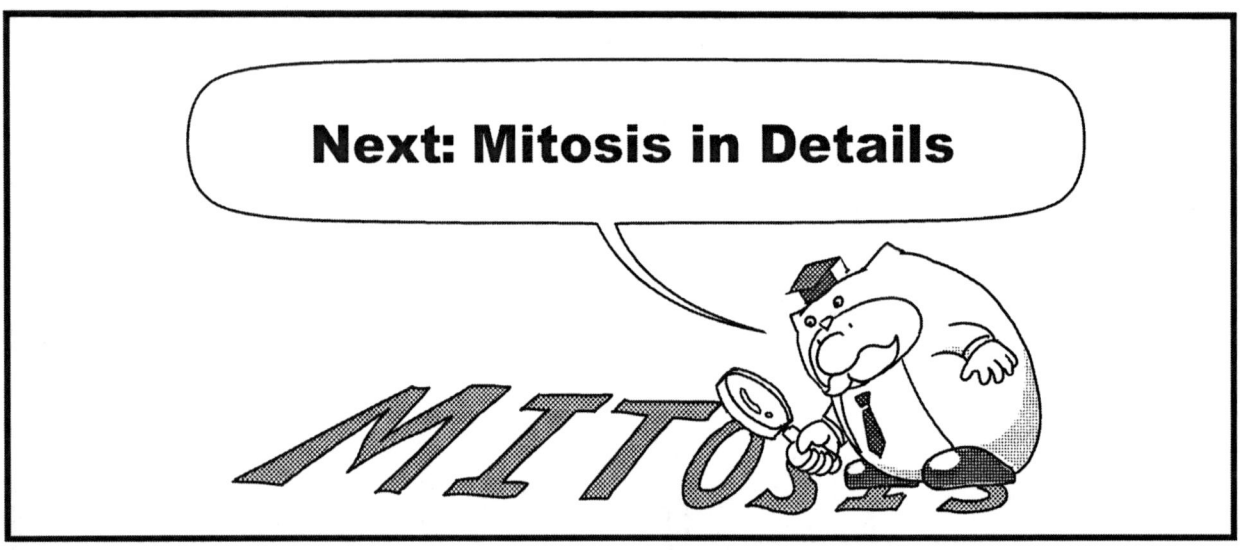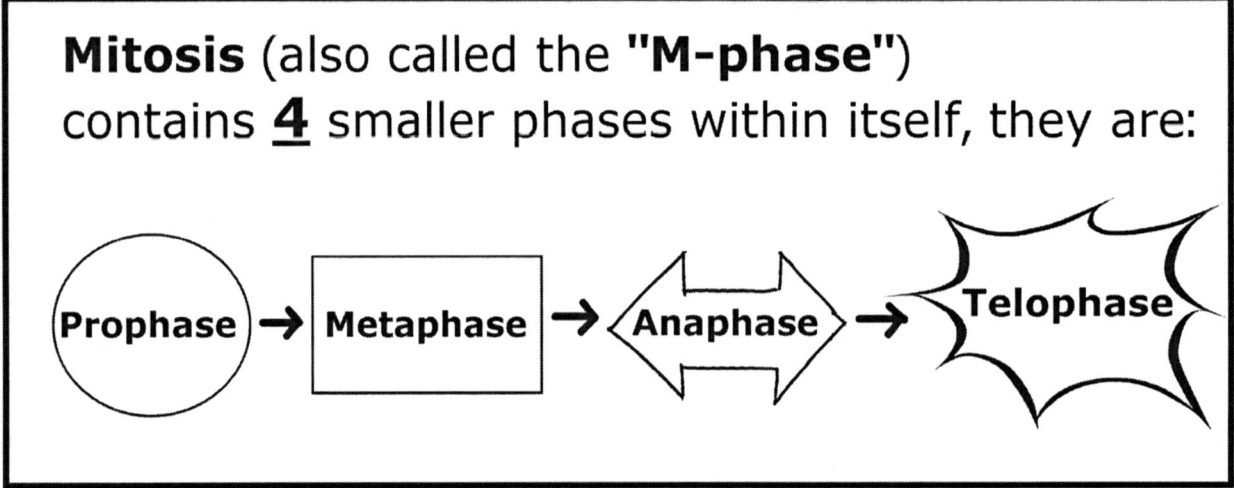

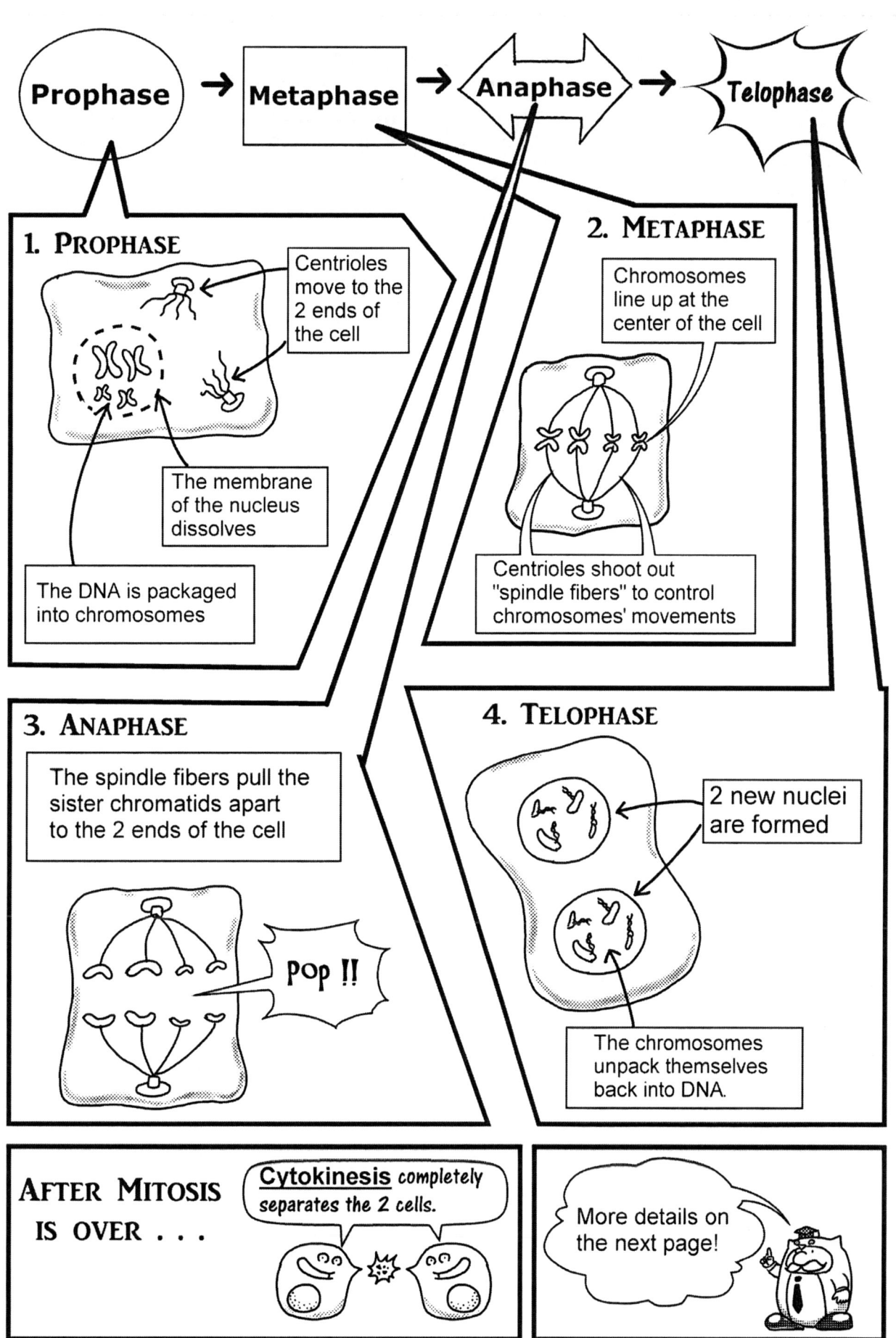

Prophase in Detail:

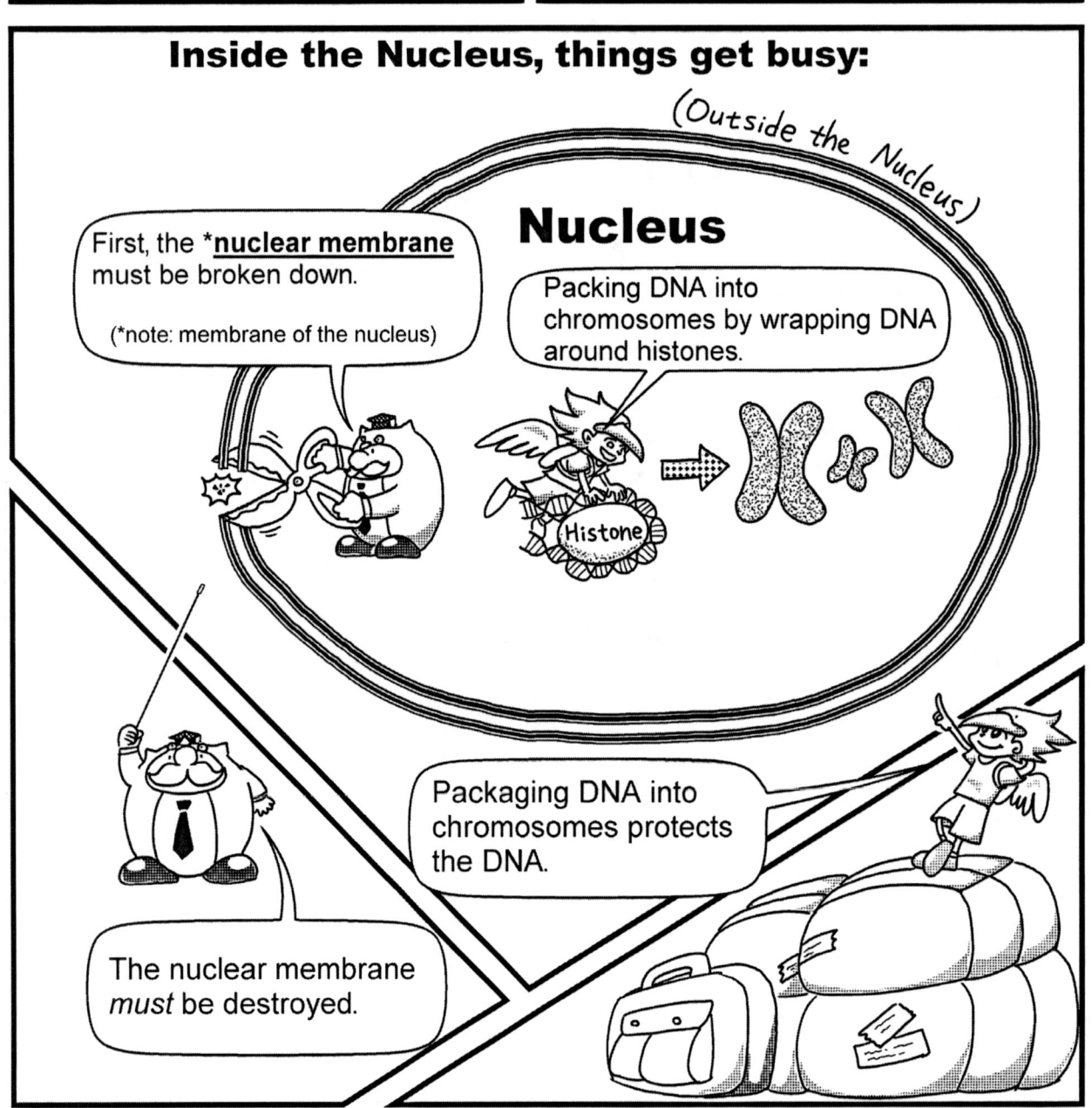

Metaphase in Detail:

Centrioles use spindle fibers to move chromosomes to the center of the cell.

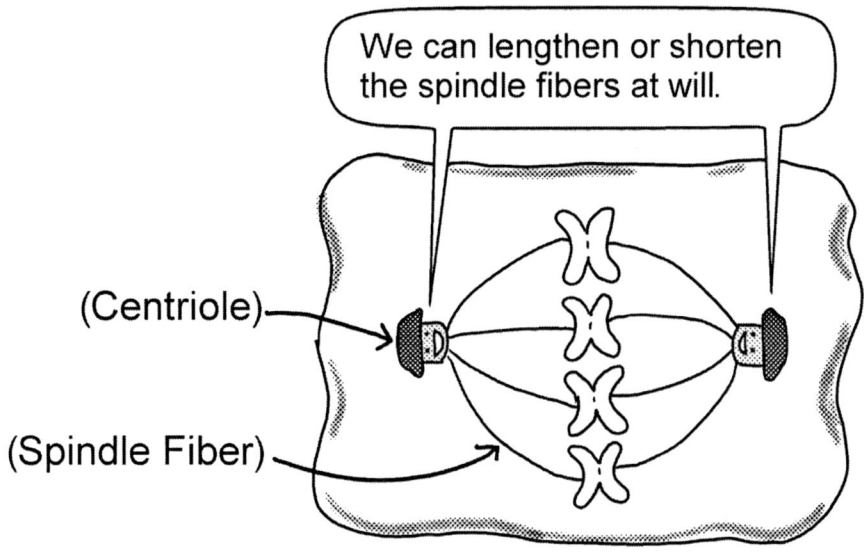

Anaphase in Detail:

Spindle fibers shorten, which pulls the sister chromatids apart and send them to the 2 opposite ends of the cell.

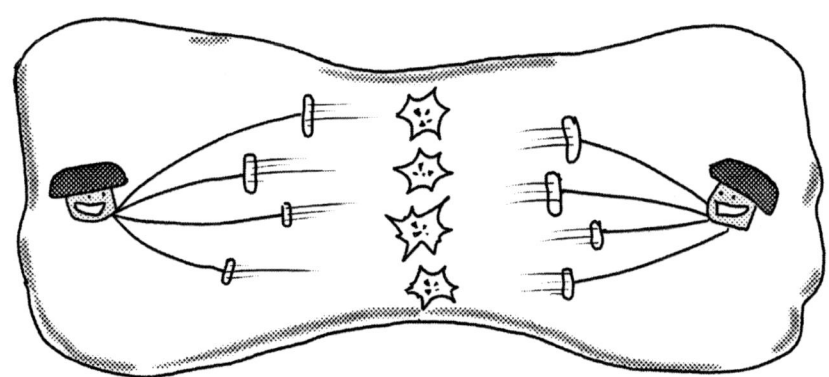

During this phase, the cell also assumes an elongated shape.

Telophase in Detail:

The previously-destroyed nuclear membrane reappears!

It wraps around the 2 groups of chromosomes and forms 2 new *nuclei.

(*nuclei=plural form of the word "nucleus")

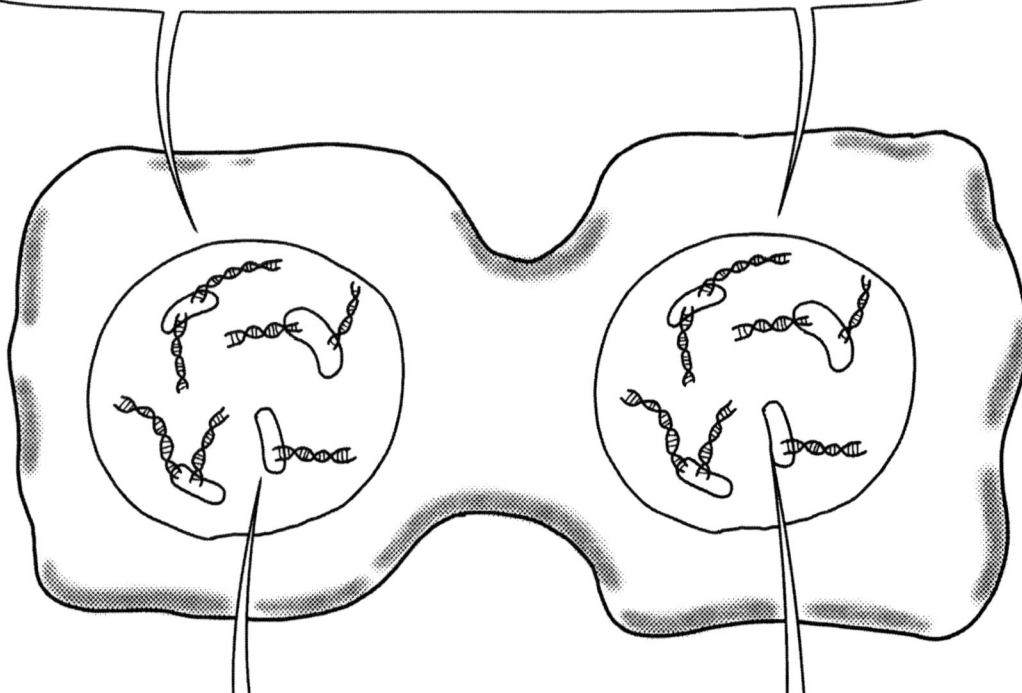

The chromosomes also start to unpack themselves into strands of DNA.

BY THE WAY . . .

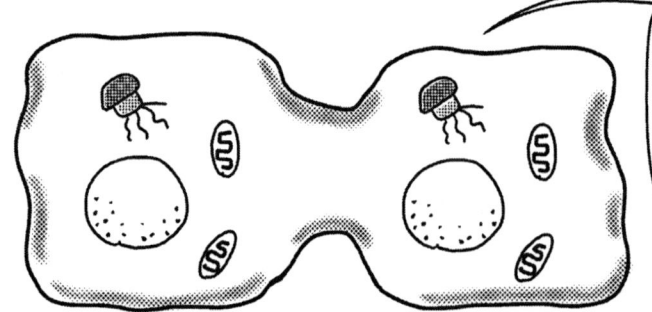

The other cell structures and nutrients/resources will also be fairly (sort of) distributed to both sides.

Chapter 14: Elements in Living Things

Science Standard: 8.6.b
Students know that living organisms are made of molecules consisting largely of carbon, hydrogen, nitrogen, oxygen, phosphorus, and sulfur.

NEW VOCABS

* **Atom:** The basic unit that makes up all matter. It is very small.

* **Bond** [between atoms]: The act of connecting 2 individual atoms together, making them stuck with each other.

* **Covalent Bond:** A bond that is established by 2 atoms sharing electron(s) with each other.

* **Electron:** It is part of an atom. It is a tiny particle that flies around the nucleus. It has a negative electrical charge.

* **Ionic Bond:** A bond that is established by 1 atom transferring 1 or more of its electron to another atom.

* **Ionic Compound:** A product that is formed by a group of atoms linked together via ionic bonds.

* **Matter:** Anything that has mass and takes up space.

* **Molecule:** A product that is formed by a group of atoms linked together via covalent bonds.

* **Nucleus** [of an atom]: The core of an atom. It accounts for most of an atom's mass. It has a positive electrical charge.

page 125

Unit 2: Chapter 14

"In nature, there are only 2 main categories of existences."

Matter & Energy

What is Matter?

"Matter" is anything that:
(1) has mass (you can weigh it)
(2) and takes up space.

Yup, all matter takes up space

Do you know?

Even "air" is matter.

"I have mass, and I DO take up space!"

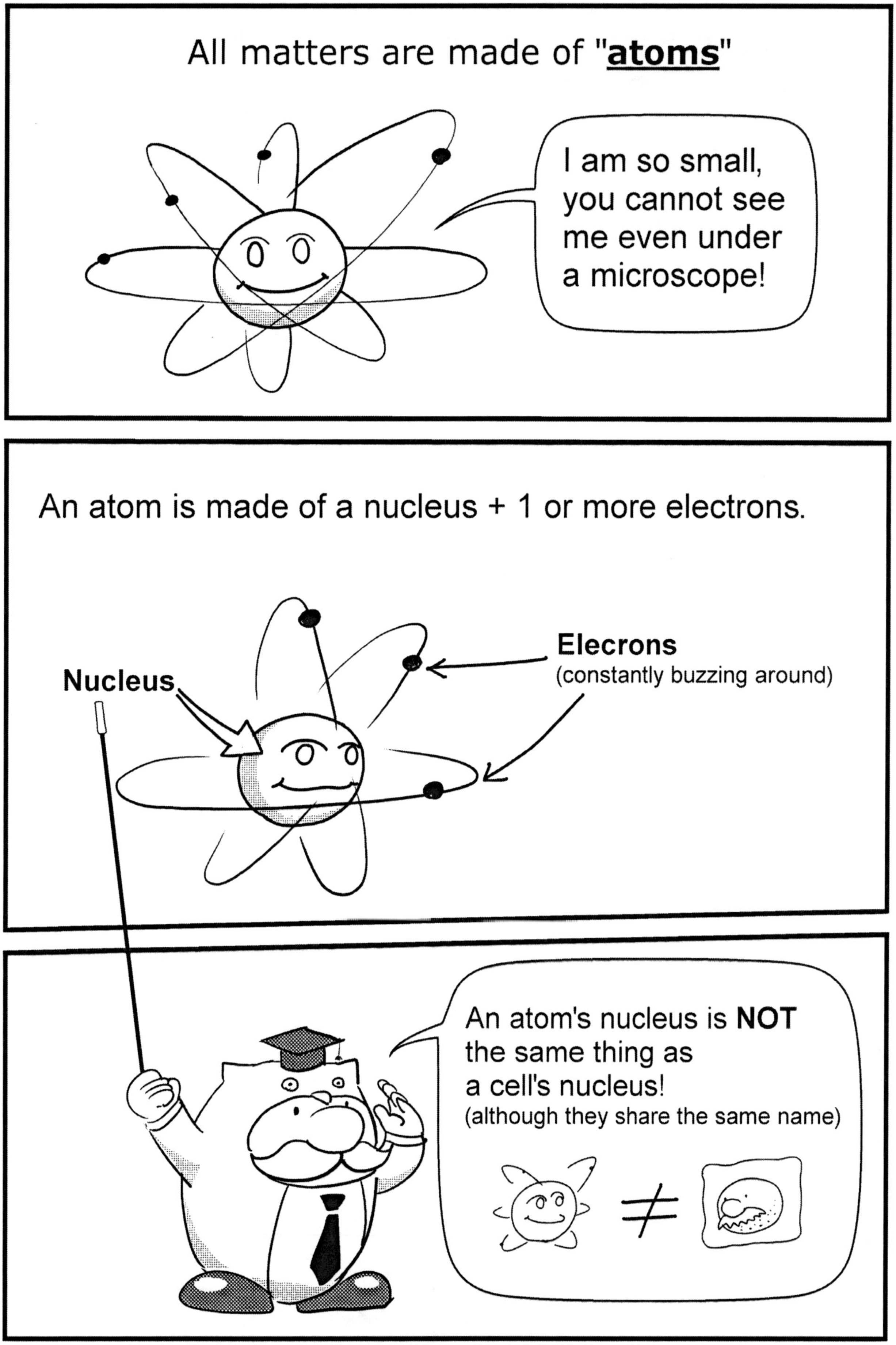

All matters are made of "**atoms**"

I am so small, you cannot see me even under a microscope!

An atom is made of a nucleus + 1 or more electrons.

Nucleus

Elecrons (constantly buzzing around)

An atom's nucleus is **NOT** the same thing as a cell's nucleus! (although they share the same name)

Sometimes, 2 atoms can *"get stuck"* with each other. We call this "**bonding**".

There are 2 types of Bonds:

(1) Ionic Bond:
   Formed by **"electron transfer"**

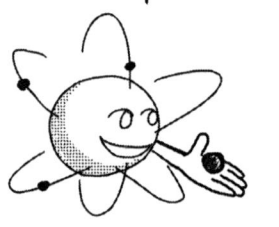 "Take an electron. It's my gift to you!"

 "Thank you!"

When 1 atom gives electron(s) to another atom, it creates an attractive force that keeps the 2 atoms bonded.

(2) Covalent Bond:
   Formed by **"electron sharing"**

"Sharing is caring~"

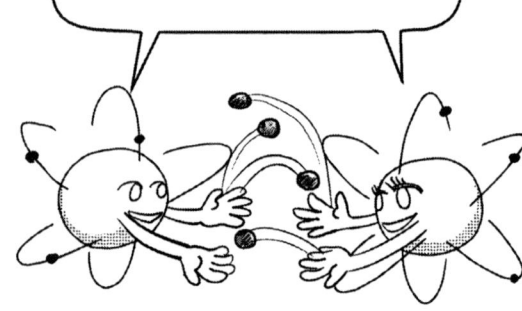

Both atoms lend out an equal number of electrons to create a **shared electron** reservoir. And both atoms act *as if* all the electrons in the reservoir were its **own.**

 "Cool ~"

When a group of atoms are linked by ionic bonds, we call the group of atoms an **"Ionic Compound."**

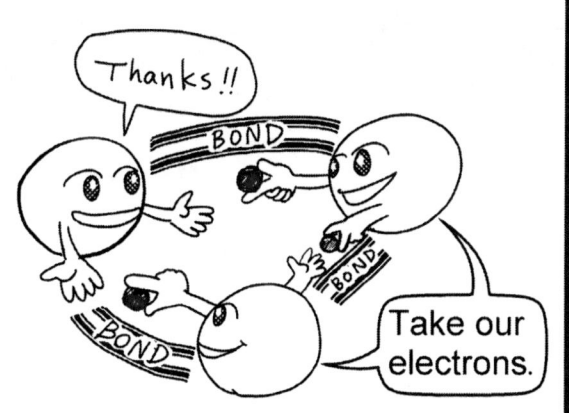

When a group of atoms are linked by covalent bonds, we call the group of atoms a **"Molecule."**

The human body is made of many different types of atoms.

Most of these atoms are found in **"molecules"**, which are atoms linked together by **Covalent Bonds.**

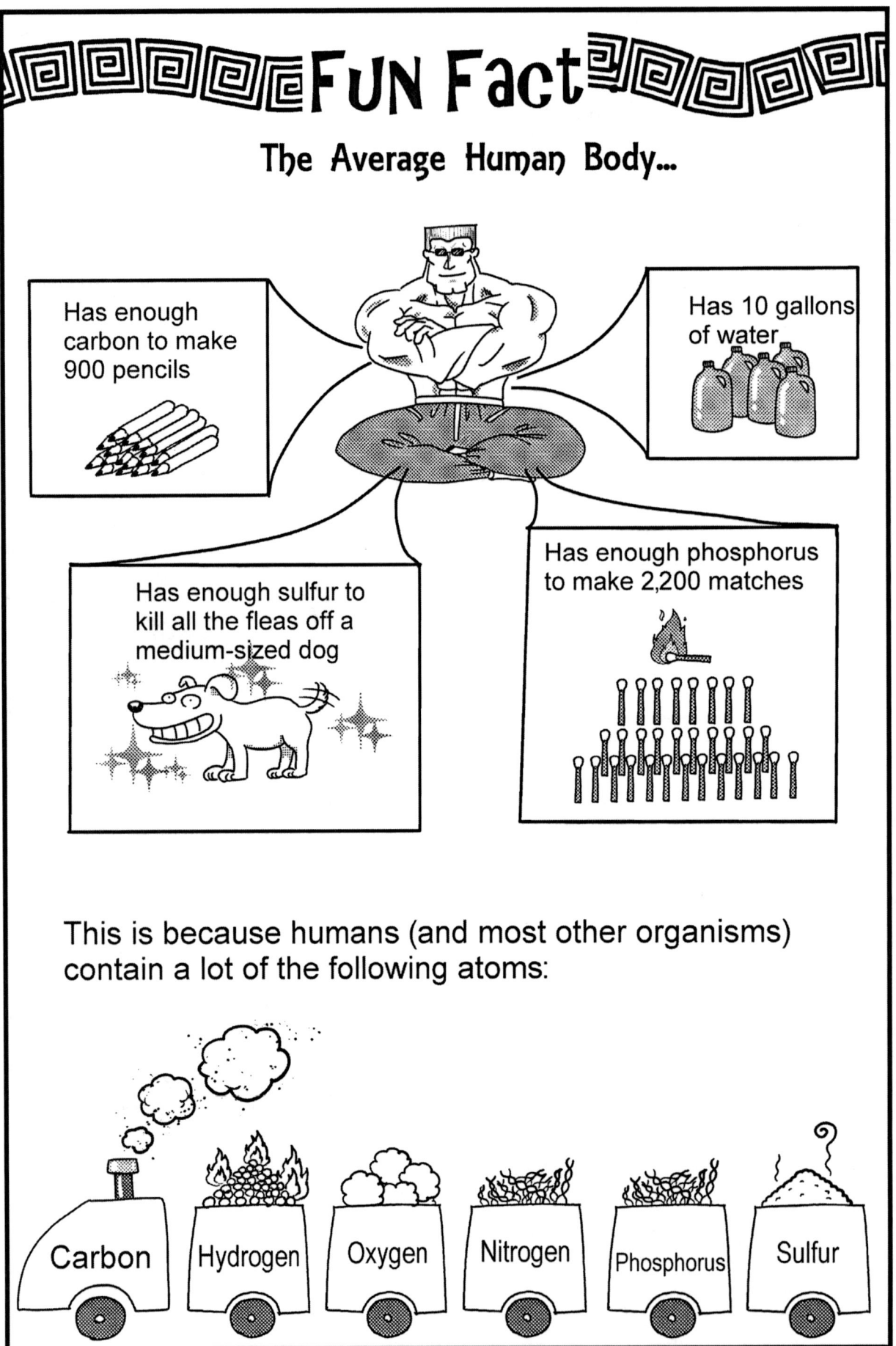

Carbon is very important because it is used to make...

FATS

PROTEIN

SUGAR

Note: Carbon is also the #1 most abundant element in most organisms (including humans).

Hydrogen and oxygen are also used in many things, such as glucose, fats (...etc..) and... water! (Water = $\underline{H}_2\underline{O}$)

Nitrogen is used to make:

PROTEIN

D.N.A.

Phosphorus is used to make:

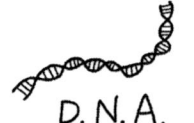
D.N.A.

Sulfur is used to make:

PROTEIN

FUN FACTS #1

The human body is 70% water.

FUN FACTS #2

Feces (poop) and urine (pee) contain a lot of nitrogen useful to plants. So they are often used as fertilizers.

FUN FACTS #3

When eggs rot, the sulfur in its proteins will come out into the air, which produces the "rotten egg smell."

Unit 2: Chapter 14

Chapter 15: Carbon Compounds

Science Standard: 8.6.c
Students know that living organisms have many different kinds of molecules, including small ones, such as water and salt, and very large ones, such as carbohydrates, fats, proteins, and DNA.

NEW VOCABS

* **Amino Acid:** A monomer that makes up the polymer protein.

* **Cellulose:** A type of polysaccharide. Its more common name is "plant fiber." Herbivores such as cows and horses can break it down into monomers and absorb it for nutrients.

* **DNA:** Deoxyribonucleic Acid. It is a macromolecule made of the monomer nucleic acid. It carries genetic information.

* **Fatty Acids:** A monomer that makes up the polymer lipid. It can be used as a nutrient to provide energy for an organism..

* **Lipid:** A macromolecule that consists mostly of fatty acids. There are 2 main types of lipids: Fat and cholesterol. We tend to consider lipid as a polymer *(But some scientists disagree).*

* **Macromolecule:** A big molecule that consists of many atoms.

* **Micromolecule:** A small molecule that consists of fewer atoms.

- **Monomer:** A type of micromolecule that can join with other monomers of the same kind to form a macromolecule called a "polymer".

- **Monosaccharide:** Also known as simple carbohydrates. It is a monomer. Examples are glucose and fructose.
  It tends to have a sweet taste.

- **Nucleotide:** A monomer that makes up the polymer "nucleic acid."

- **Organic Chemistry:** The study of molecules that contain carbon(s).

- **Polymer:** A macromolecule that is made of multiple monomers in a consistent, repeating pattern.

- **Polysaccharide:** Also known as complex carbohydrates. It is a polymer that consists of monosaccharides.
  Examples are starch and cellulose.

- **Protein:** A polymer that consists of the monomer amino acids. An essential molecule to organisms.

- **RNA:** Ribonucleic Acid. It is a macromolecule made of the monomer nucleic acid. Similar to DNA, it carries genetic information.

- **Starch:** A type of polysaccharide. It is found in wheat, rice, and potatoes…etc. It can be broken down into monosaccharide and be absorbed by our body as nutrients.

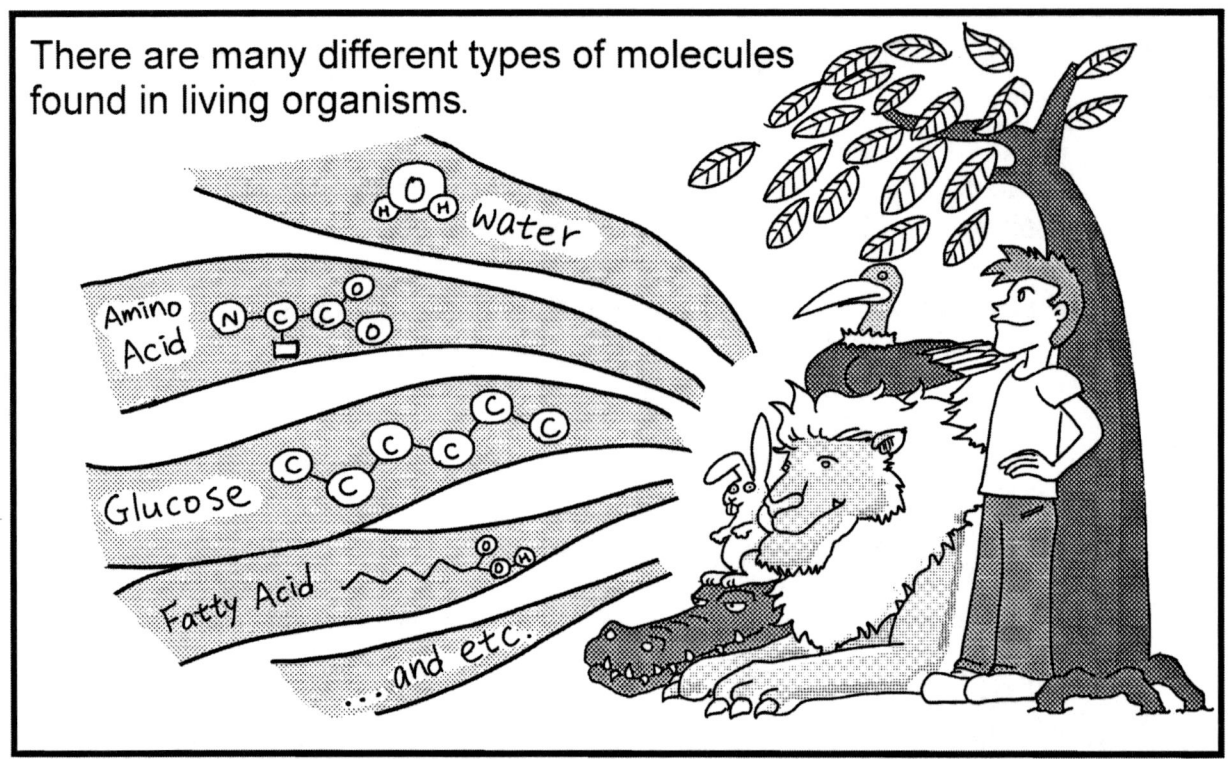
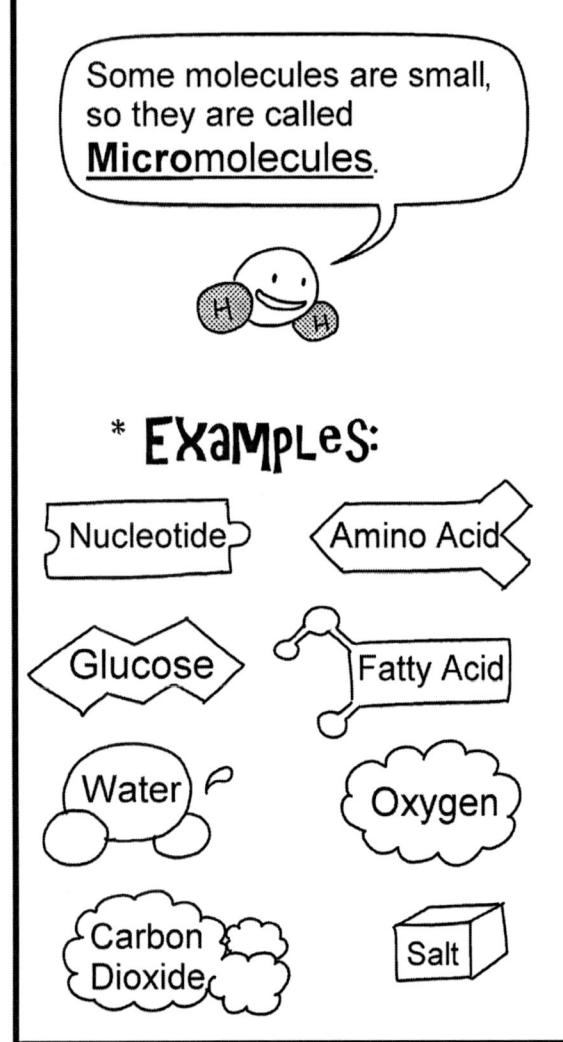
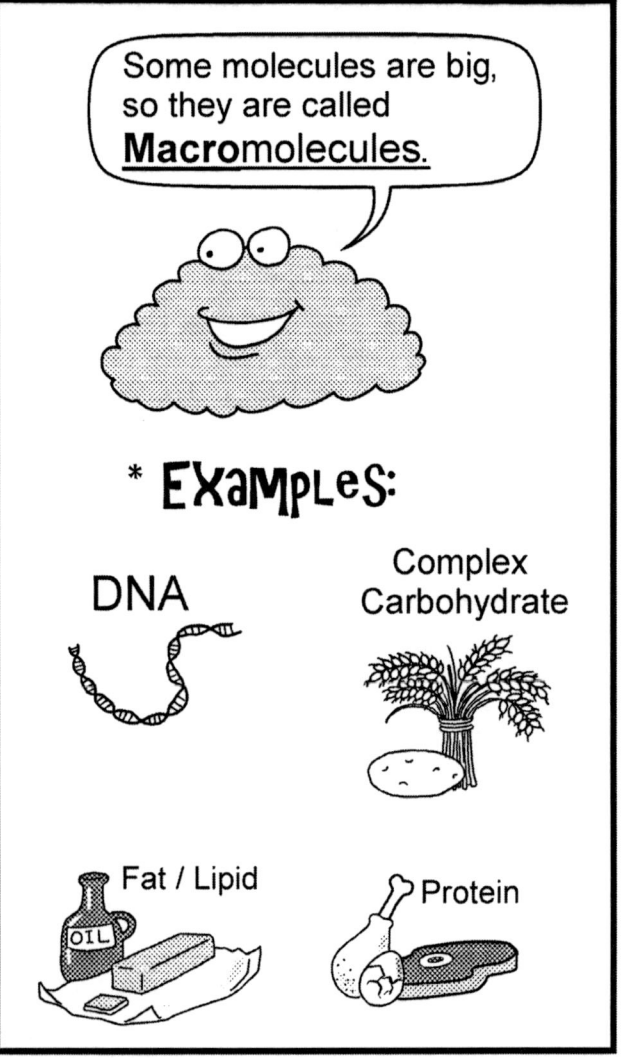

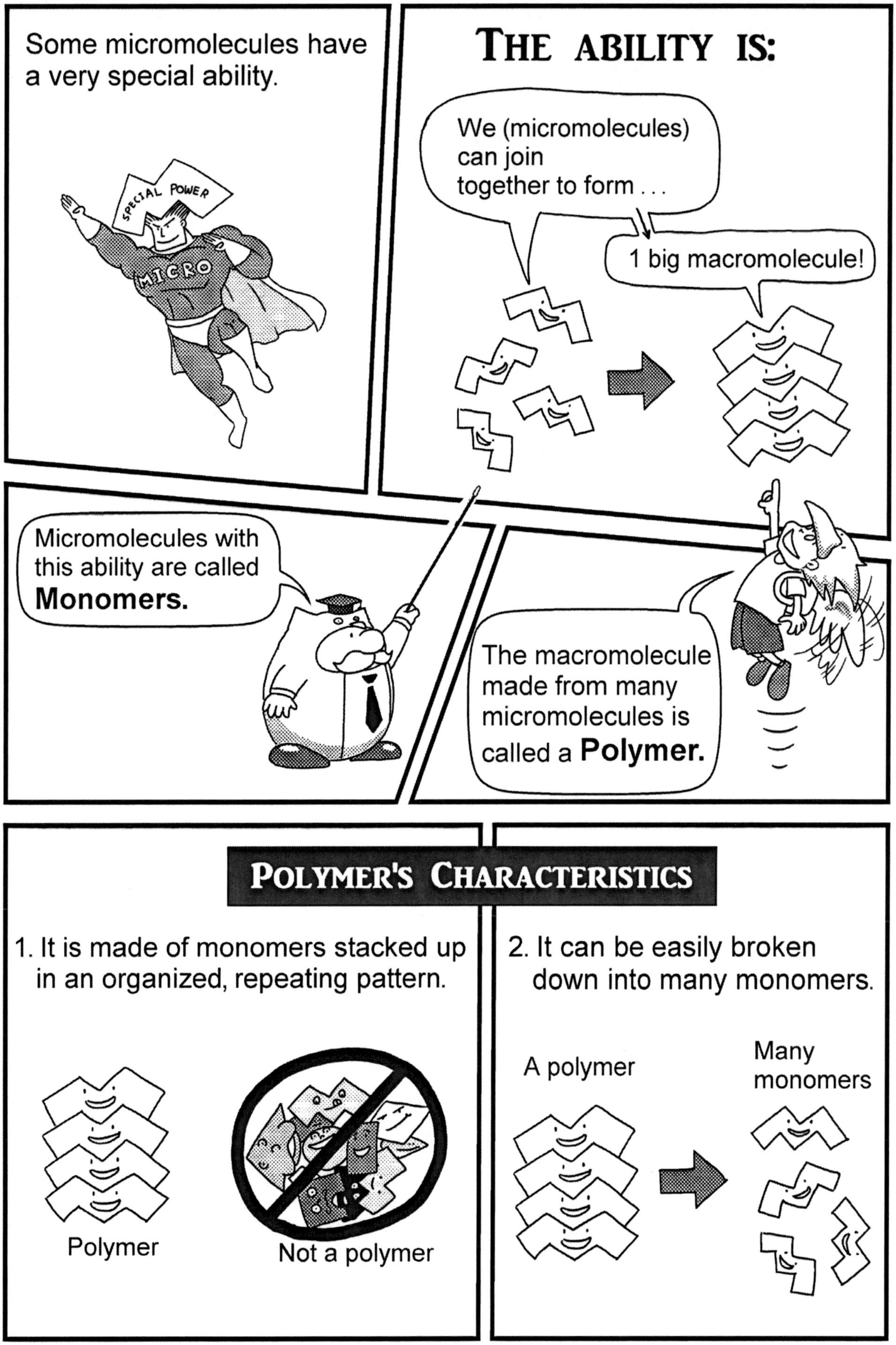

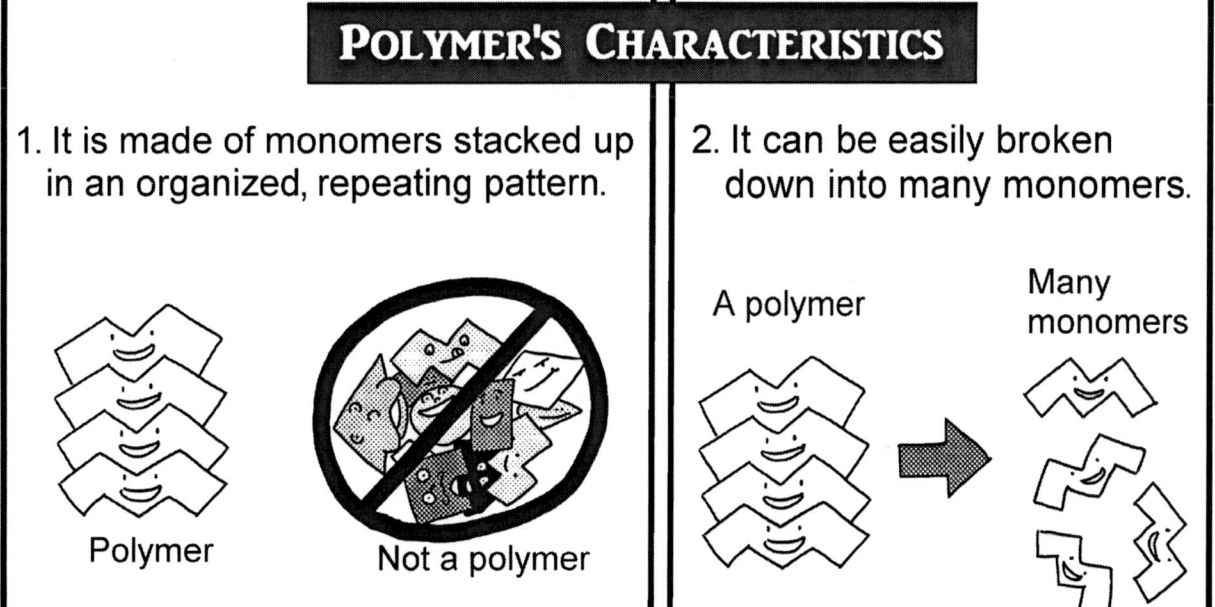

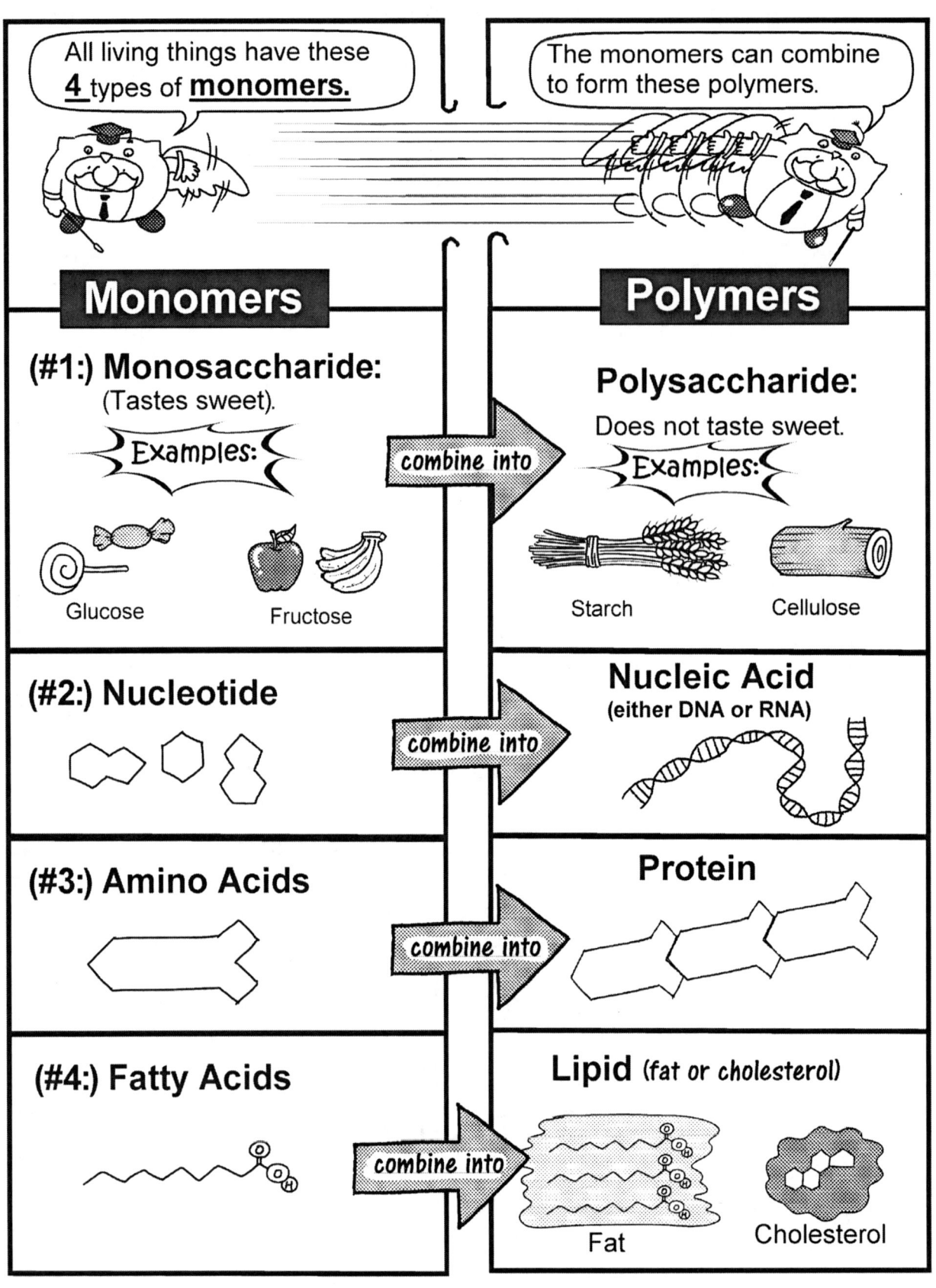

* **Note:** Some scientists do not count fatty acids are monomers because they do not form lipids in a *highly* organized, repeating pattern.

 Vocabularies Associated with monosaccharides and polysaccharides

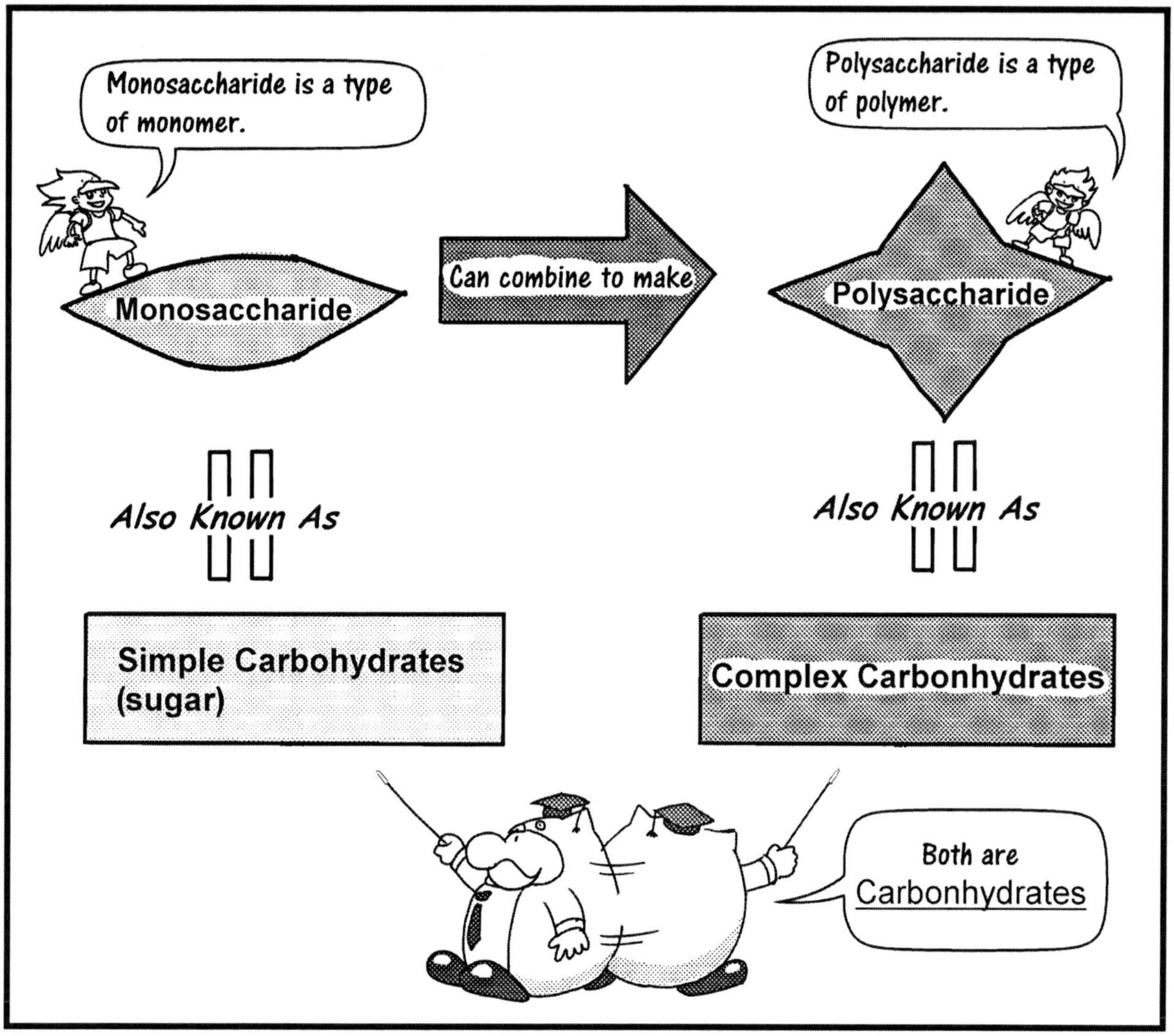

"Monosaccharide is a type of monomer."

Monosaccharide — Can combine to make → **Polysaccharide**

"Polysaccharide is a type of polymer."

Also Known As = **Simple Carbohydrates (sugar)**

Also Known As = **Complex Carbonhydrates**

Both are **Carbonhydrates**

More information on Polysaccharides:
There are 2 types of polysaccharides: Starch and Cellulose

Starch:
Has a "powdery" texture and can be found in:

Cellulose:
Has a "fibrous" texture and can be found in:

page 137

Unit 2: Chapter 15

Humans can use **starch** as a nutrient because our body can break it down into **monosaccharides**, which can be absorbed by us.

But **Cellulose *cannot be broken*** down into monosaccharides by humans. So, it holds no nutritional value for humans.

However, we still eat cellulose because it ***helps clean*** our digestive system as it passes through our intestines.

By the way, ***herbivore's*** (i.e. cow's) digestive systems have special bacteria that can break cellulose down into monosaccharides.

So they can actually get nutrients out of cellulose

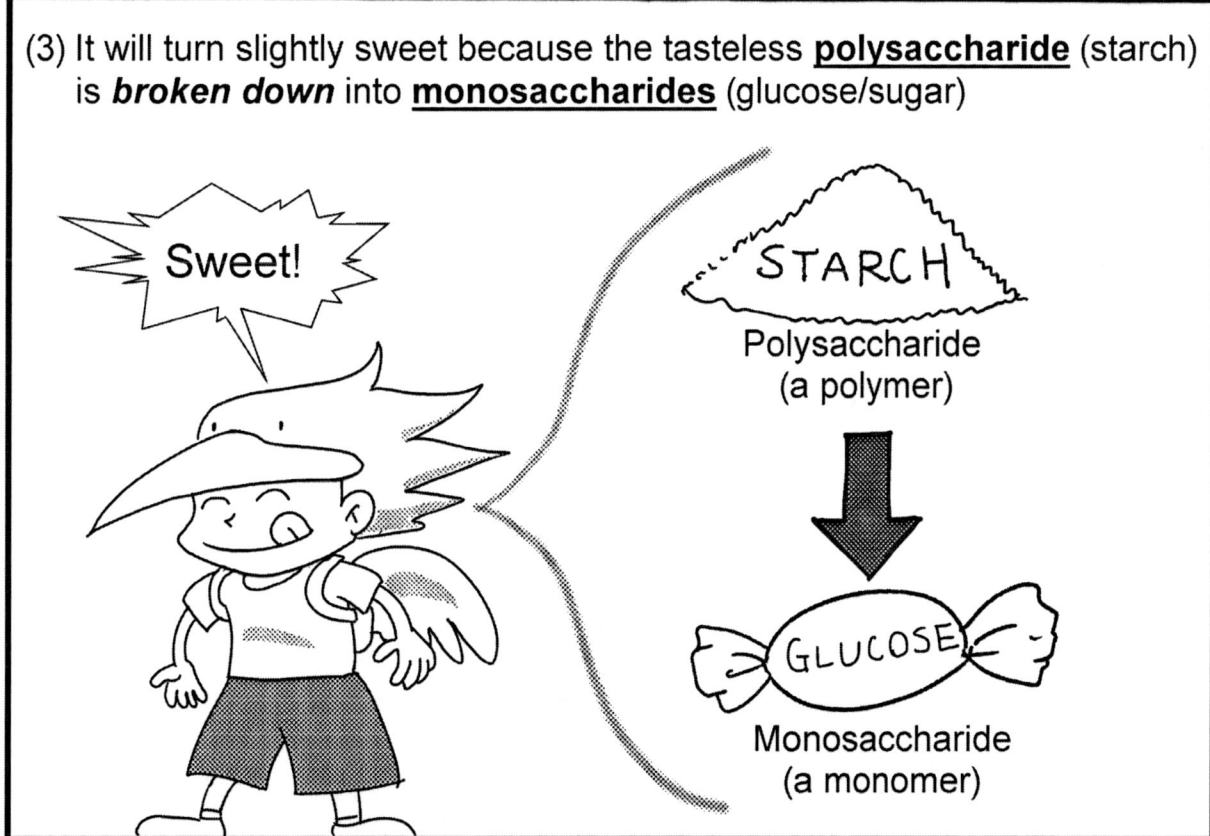

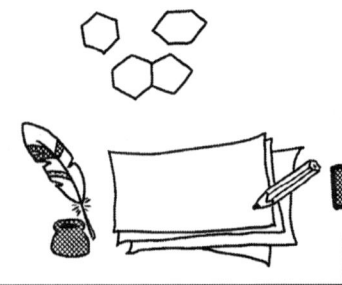

Because all the food we eat are *living* things, we also end up eating a lot of their **DNA**

Our body *breaks* down the DNA into **nucleotides**. Then, we use them to build our *own* DNA!

We need new DNA to make new cells!

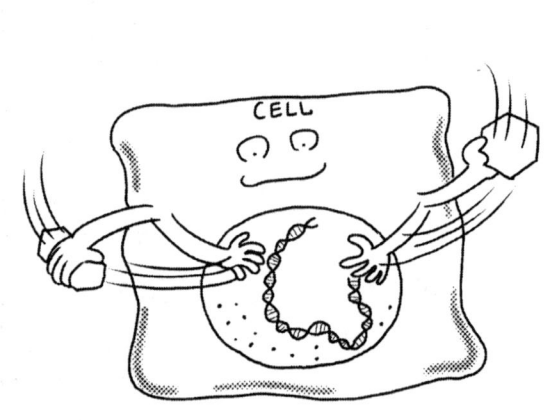

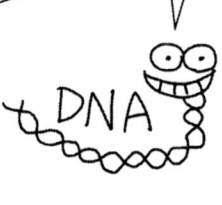

They both carry genetic info, more details in Unit 3: Chapter 25.

If a cell is like a city...

Nucleotides are like the pen and paper used to make the Manual.

DNA is like the city's User's Manual.

Our Body builds a new city (a new cell). The city's is made of **amino acids.**

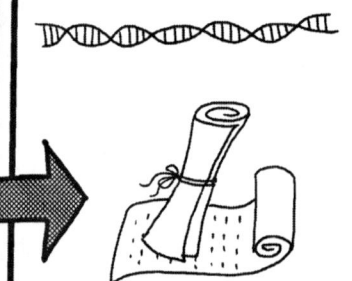

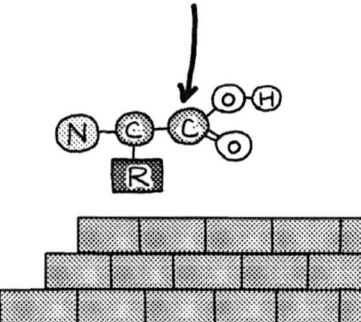

page 140

Unit 2: Chapter 15

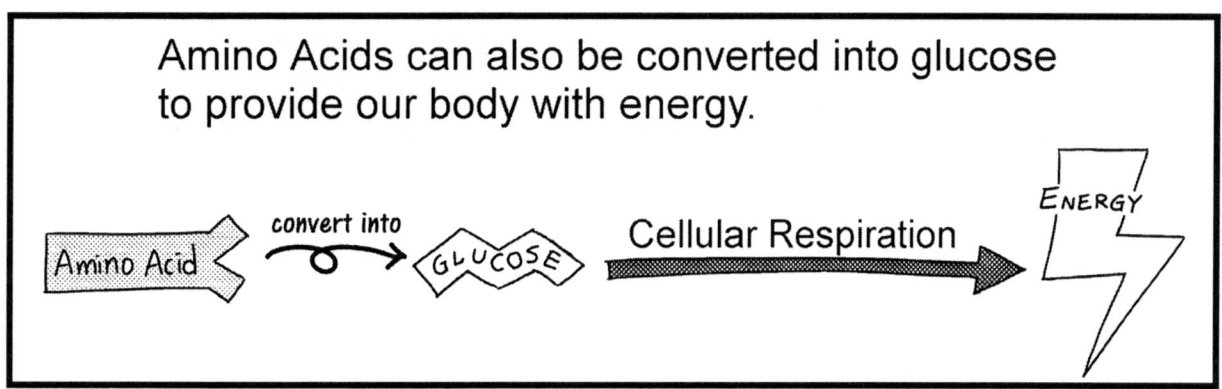
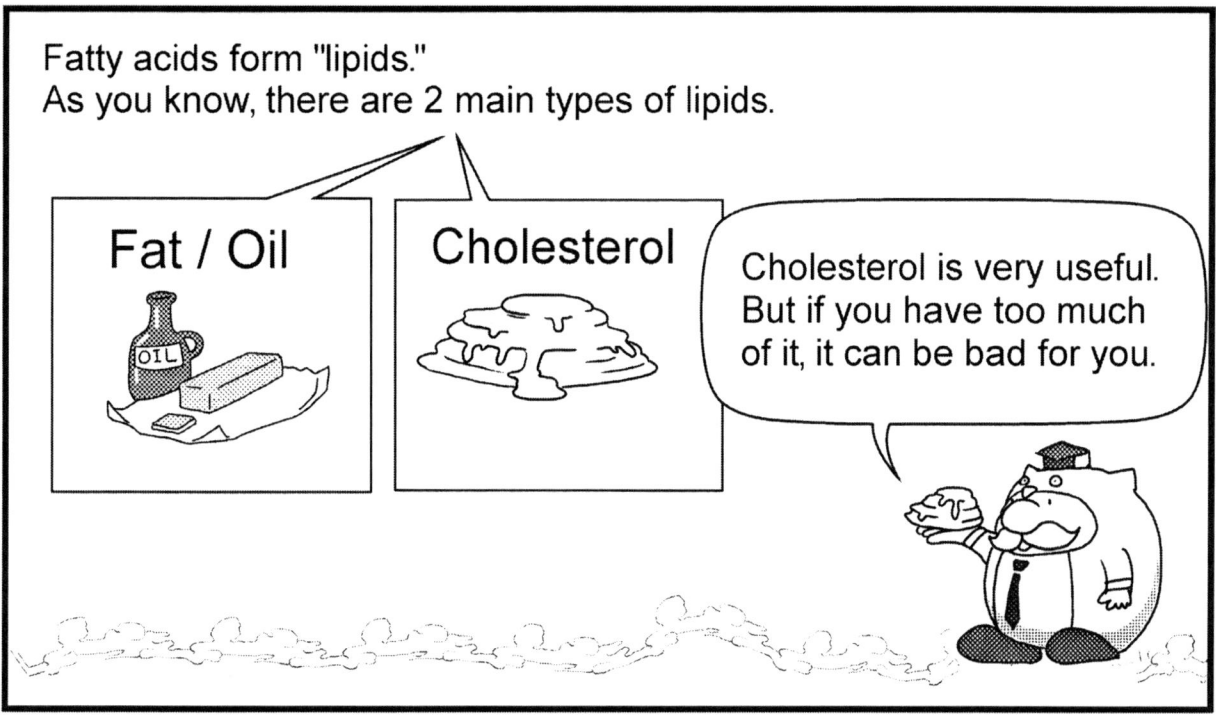

Chapter 16: Cell Membrane and Diffusion

Science Standard: Bio/LS . 1 . a
Students know cells are enclosed within semipermeable membranes that regulate their interaction with their surroundings.

NEW VOCABS

* **Active Transport:** A process in which transport proteins pump molecules across the cell membrane using the energy from ATP.

* **Concentration:** The ratio of solutes / solvent in a solution.

* **Concentration Gradient:** The difference in concentration between 2 regions across the cell membrane.

* **Diffusion:** The movement of solutes from a place of higher concentration to a place of lower concentration.

* **Endocytosis:** The process of a cell "swallowing" an object by movements of its cell membrane.

* **Exocytosis:** The process of a cell "spitting out" something it wants to get rid of (such as wastes).

* **Facilitated Diffusion:** A type of diffusion that requires transport protein to happen.

* **Hydrophilic:** This term describes the tendency of a molecule preferring to interact with a polar molecule.

* **Hydrophobic:** This term describes the tendency of a molecule preferring to interact with a nonpolar molecule.

* **Hypertonic:** A situation where the concentration inside a cell is greater than that of the outside.

* **Hypotonic:** A situation where the concentration inside a cell is lower than that of the outside.

* **Isotonic:** A situation where the concentration inside a cell is the same as that of the outside.

* **Osmosis:** The movement of water from a place of lower concentration to a place of higher concentration.

* **Phagocytosis:** A large-scale endocytosis when a cell takes in a particle by wrapping itself around it.

* **Phospholipid bilayer:** This term describes the way the cell membrane (and nuclear membrane) are structured —they are made of 2 layers of phospholipids.

* **Pinocytosis:** A small-scale endocytosis when a cell consumes a particle by making a dent in its membrane.

* **Selectively Permeable:** It is an adjective used to describe a membrane that allows only certain substances to pass through.

* **Solute:** The material that is to be dissolved by the solvent.

* **Solution:** The mixture of solutes and solvent.

* **Solvent:** The substance that is to dissolve the solute. In middle/high school science labs, solvent usually means water.

* **Transport Protein:** A gate made of protein embedded in the cell membrane. It allows certain molecules to pass through the cell membrane. Some transport proteins are equipped with the ability to actively pump molecules across the membrane. (Some substances, such as sodium, can only get in/out of the cell through transport proteins.

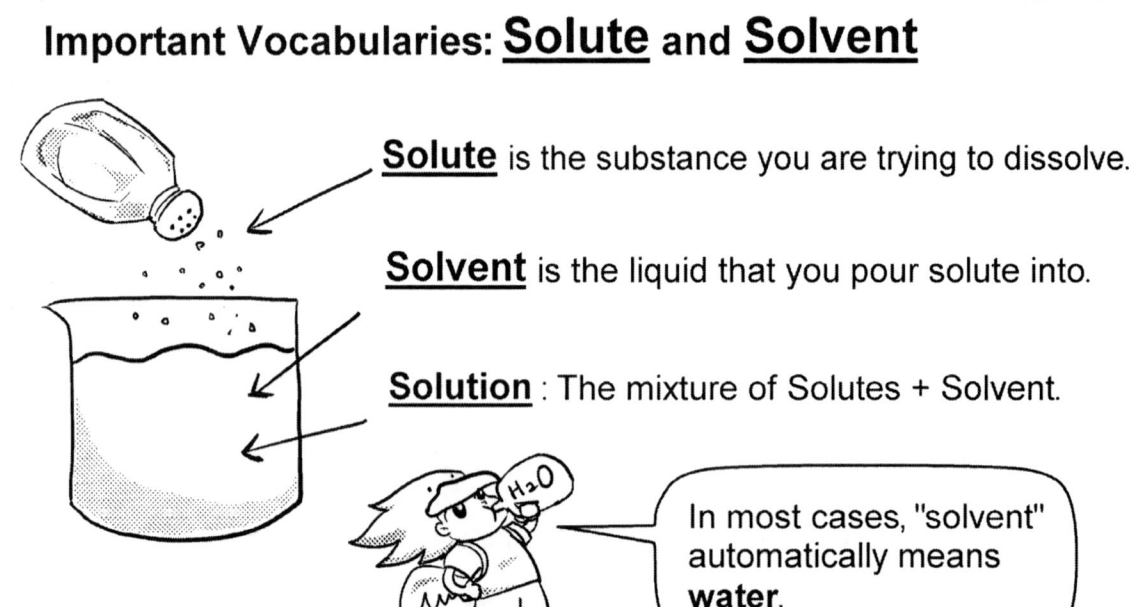

Concentration:
The ratio of solute vs. solvent in a solution.

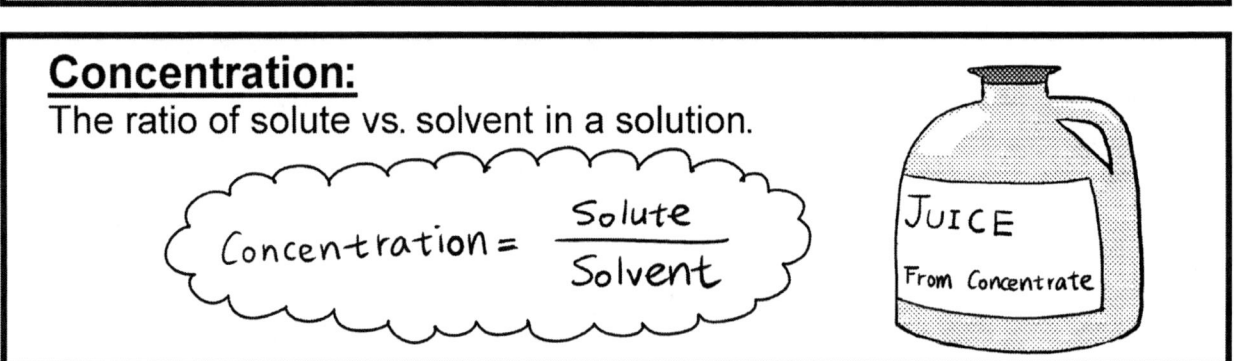

$$\text{Concentration} = \frac{\text{Solute}}{\text{Solvent}}$$

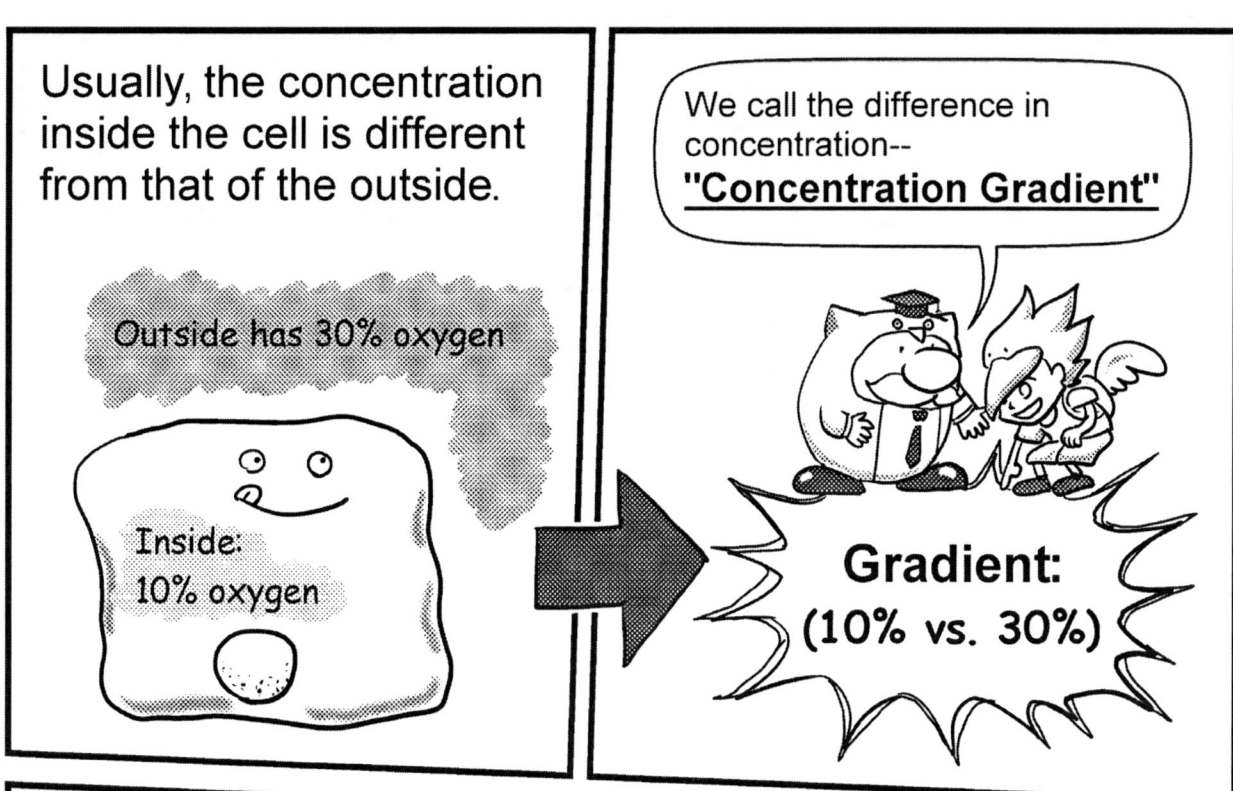

Usually, the concentration inside the cell is different from that of the outside.

We call the difference in concentration-- **"Concentration Gradient"**

When there is a concentration gradient, **"Diffusion"** happens.

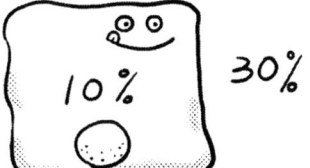

"Diffusion": The movement of solutes from a place of higher concentration to a place of lower concentration.

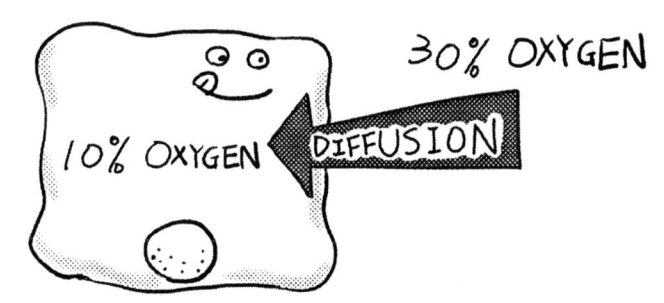

Diffusion Example: The movement of perfume molecules moving across a concentration gradient.

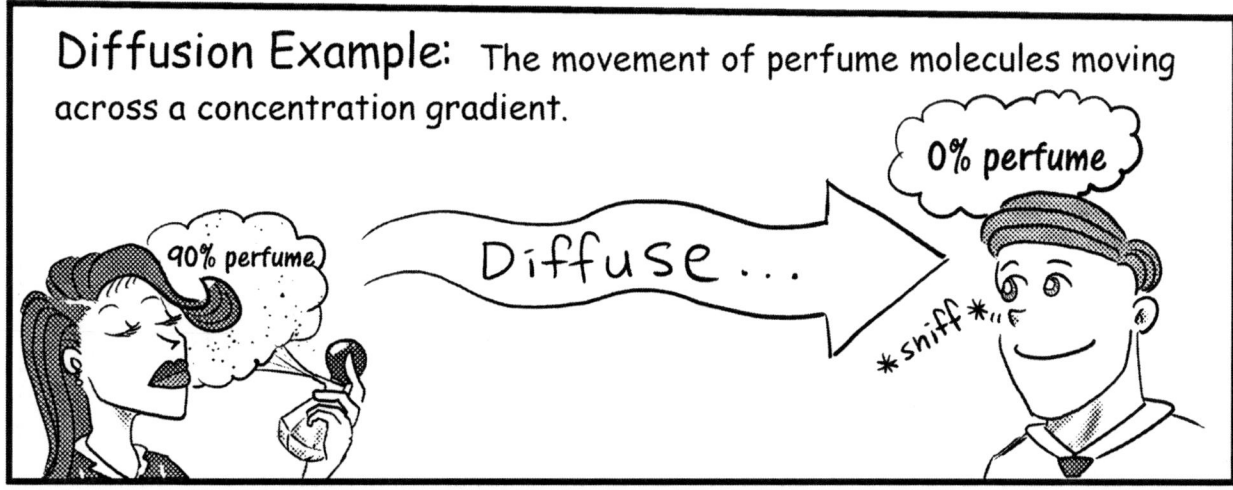

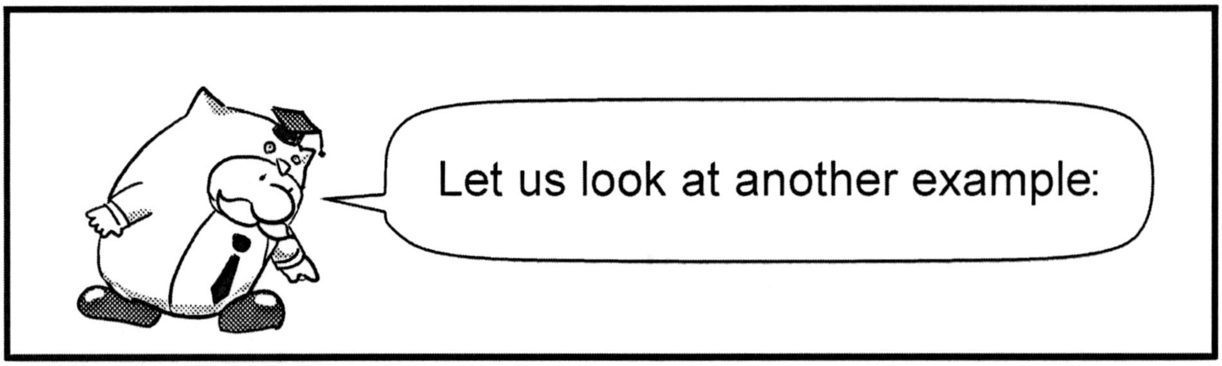

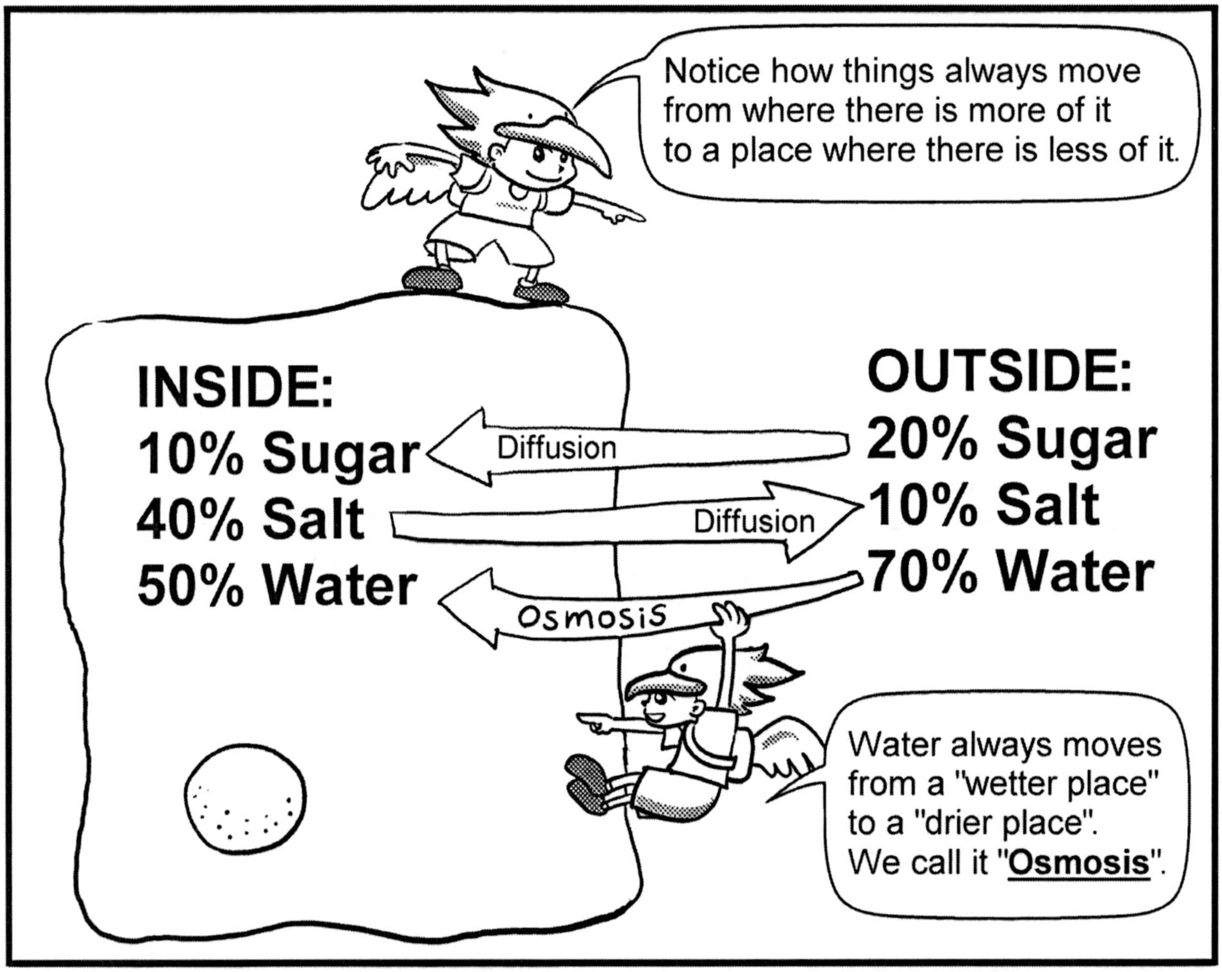

Review: Things Naturally Balance Out

Where there is a *lower* concentration, that means it has:

Less% Solutes ← Solutes will **diffuse** this way

&

Higher% Water → Water will **osmosis** this way

Where there is a *higher* concentration, that means it has:

Higher% Solutes

&

Less% Water

Some Real-Life Examples of Osmosis:

Example #1:
Put a dry rag into a bucket of water, and the water will move into the dry rag.

Example #2:
Pour salt (dry) on a slug (which is wet), and water will move from the slug into the salt.

This can kill us!

This is also why honey (high% sugar solute) does not spoil--Bacteria tend to dry up and die (from osmosis)!

page 147

Unit 2: Chapter 16

About: Hypertonic, Hypotonic, and Isotonic

Hypertonic means the inside of a cell has a *higher concentration* of solutes than the outside.

Outside: Low% solutes, High% water.

Inside: High% solutes, Low% water.

Hypotonic means the inside of a cell has a *lower concentration* of solutes than the outside.

Outside: High% solutes, Low% water.

Inside: Low% solutes, High% water.

Isotonic means the inside and outside of the cell has the same concentration.

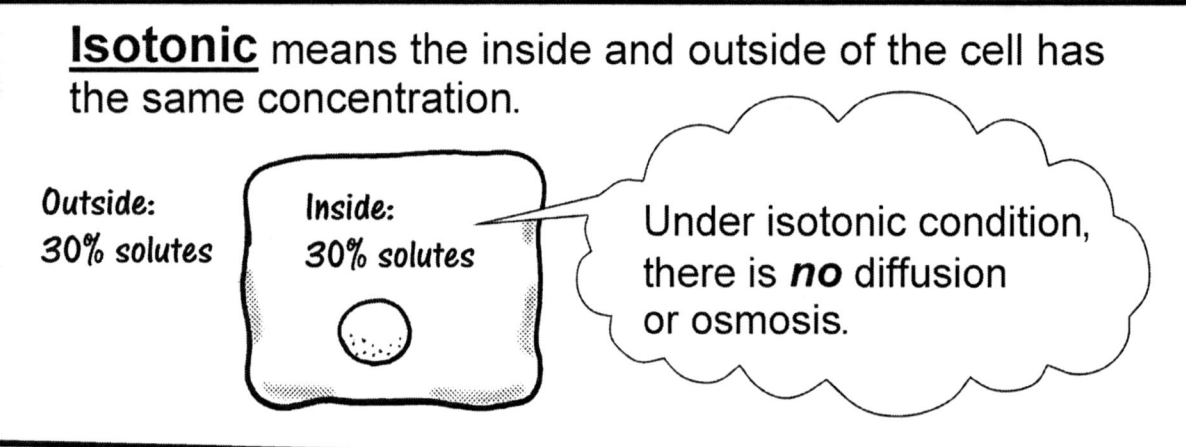

Outside: 30% solutes

Inside: 30% solutes

Under isotonic condition, there is **no** diffusion or osmosis.

Fun Fact:

When you put a red blood cell in water, the cell's hypertonic condition will cause water to osmosis into the cell, ballooning it up, sometimes even causing it to explode!

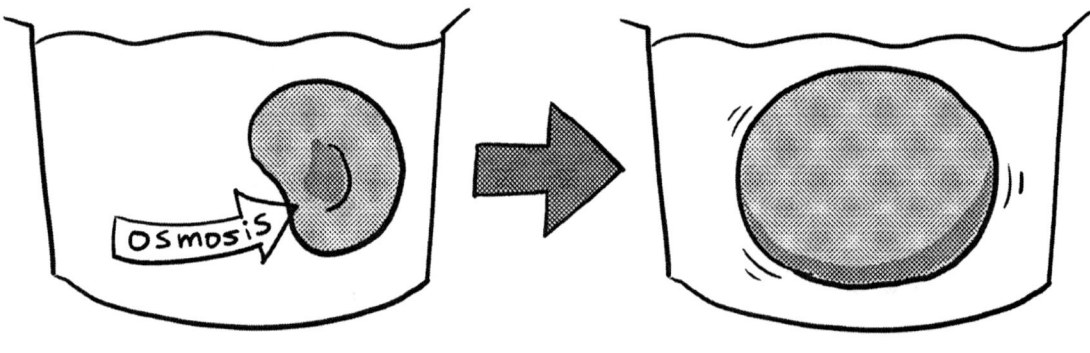

Review:
We have already learned that a nucleus' surface is made of the **"Nuclear Membrane"**.

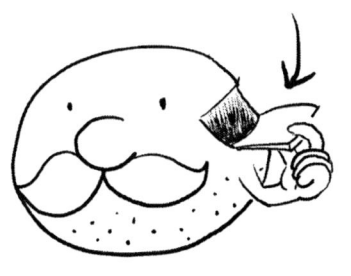

A cell's "skin" is called the **"Cell Membrane"**. It is made of the *same* substance as the *nuclear membrane*.

Cell membranes (and nuclear membranes) are **selectively-permeable**, which means that some things can pass through it... some cannot.

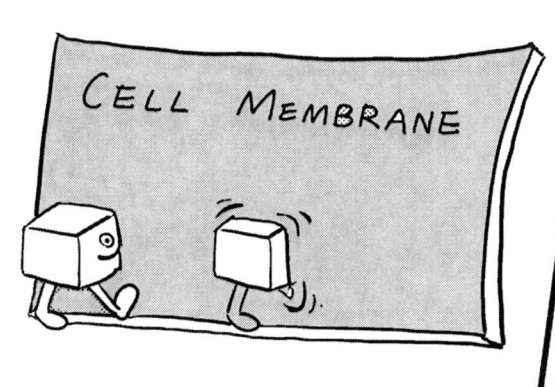
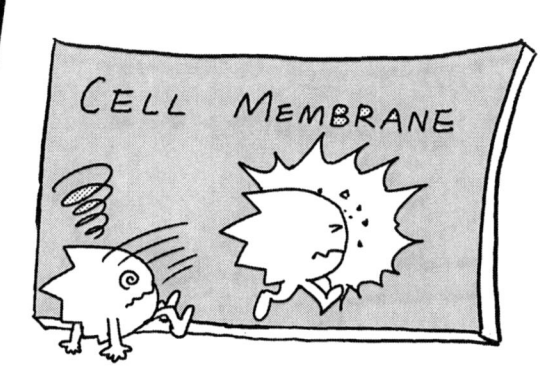

Substances that *easily* pass through the membrane:

Substances that can only pass through *certain parts* of the membrane:

Substances that *can never* pass through the membrane:

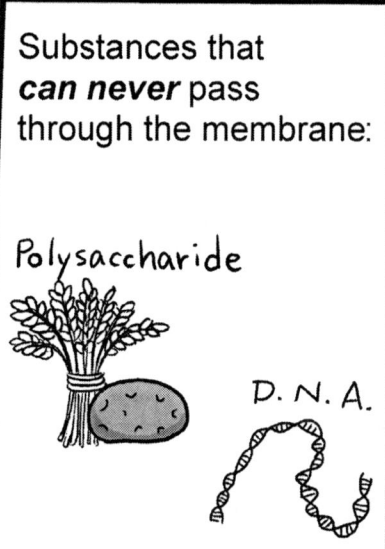

page 149

Unit 2: Chapter 16

Both the Cell Membrane and Nuclear Membrane are arranged in a structure called "**Phospholipid Bilayer**"

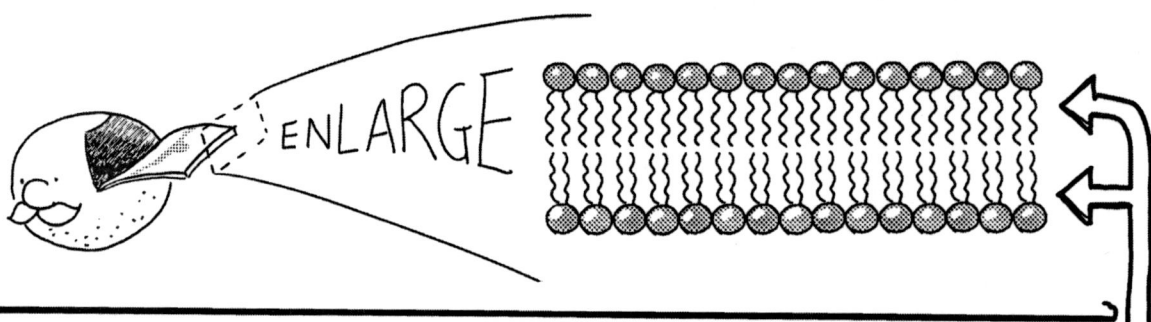

Why call it "Phospholipid Bilayer"?

The word **"bilayer"** means **"2-layers"** because...look! There are 2 layers!

The word **"Phospholipid"** stands for the small molecules that make up the membrane.

Phosphate head

Lipid tail

The phosphate head is <u>hydrophilic</u>, which means it likes water.

The tail is <u>hydrophobic</u>, which means it hates water.

A *good way* to remember this is to think of phospholipid as a *dog*:

Hydrophilic Head (water-loving)

Hydrophobic Tail (Water-hating)

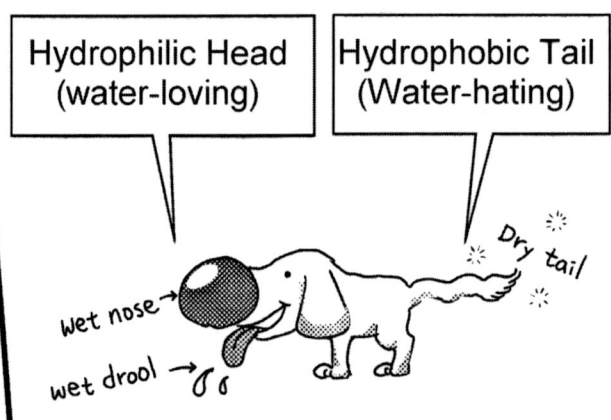

page 150

Unit 2: Chapter 16

On the other hand, **Nonpolar** molecules do not have a (+) or (−) end.

Examples of Nonpolar Molecules:

Oxygen   Carbon Dioxide   Oil

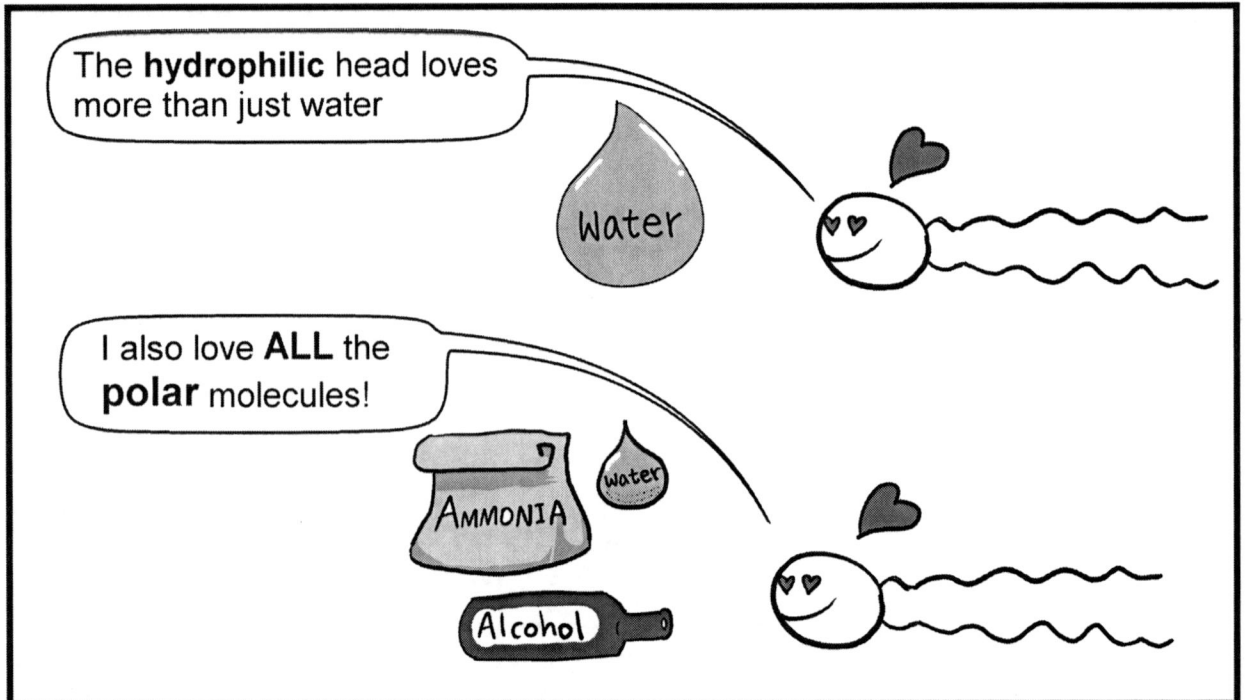

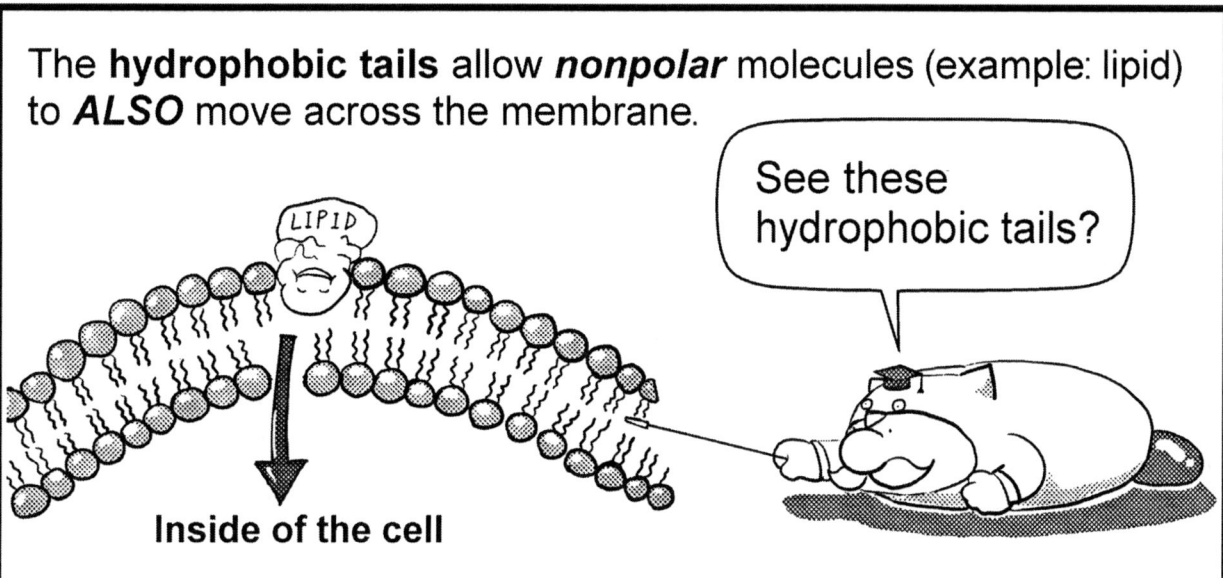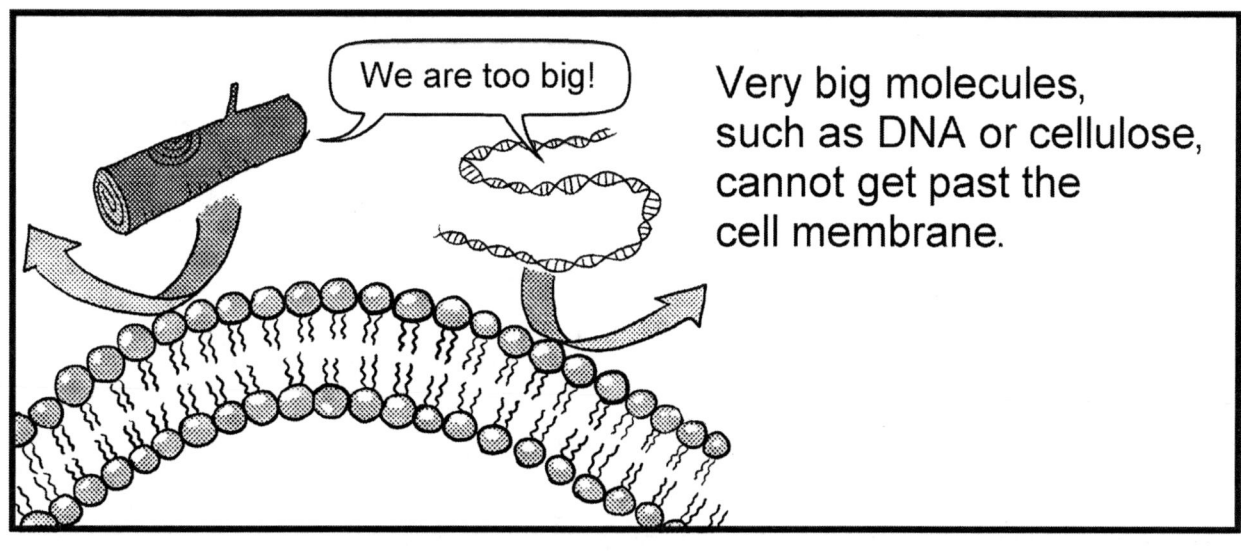

Some small molecules (such as glucose) still need help getting past the cell membrane.

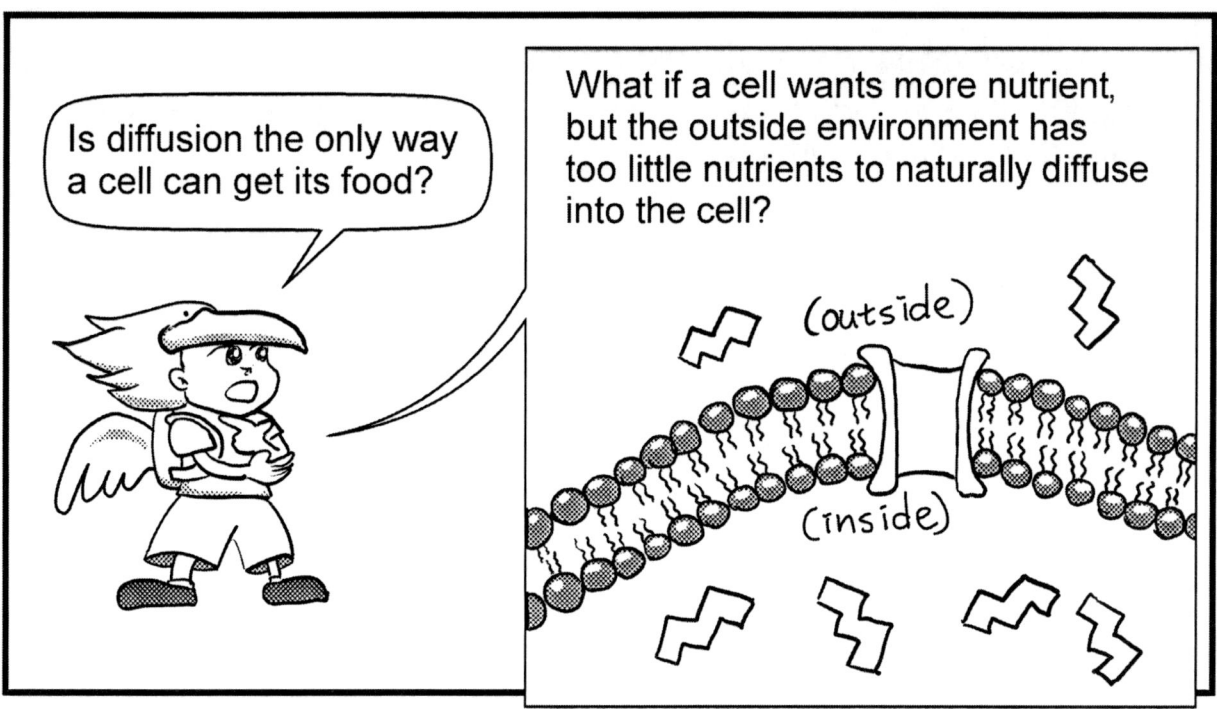

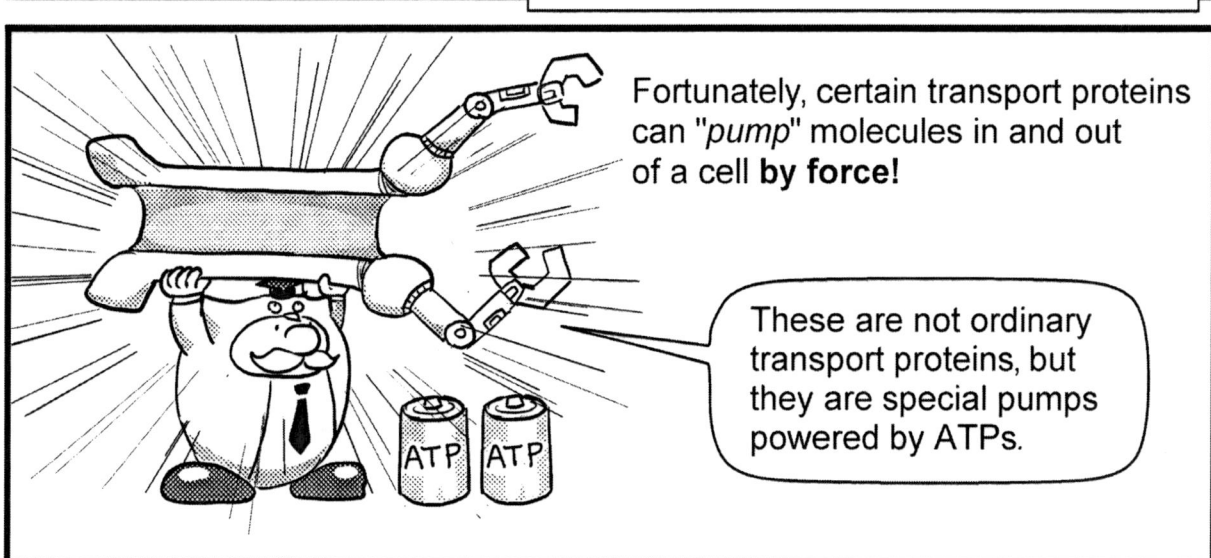

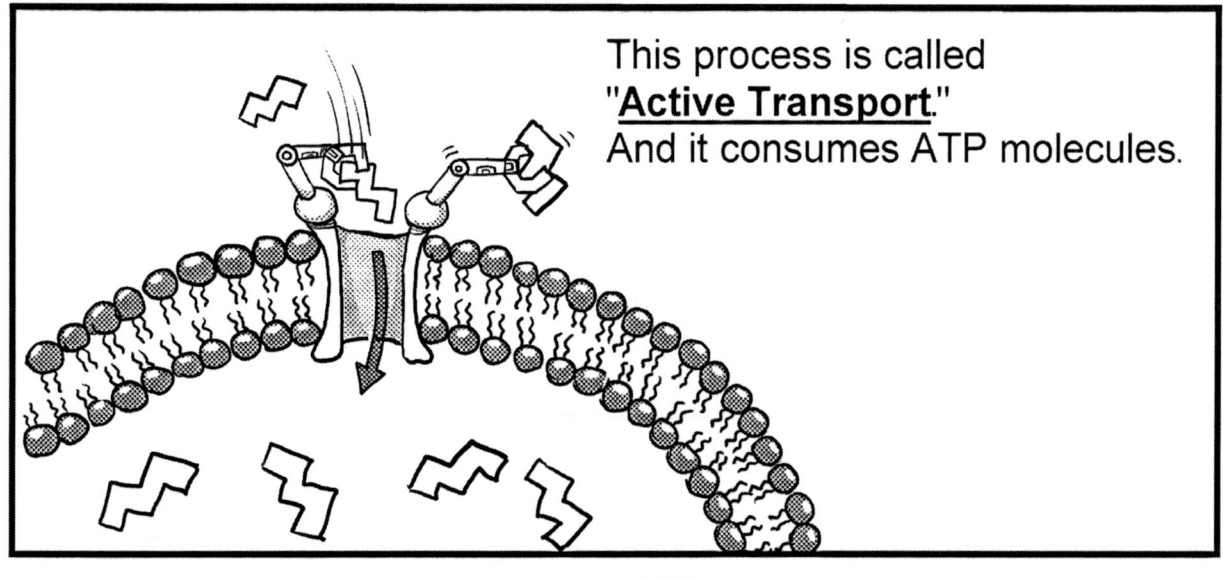

page 155

Unit 2: Chapter 16

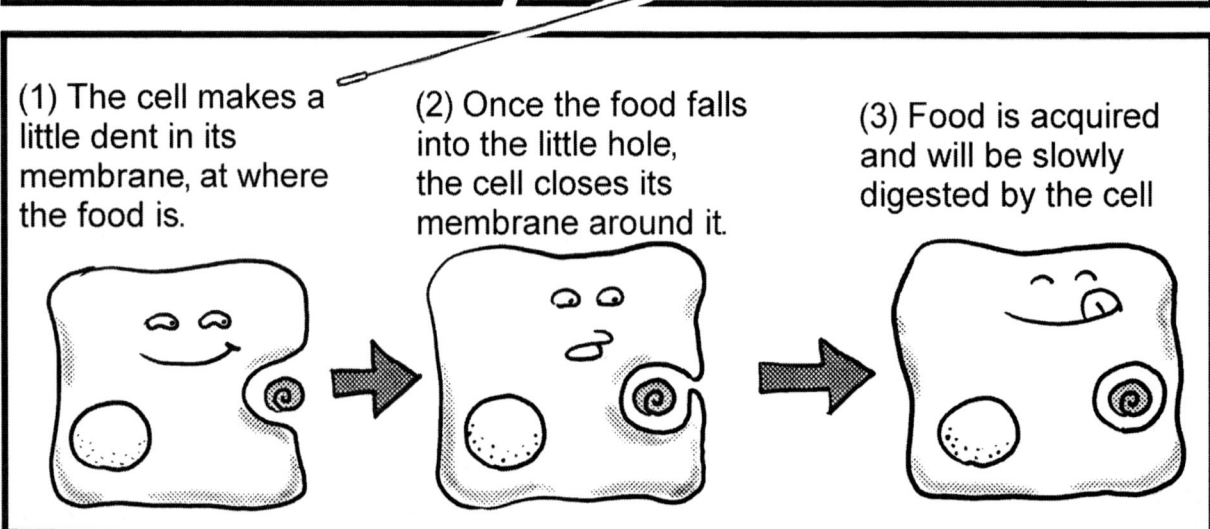

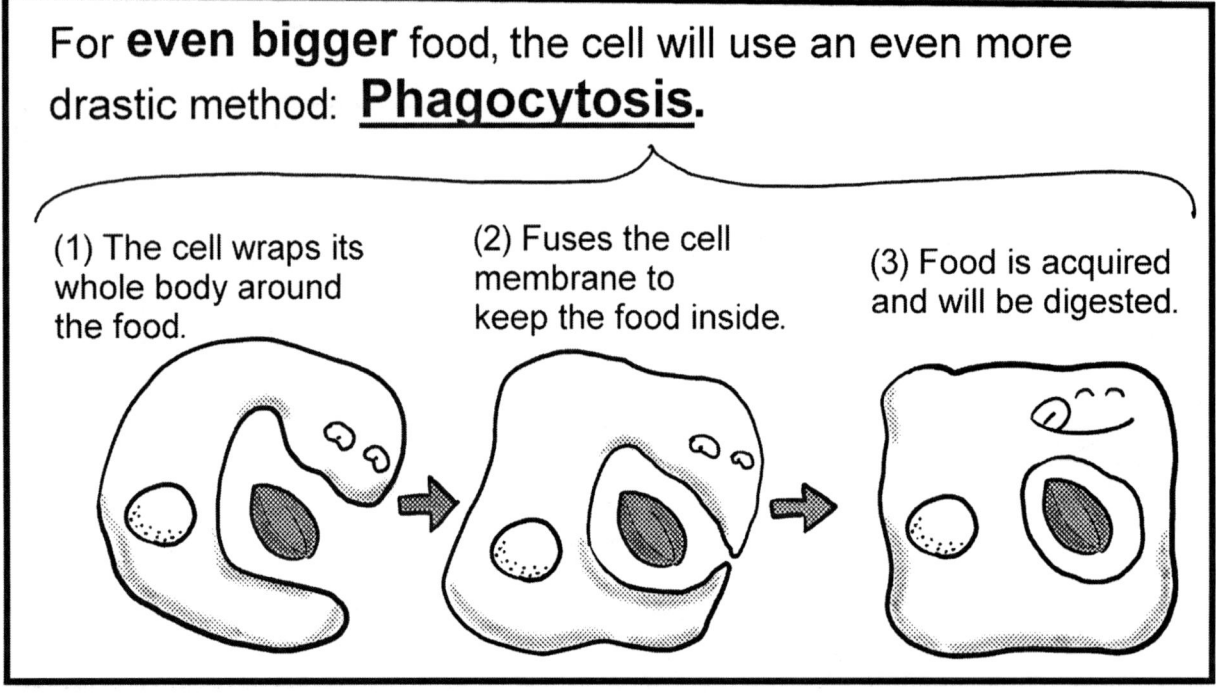

Both "pinocytosis" and "phagocytosis" are types of **"Endocytosis"**, which is a general term describing a cell swallowing things into itself.

The opposite of endocytosis is **"Exocytosis"**, which is a cell ridding of things it doesn't want (such as wastes).

EXOCYTOSIS

All the "-cytosis" consume ATP for energy. But it's worth it!

Review:
These processes consume ATP for energy...

Active Transport

Endocytosis (including pino/phago-cytosis)

Exocytosis

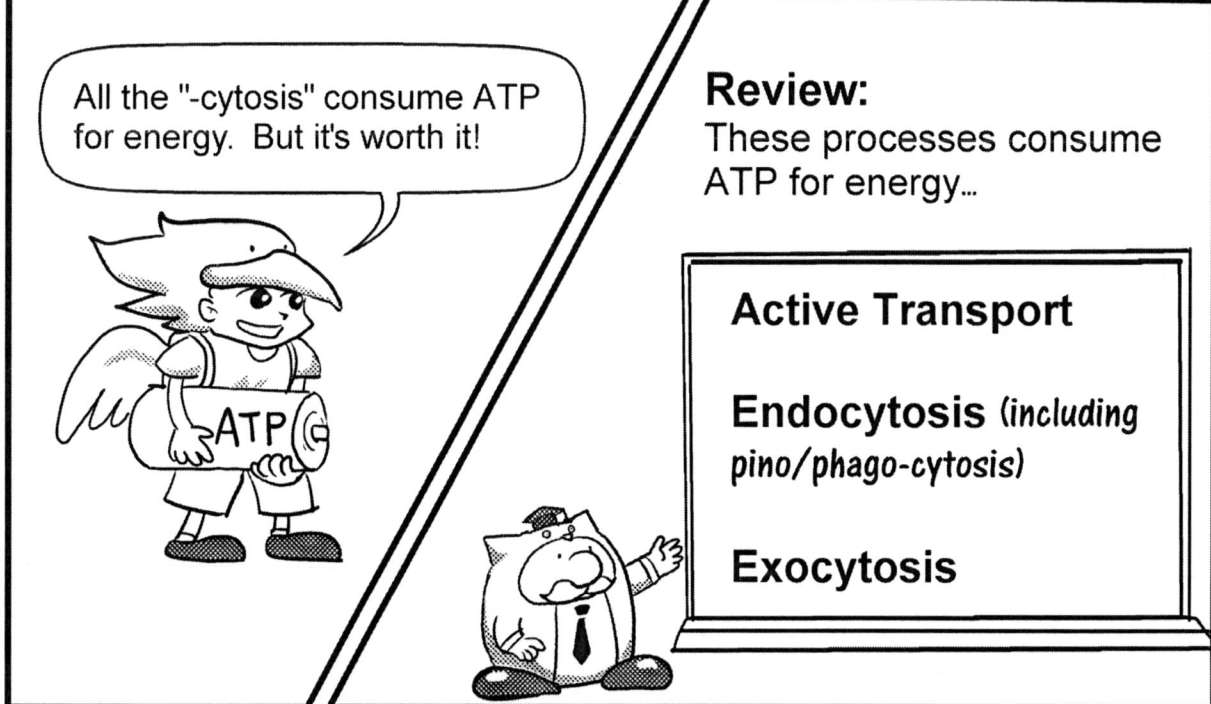

Unit 2: Chapter 16

Chapter 17: Virus, Prokaryote, and Eukaryote

Science Standard: Bio/LS . 1 . c
Students know how prokaryotic cells, eukaryotic cells (including those from plants and animals), and viruses differ in complexity and general structure.

NEW VOCABS

* **Eukaryotes:** Any cell that has a nucleus.

* **Prokaryotes:** Any cell that does not have a nucleus. In other words, it does not have a nuclear membrane to wrap around its genetic information (DNA), leaving its DNA naked and exposed.

* **Virus:** A virus is not a cell—thus, it is not considered as a living thing. It is made of a protein coat that wraps around its genetic information (RNA/DNA). It is an infective agent, meaning that it can spread from 1 living cell to another.

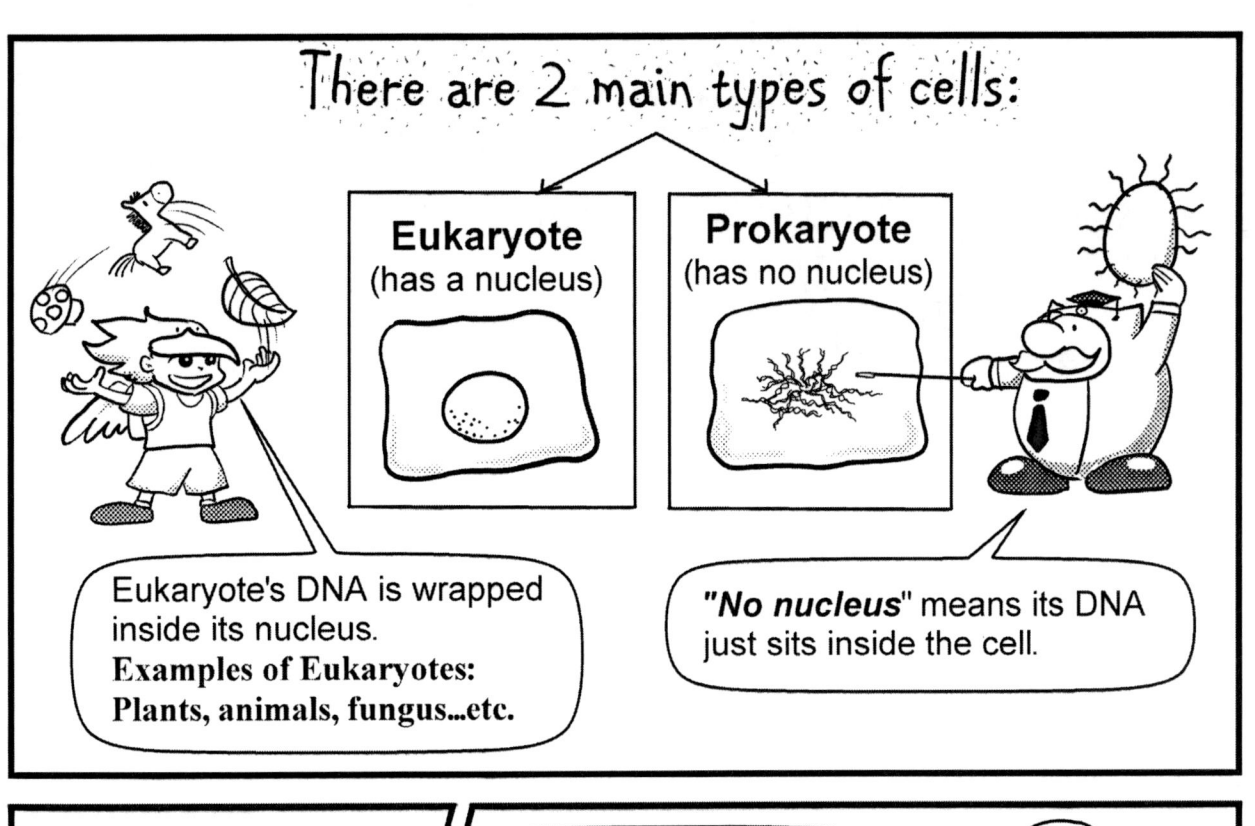

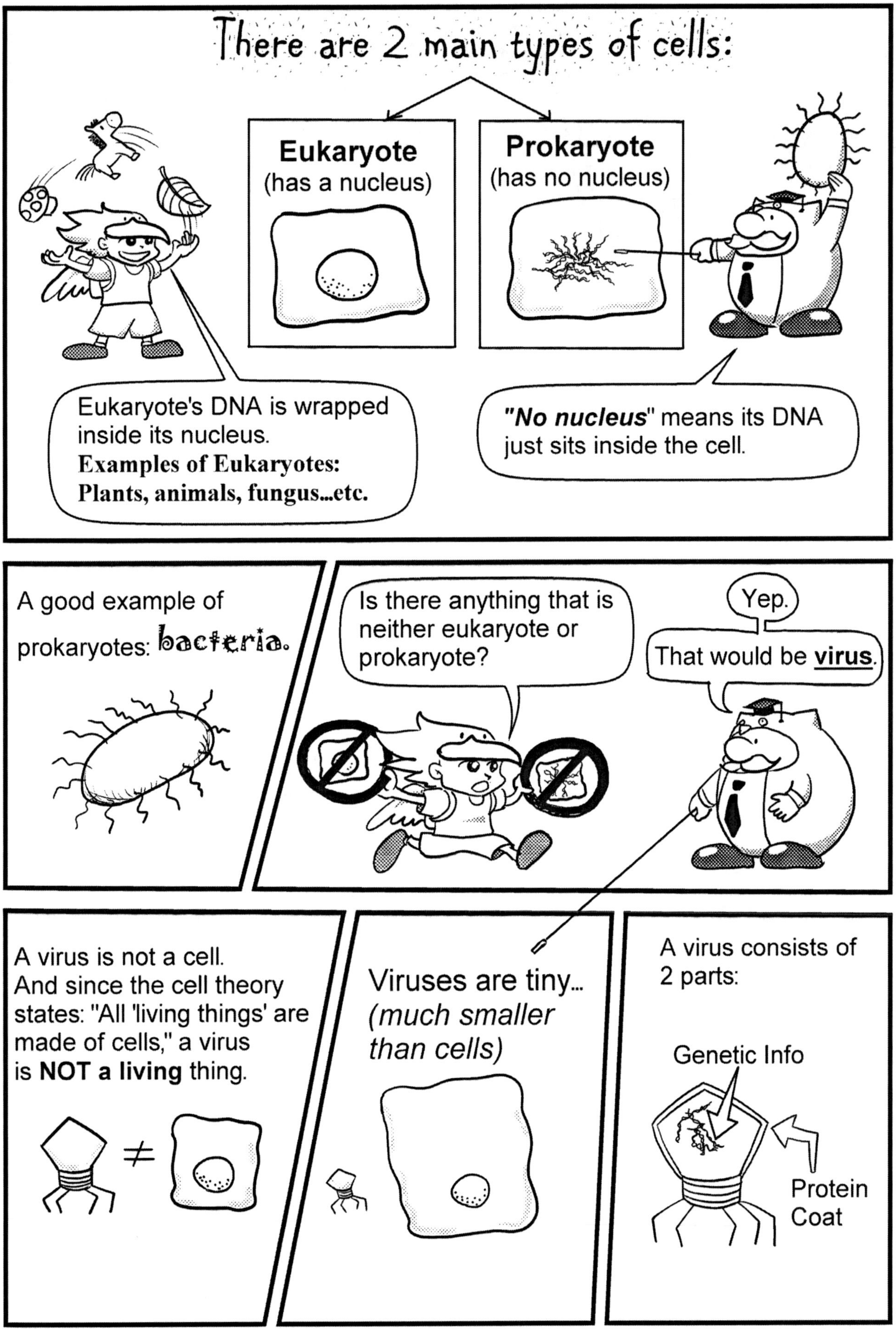

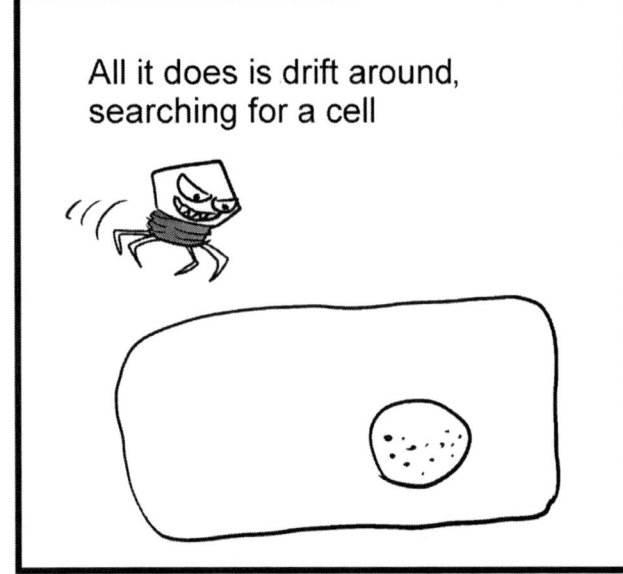

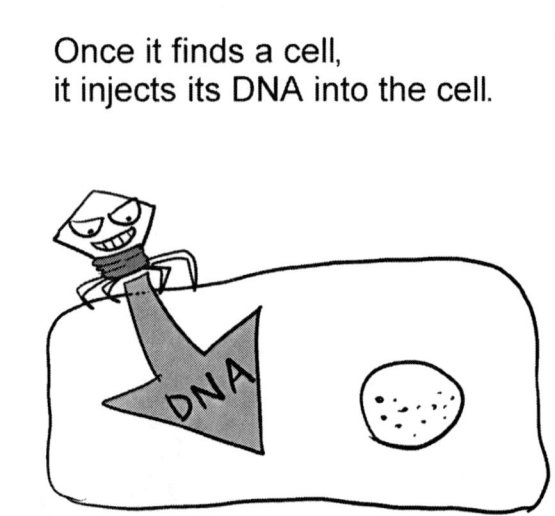

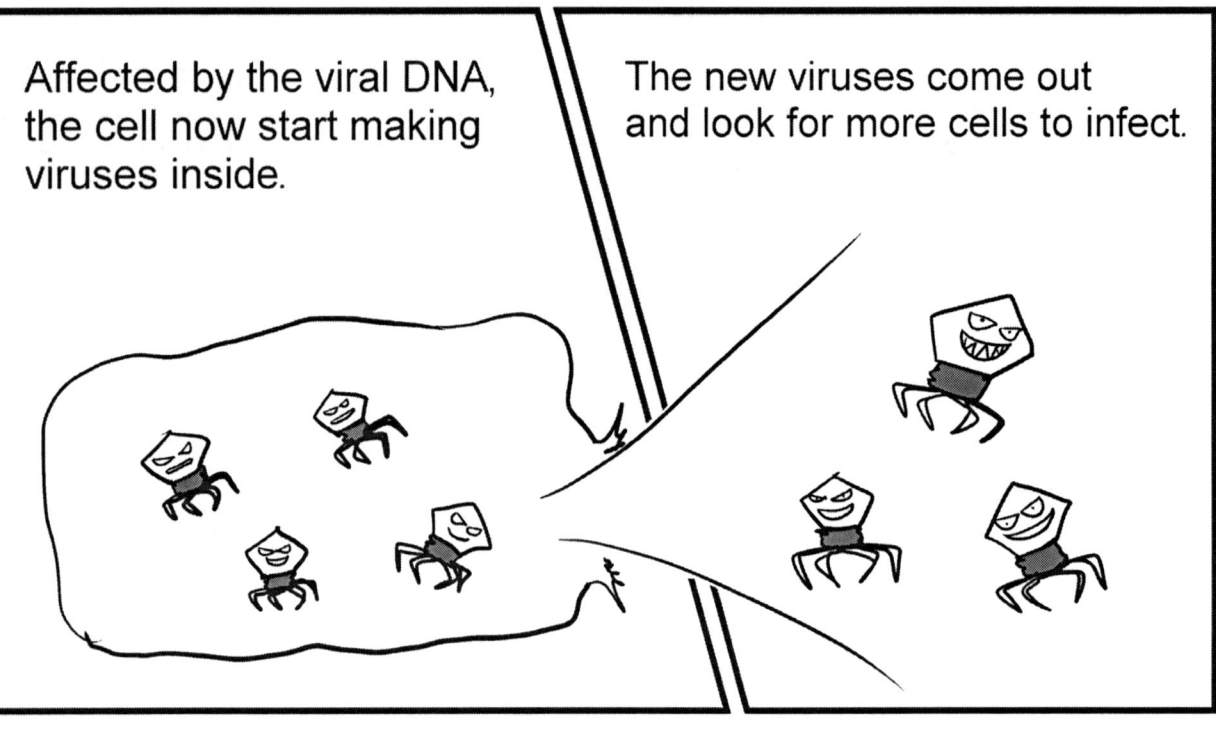

page 160

Unit 2: Chapter 17

Unit 3: Genetics

Chapter 18: Sexual Reproduction

Science Standard: 7.2.a
Students know the differences between the life cycles and reproduction methods of sexual and asexual organisms.

NEW VOCABS

* **Asexual Reproduction:** A type of reproduction in which only 1 parent is involved. This process usually involves mitosis.

* **Fertilization:** The process of a sperm fusing (joining) with an egg.

* **Offspring:** Children/young/descendants of a particular parent.

* **Reproduction:** The act of producing offspring.

* **Sexual Reproduction:** A type of reproduction in which a sperm and an egg are fused (joined) into an offspring.

In nature, there are 2 main styles of *__reproduction__.

Method 1: Asexual Reproduction

Method 2: Sexual Reproduction

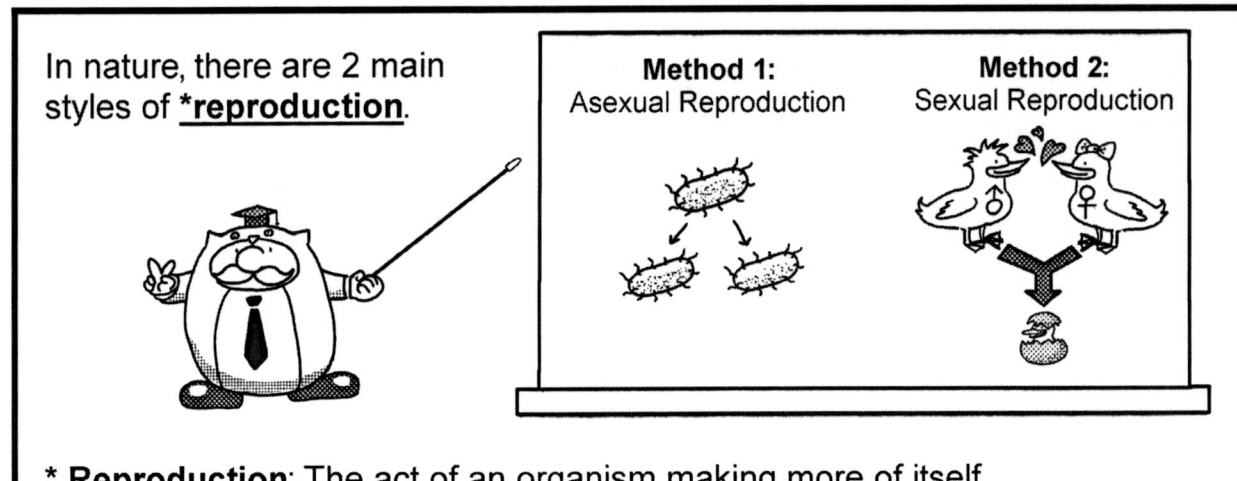

* **Reproduction**: The act of an organism making more of itself.

In **Asexdual** Reproduction, 1 organism simply makes identical copies of itself by *Mitosis*.

"I'm called 'the parent cell' because I am both the mom and the dad."

"And we look exactly like our parent!"

In **Sexual** Reproduction, a male sex cell (also called sperm) combines with a female sex cell (also called egg) to form a new organism.

(sperm) (egg) "This joining process is called **'fertilization.'**"

The Good Thing about Sexual Reproduction:
It creates offsprings ("children") that possess the traits of both the father and the mother..

Dad: Big Ears   **Mom:** Long Tail

Give Birth to

Child: Big Ear + Long Tail

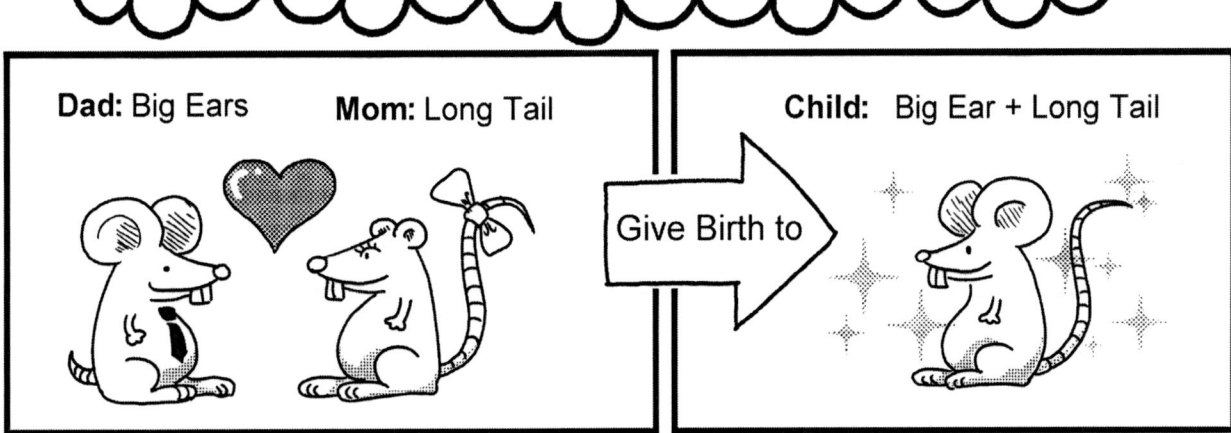

page 163

Unit 3: Chapter 18

In other words, sexual reproduction increases varieties in a population.

For example, the human population has a lot of varieties thanks to sexual reproduction.

Because humans reproduce sexually, we see a lot of varieties in individuals!

Chapter 19: Dominant and Recessive Traits

Science Standard: 7.2.c
Students know an inherited trait can be determined by one or more genes.
Science Standard: 7.2.d
Students know plant and animal cells contain many thousands of different genes and typically have two copies of every gene. The two copies (alleles) of the gene may or may not be identical, and one may be dominant in determining the phenotype while the other is recessive.

NEW VOCABS

* **Alleles:** Varieties of a gene.

* **Carrier:** Someone whose recessive allele is masked (hidden) by the dominant allele and not expressed (shown) in its phenotype.

* **Codominance:** A situation when 2 different alleles of a gene both dominate and express themselves in the organism's phenotype.

* **Dominant allele:** The allele that "dominates" over the other allele.

* **Genotype:** The type of alleles an organism has.

* **Heterozygous:** A situation when the 2 alleles of your gene are different.

* **Homozygous:** A situation when the 2 alleles you have for a gene are completely identical.

* **Incomplete Dominance:** A situation when 2 different alleles of a gene cause a phenotype that is an intermediate (in-between) of the 2 alleles.

* **Multiple Allele:** A situation in which there are 3 or more types of alleles for a gene.

* **Phenotype:** The outward phsyical characteristics of an organism, as determiend by its genotype.

* **Polygenic Trait:** A trait that is encoded (determined) by more than 1 gene.

* **Recessive allele:** The allele that "is dominated by" the other allele.

* **Traits:** Outwardly-showing characteristics of an individual organism.

We have learned that different genes control different traits. **Traits** are: "observable characteristics of an individual."

Examples of Traits:
* Brown Hair
* Small Nose
* Thin
* Cheerful
* Loves Reading

However: Qualities that do not come from genes do not count as "traits."

NOT a Trait: Dying her hair to a different color.

Also NOT a trait: Changing her nose by plastic surgery.

Only characteristics that come from **Genes** can count as **Traits**.

Remember:
In previous chapters, we learned that all organisms have **2 sets of chromosomes.**

"Two is better than one!"

Because chromosomes are in pairs. **Genes** are in **pairs**, too!

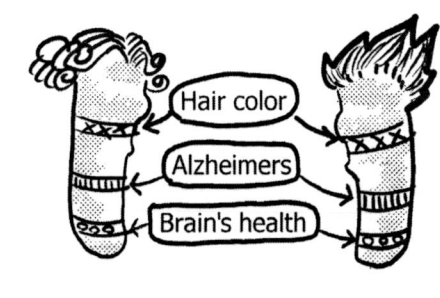

Hair color
Alzheimers
Brain's health

page 167

Unit 3: Chapter 19

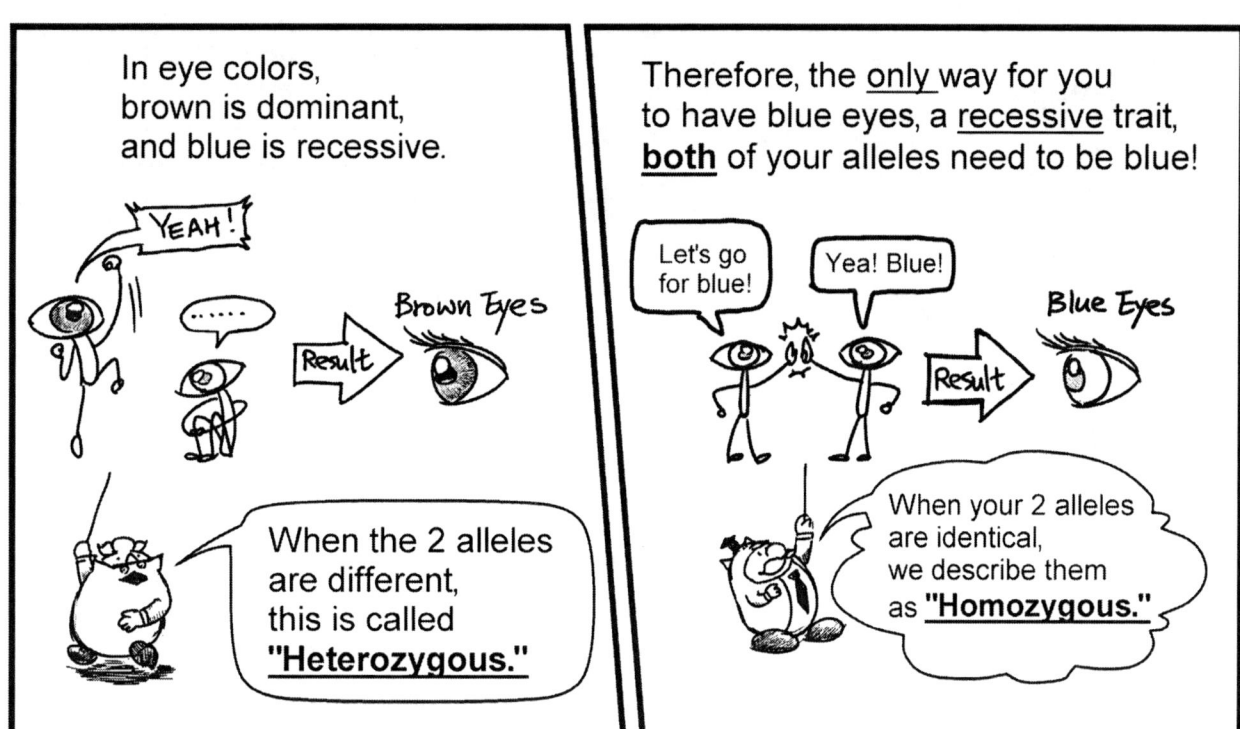

Examples of Dominant/Recessive Alleles in Humans:

Detached Earlobe (Dominant)

Attached Earlobe (Recessive)

Widow's Peak (Dominant)

No Peak (Recessive)

Cleft Chin (Dominant)

No Cleft (Recessive)

Curly Hair (Dominant)

Straight Hair (Recessive)

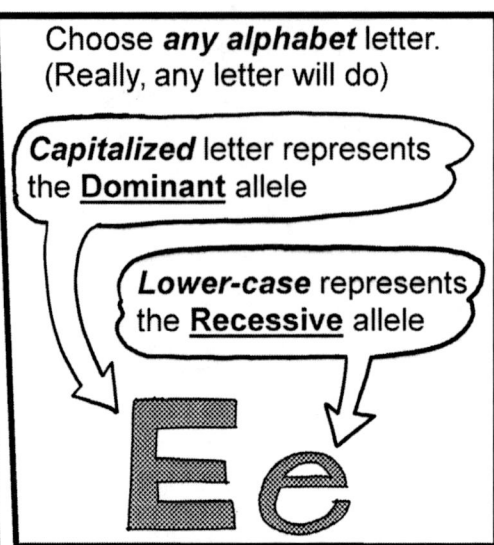
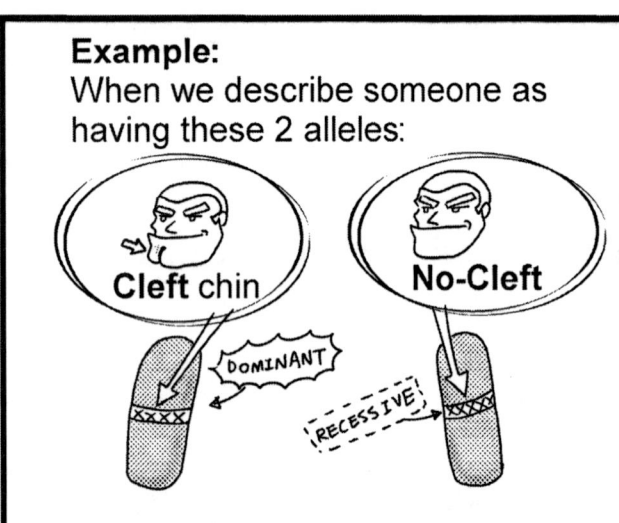
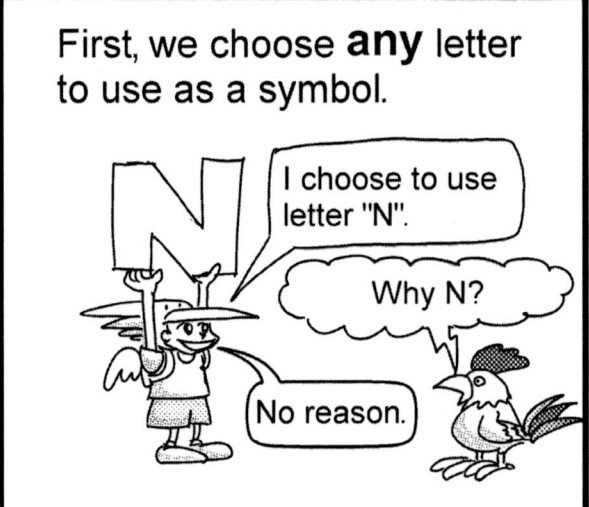
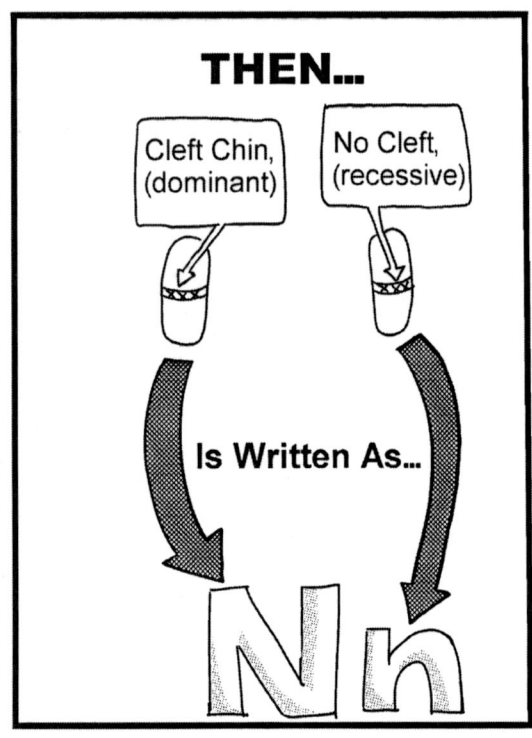

Question:
So, if your alleles are **Nn**, what chin will you end up with?

Answer:
Cleft Chin--because your dominant "N" beats your recessive "n".)

NEW VOCABS

The letters **"Nn"** describes what genes you have, we call this your **"genotype"**.

The word **"cleft chin"** describes the trait caused by the gene, we call it your **"phenotype"**.

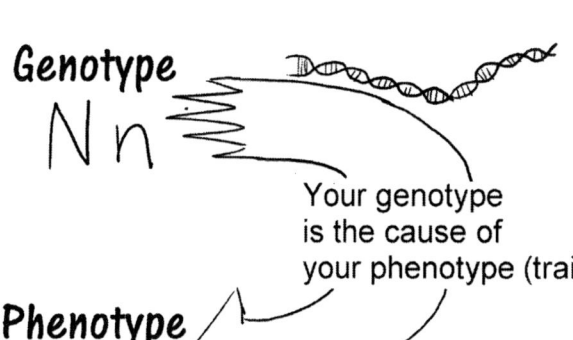

Your genotype is the cause of your phenotype (trait).

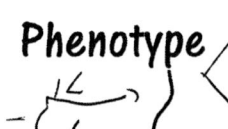

Review:

Genotype:
Nn:

N beats n, producing **phenotype: "Cleft Chin"**.

Genotype:
NN:

2 N's will produce **phenotype: "Cleft Chin"**.

Genotype:
nn:

2 n's will produce **phenotype: "No-Cleft"** Chin.

THERFORE...

2 people with the **same genotype** will always have the **same phenotype**.

Same Genotype   **Same Phenotype**

However:

Just because 2 people have the same **phenotype**, it does not mean they have the same **genotype**!

Because...

My genotype could be NN or nn

My genotype could be NN or nn

But if 2 people have the **same Recessive Phenotype**. Then you *can tell* always know their genotype *must be* nn.

Only the genotype "nn" can give me this chin.

Exactly! There is no other possibility!

page 172

Unit 3: Chapter 19

Remember:
A recessive's allele is *inactive* in the presence of a dominant allele!

"This "n" has no effect!"

This brings us to the topic of **"Carriers"**.

Carrier:
Someone whose recessive allele is hidden by the power of a dominant allele, so we do not see any recessive phenotype on him..

"I have 1 "n""

"But I still have a cleft chin."

Some genetic diseases are recessive. This means that *anyone* could be a carrier yet still look healthy.

Example of Recessive Disease:
"Albinism" (pale hair and skin)

Genotype: Aa
(healthy carrier)

Genoptype: aa
(albino disease)

You need to examine someone's **Genotype** to know whether it is a carrier.

So far, the examples we have listed are simple cases that only involve 2 varieties of alleles.

page 173

Unit 3: Chapter 19

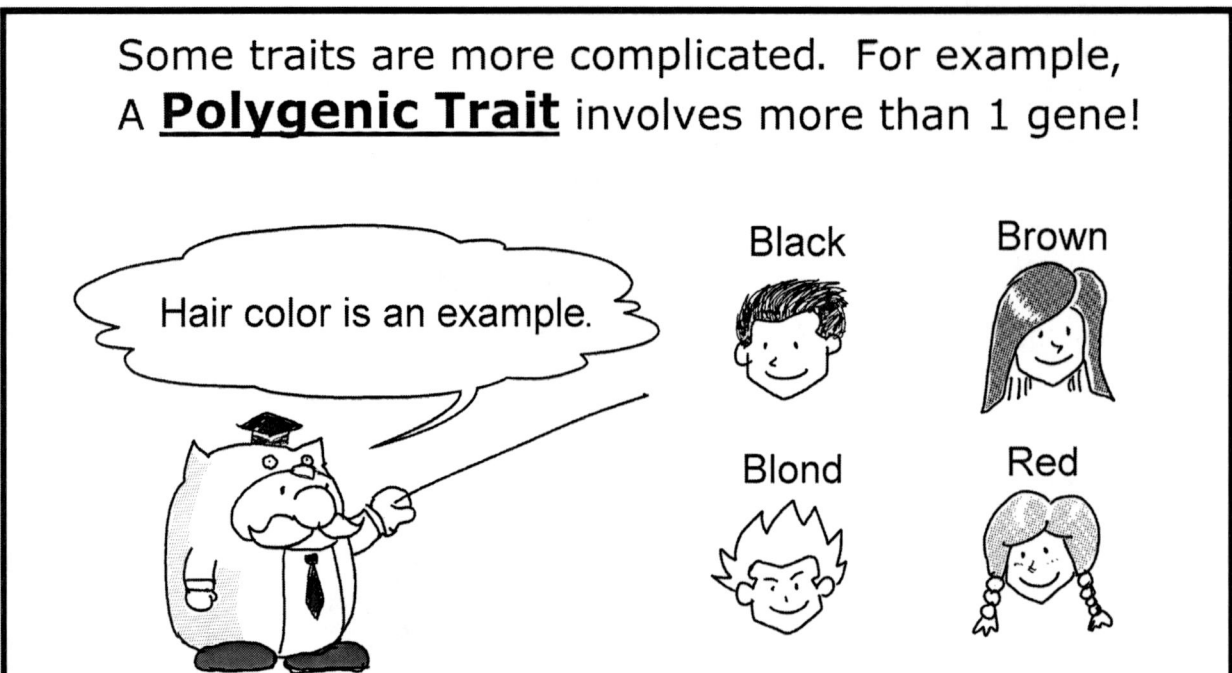

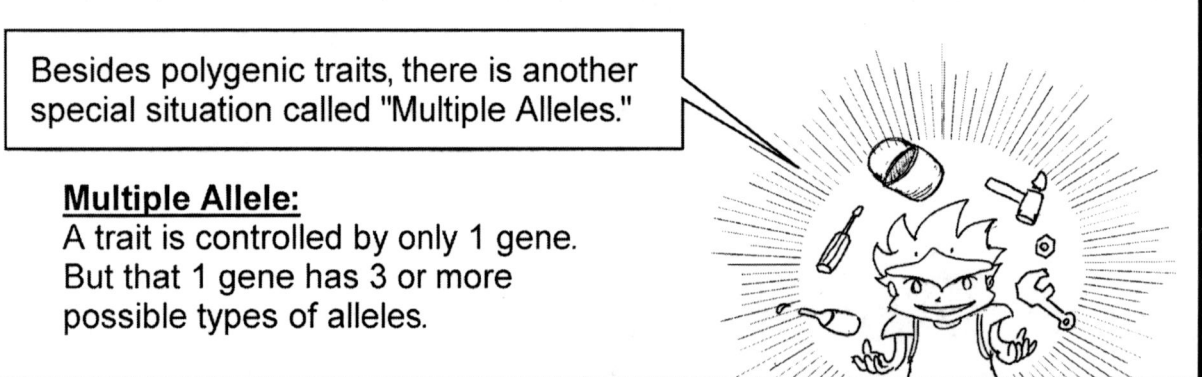

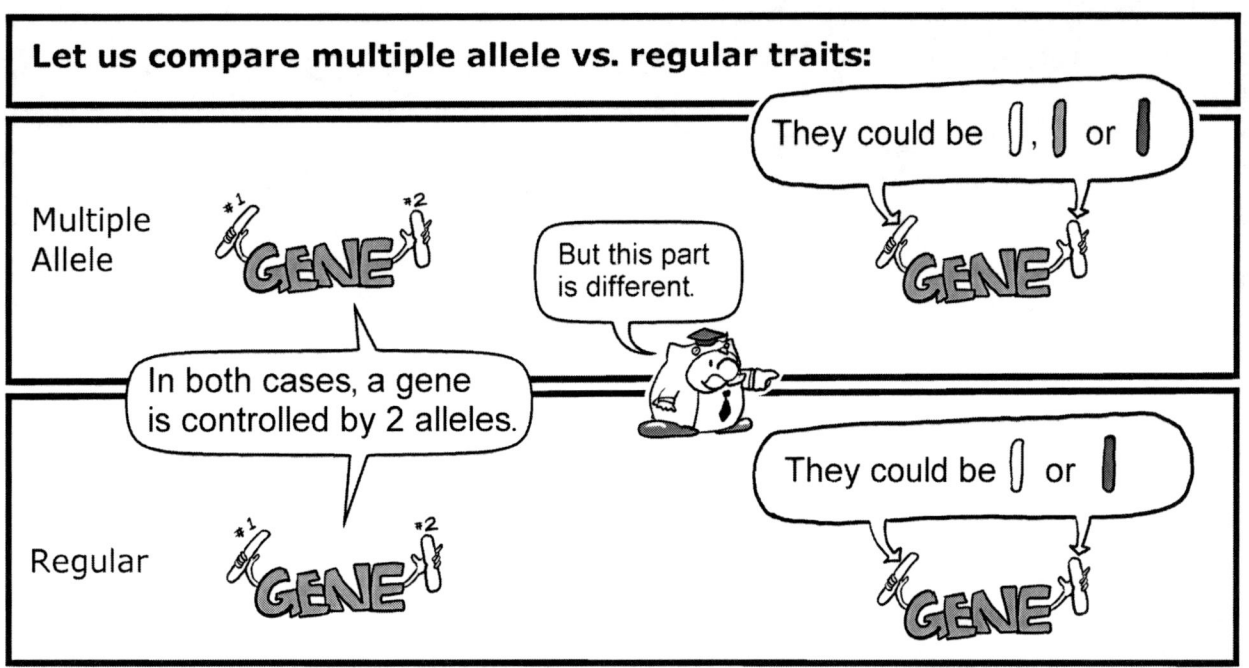

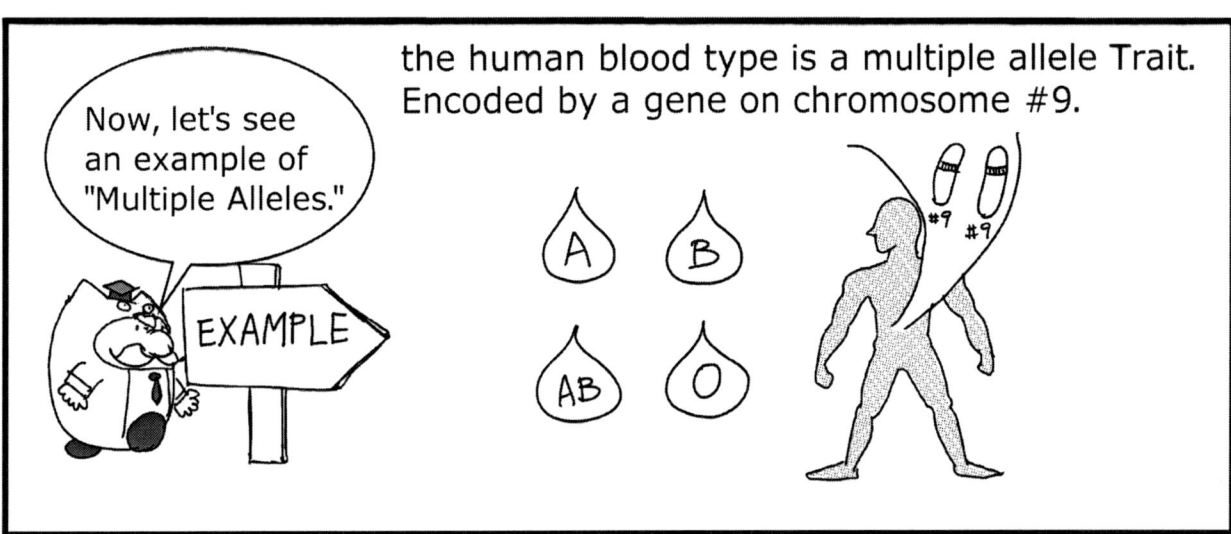

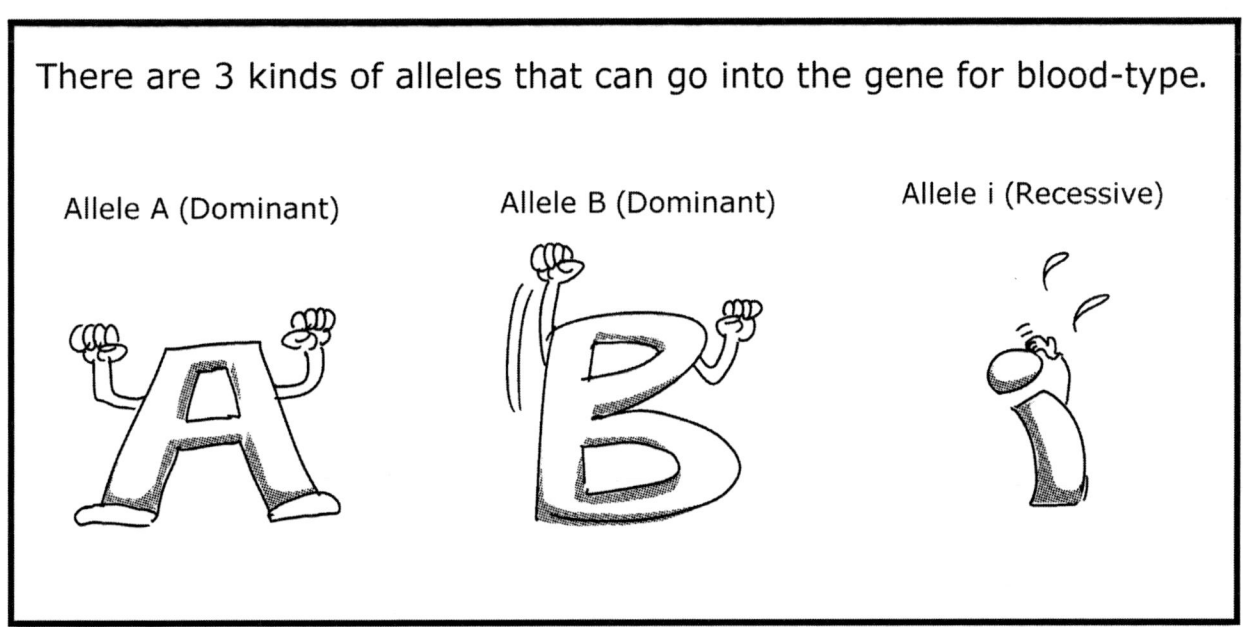

When both of your alleles are A . . . you have **Type A** blood.

If you have A and i alleles, you'll still have **Type A** blood.

  (Recessive)

When both of your alleles are B . . . you get **Type B** blood.

If you have B and i alleles . . . you'll still get **Type B** blood.

 RECESSIVE

When both of your alleles are i . . . you get **Type O** blood.

When you have 1 allele A and 1 allele B . . . you get **Type AB** blood.

  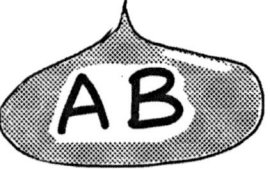

The human blood type is an example of how Multiple Alleles can create a wide variety of traits.

   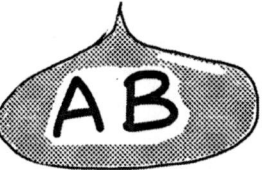

Is there a trait that is **BOTH** Multiple Alleles **AND** Polygenic?

Yup! "**Eye color**" is a trait that is both Mulitple-Alleles and Polygenic.

it is affected by more than one gene (**polygenic**).

And some of the genes have 3 or more potential varieties of alleles (**multiple allele**).

Both Polygenic AND Multiple Alleles? No wonder there are so many varieties of eye color.

The cleft vs. no-cleft chin's example is a very simple case of dominance.

A simple Win/Lose situation.

However, not all cases are this simple.

Special Case #1: Co-dominance:

Definition: BOTH alleles' traits are shown in someone's phenotype!

EXAMPLE: Blood Type AB

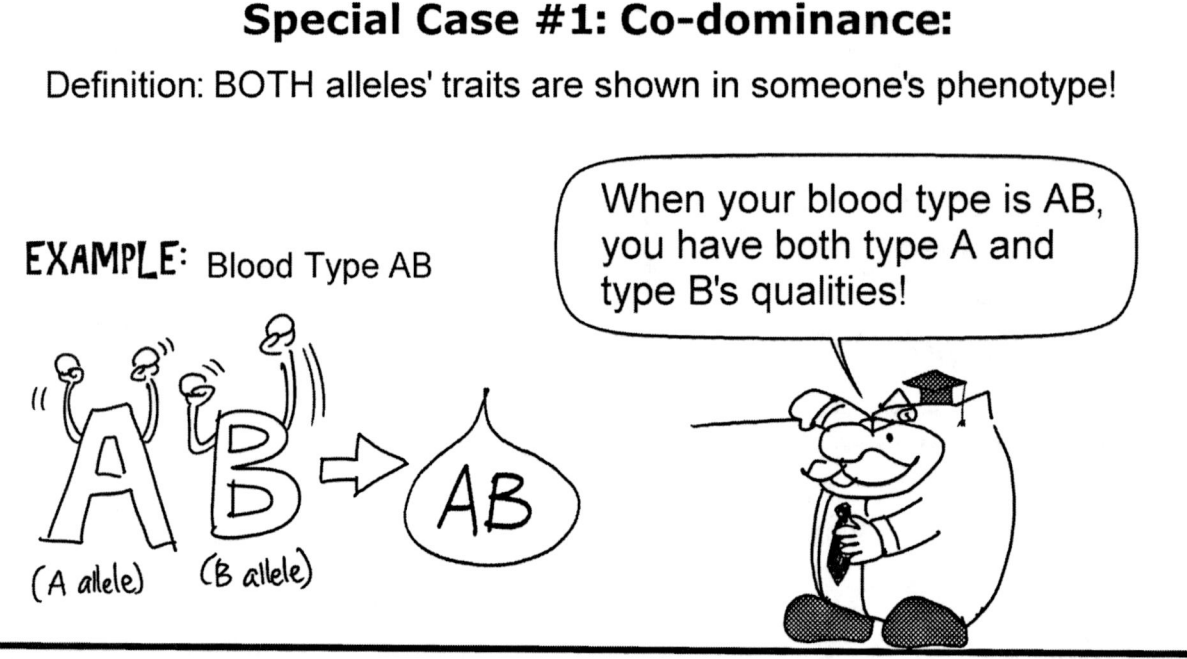

When your blood type is AB, you have both type A and type B's qualities!

Special Case #2: Incomplete Dominance

Definition: This is a situation when neither alleles wins, resulting in a a phenotype that has an "in-between" appearance.

EXAMPLE: Certain types of flowers

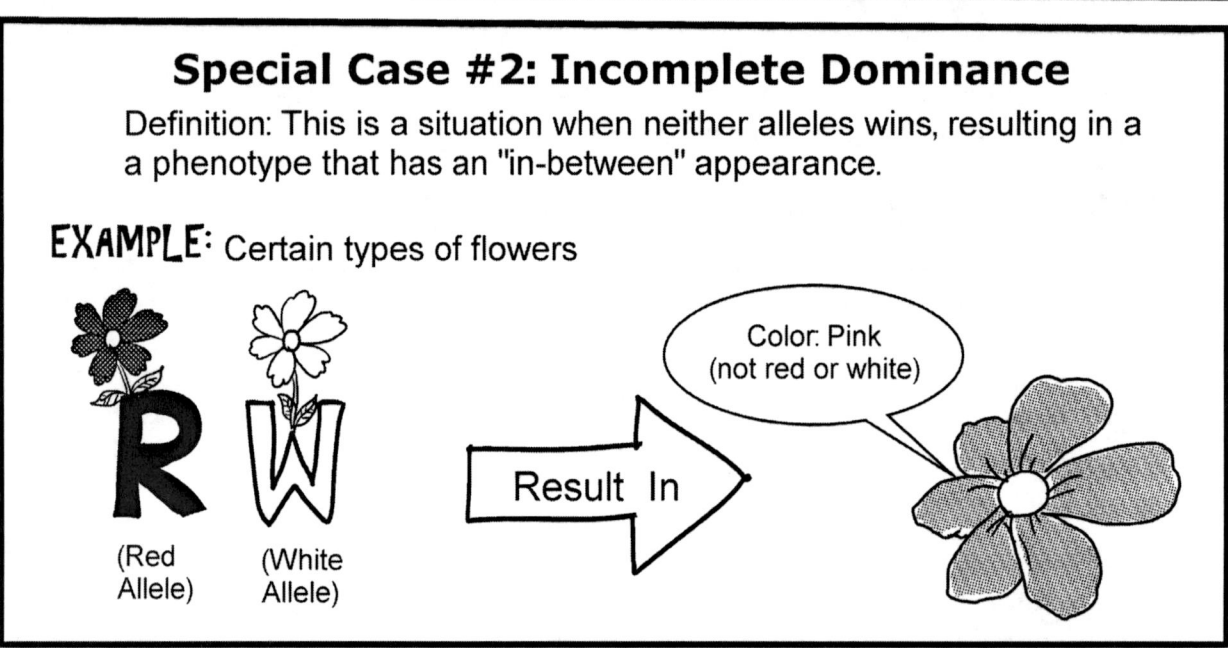

Color: Pink (not red or white)

Chapter 20: What Are "Genes" Made Of?

Science Standard: 7.2.e
Students know DNA (deoxyribonucleic acid) is the genetic material of living organisms and is located in the chromosomes of each cell.

NEW VOCABS

* **Deoxyribonucleic Acid:** Abbreciated as D.N.A. It carries genetic information for all organisms and has a "double-helix" shape.

* **Double Helix:** It is a shape that looks like a ladder twisted into a spiral shape. It is the shape of DNA.

Review on "Traits": Different organisms have different traits.

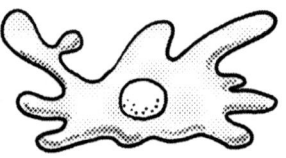

Amoeba:
- single-celled
- move by crawling
- have a nucleus

Mallard (Duck):
- flat beak
- can fly
- webbed feet

Clover:
- triple-leaf
- flower with 6 petals
- photosynthesis

You can find *different* traits even within the *same* species:

Mallard #1
- Bigger beak
- Dark wings

Mallard #2
- Bigger eyes
- Dark torso

Mallard #3
- Bigger tail
- Longer feathers

Every individual's unique traits are determined by D.N.A., found in a cell's nucleus.

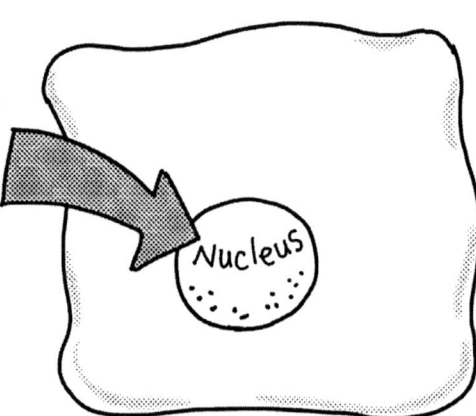

page 180
Unit 3: Chapter 20

DNA is a member of the "nucleic acid" family.

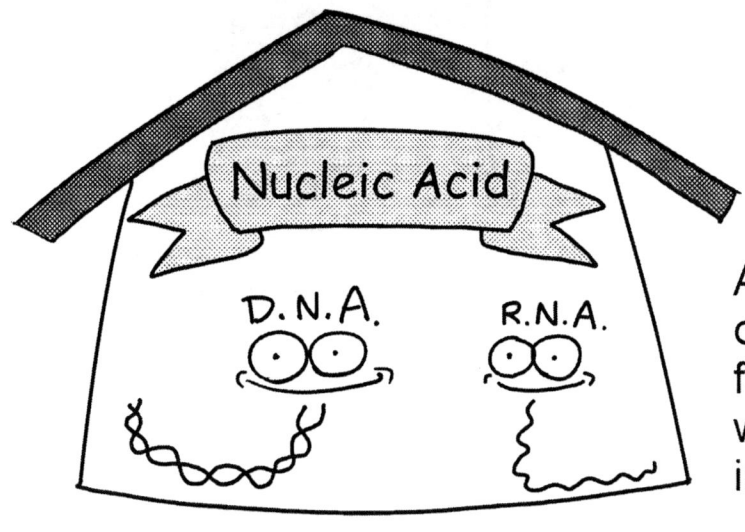

Another member of the nucleic acid family is RNA, which we will discuss in Chapter 25.

DNA's shape is ...

Not rectangular! Not triangular!

DNA's shape is a **Double Helix** shape.

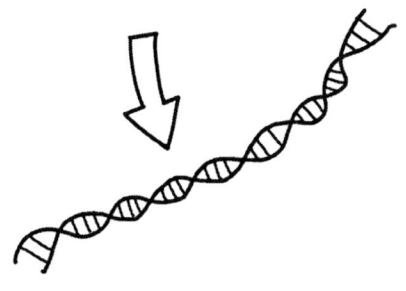

What kind of shape is a "double-helix"?

 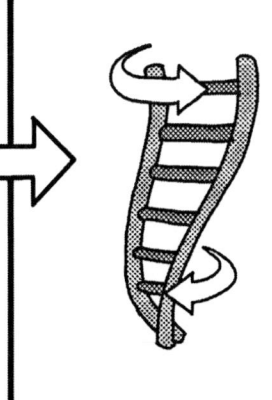 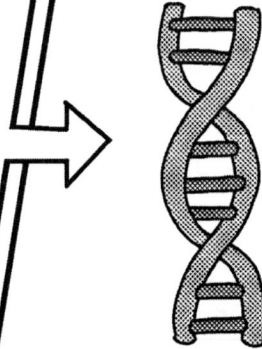

Take a ladder → Twist it! → It's a Double-Helix!

"D.N.A." is just an abbreviation, its full name is:

Deoxyribo**n**ucleic **A**cid

How does DNA get its name?

Deoxyribonucleic   Acid

The backbone is made of a sugar called **"Deoxyribose"**. It also contains a little bit of **"Phosphate"**.

The "rungs on the ladder" are made of **"Nucleotides"**.

DNA is **acidic** due to its high phosphate content.

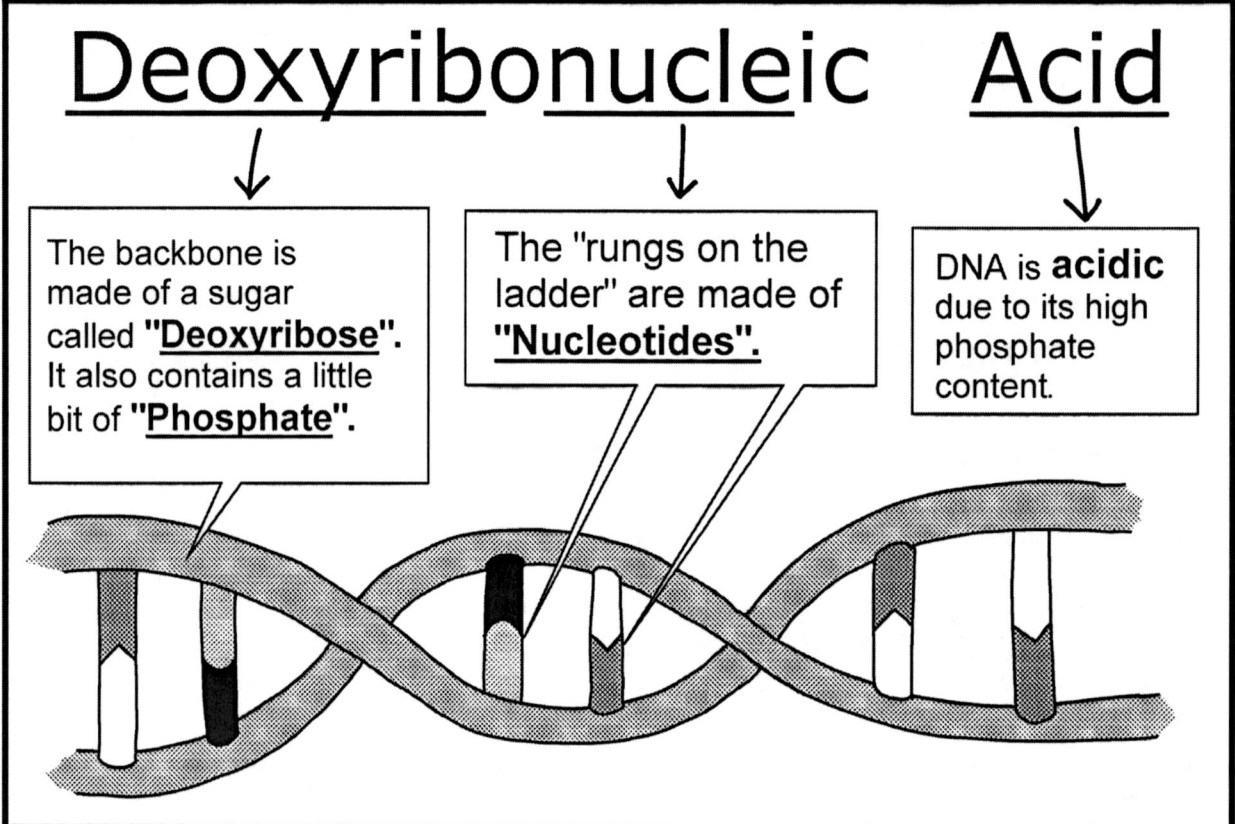

For **eukaryotic** cells (cells that have a nucleus), DNA is inside the **nucleus.**

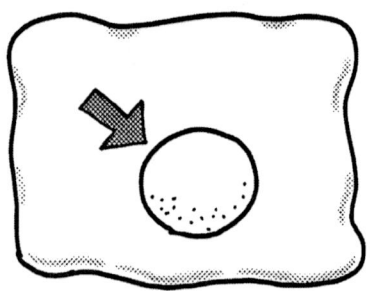

For **prokaryotic** cells (they have no nucleus), DNA is located near the **center** of a cell.

page 182
Unit 3: Chapter 20

During **cell division**, DNA is condensed and packed into packages called "**Chromosomes**".

Unorganized DNA → Packing DNA by coiling them... → And condense into a chromosome

The cell does not just randomly pack its DNA! It is very specific (and organized) in its packing method.

This means that each species produces a specific number of chromosomes during cell division.

cabbage has 18 chromosomes
Human has 46
Dog has 78
Mouse has 40

All members of the *same species* packs its DNA in the exact *same way* !

EAMPLE:

All human cells' chromosome #19 will arrange its DNA like this:

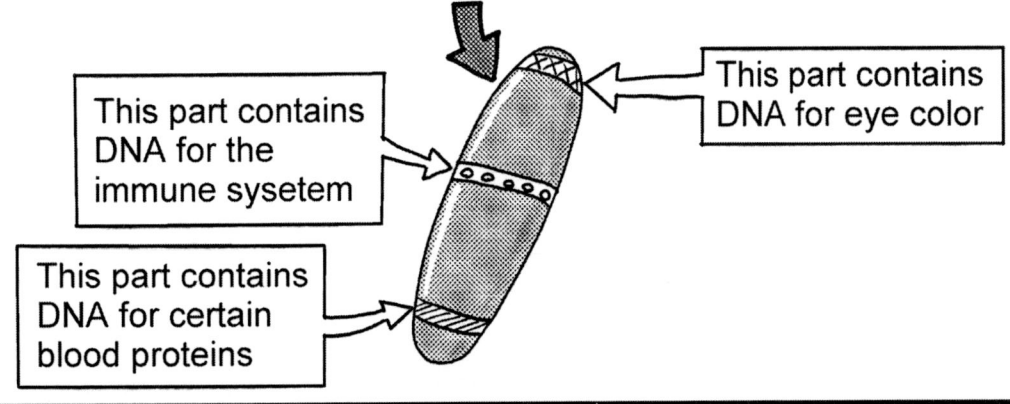

This part contains DNA for eye color

This part contains DNA for the immune sysetem

This part contains DNA for certain blood proteins

Chapter 21: Meiosis and Gametes

Science Standard Bio/LS . 2 . b :
Students know only certain cells in a multicellular organism undergo meiosis.

NEW VOCABS

* **Diploid:** This adjective describes the state at which a cell has 2 sets of chromosomes. The symbol for diploid is "2N".

* **Embryo:** An organism's offspring, in its underdeveloped state. Over time, an embryo will develop into a more complete individual.

* **Endosperm:** The part of a plant's seed that stores up nutrients for the embryo in the seed.

* **Fertilize:** The act of a sperm joining with an egg.

* **Gametes:** Commonly called "Sex cells". The male gametes are called sperms. The female gametes are called eggs.

* **Haploid:** This adjective describes the state at which a cell has only 1 set of chromosomes. The symbol for haploid is "1N".

* **Meiosis:** A type of cell division that produces gametes. It produces 4 haploid daughter cells. (This is different from mitosis, which produces 2 diploid daughter cells)

* **Polar bodies:** During the production of an egg cell, only 1 of the daughter cells becomes an egg, the rest of the daughter cells end up as "polar bodies" instead of being eggs.
Polar bodies serve no known purpose for animals.
But in plants, they turn into endosperms.

* **Somatic cells:** Any non-gamete (non-sex) cell is a somatic cell. Somatic cells make up most of the body tissues.

* **Zygote:** A fertilized egg. As soon as the sperm joins with the egg—the result is a zygote.

There are 2 types of cells in the human body:

1st Type: Somatic Cells (the typical, regular cells.)

Examples:

   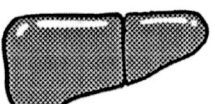

Skin   Muscle   Bone   Liver

Somatic Cells reproduce by **Mitosis**
(See Unit 2: Chapter 13, for more info on Mitosis)

"Your skin (somatic) grows back by mitosis."

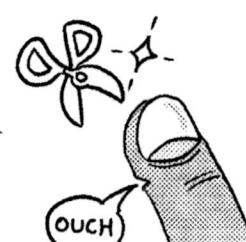 →Mitosis→  HEALED!

2nd Type: Sex Cells

Examples:

SPERM   EGG

Another name for sex cells is: **"Gametes"**.

Sperm = male gamete
Egg = female gamete

Sex cells are **NOT** made by mitosis. Rather, they are made by a different process, called **Meiosis.**

 What is Meiosis?

Meiosis is a type of cell division (different from Mitosis).

Compare: Mitosis vs. Meiosis

Mitosis:

 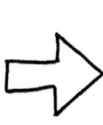  

This is a normal cell. Its DNA is described as being "**2N**", which means it has **2 sets** of chromosomes.

Right before mitosis, the cell double its DNA, giving it **4N** amount of DNA.

Then, **Mitosis** splits the cell into two **2N** daughter cells.

Meiosis:

A normal cell with 2N DNA

Double up into 4N right before cell division.

Split into 4 daughter cells with 1N DNA!

2N is also called "**Diploid**".

1N is also called "**Haploid**".

page 187
Unit 3: Chapter 21

Meiosis Step-by-Step, in Details:

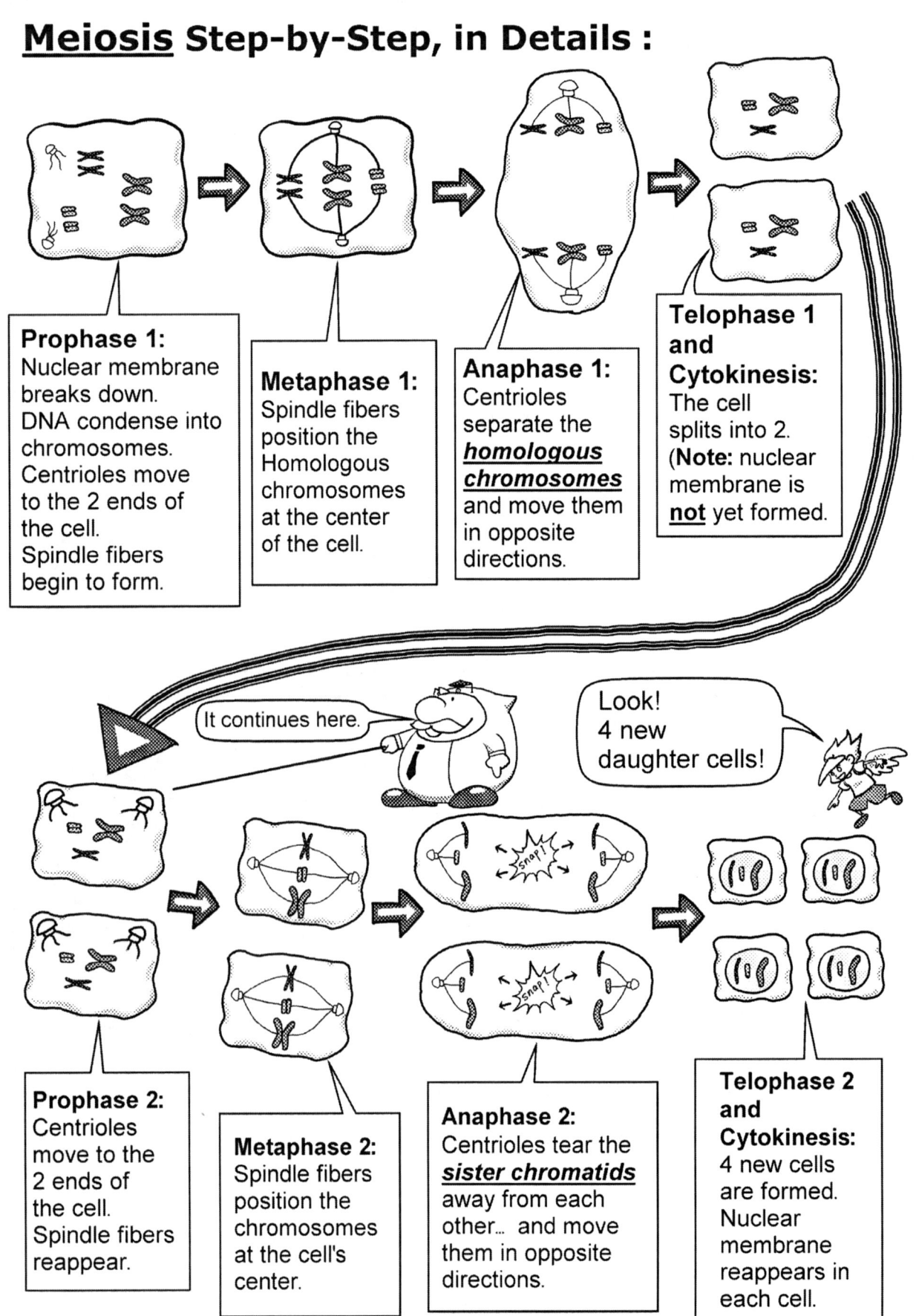

Prophase 1: Nuclear membrane breaks down. DNA condense into chromosomes. Centrioles move to the 2 ends of the cell. Spindle fibers begin to form.

Metaphase 1: Spindle fibers position the Homologous chromosomes at the center of the cell.

Anaphase 1: Centrioles separate the **_homologous chromosomes_** and move them in opposite directions.

Telophase 1 and Cytokinesis: The cell splits into 2. (**Note:** nuclear membrane is **not** yet formed.

It continues here.

Look! 4 new daughter cells!

Prophase 2: Centrioles move to the 2 ends of the cell. Spindle fibers reappear.

Metaphase 2: Spindle fibers position the chromosomes at the cell's center.

Anaphase 2: Centrioles tear the **_sister chromatids_** away from each other... and move them in opposite directions.

Telophase 2 and Cytokinesis: 4 new cells are formed. Nuclear membrane reappears in each cell.

page 188
Unit 3: Chapter 21

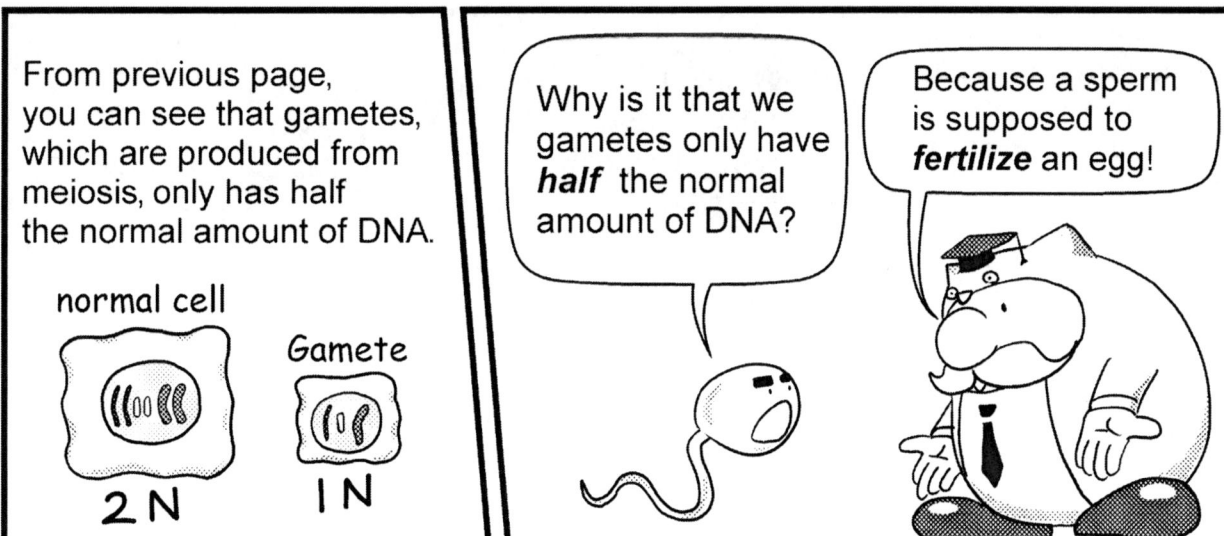

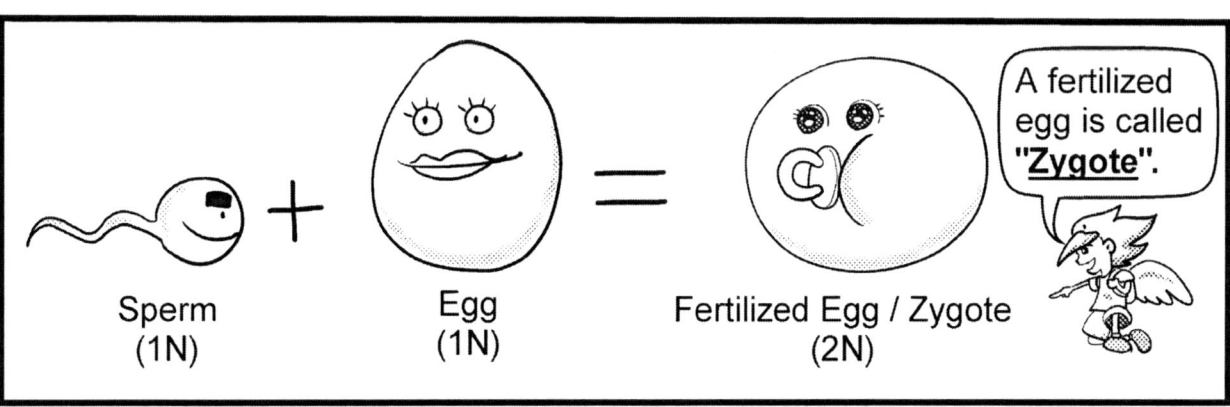

Additional Info: The production of Eggs and Sperms.

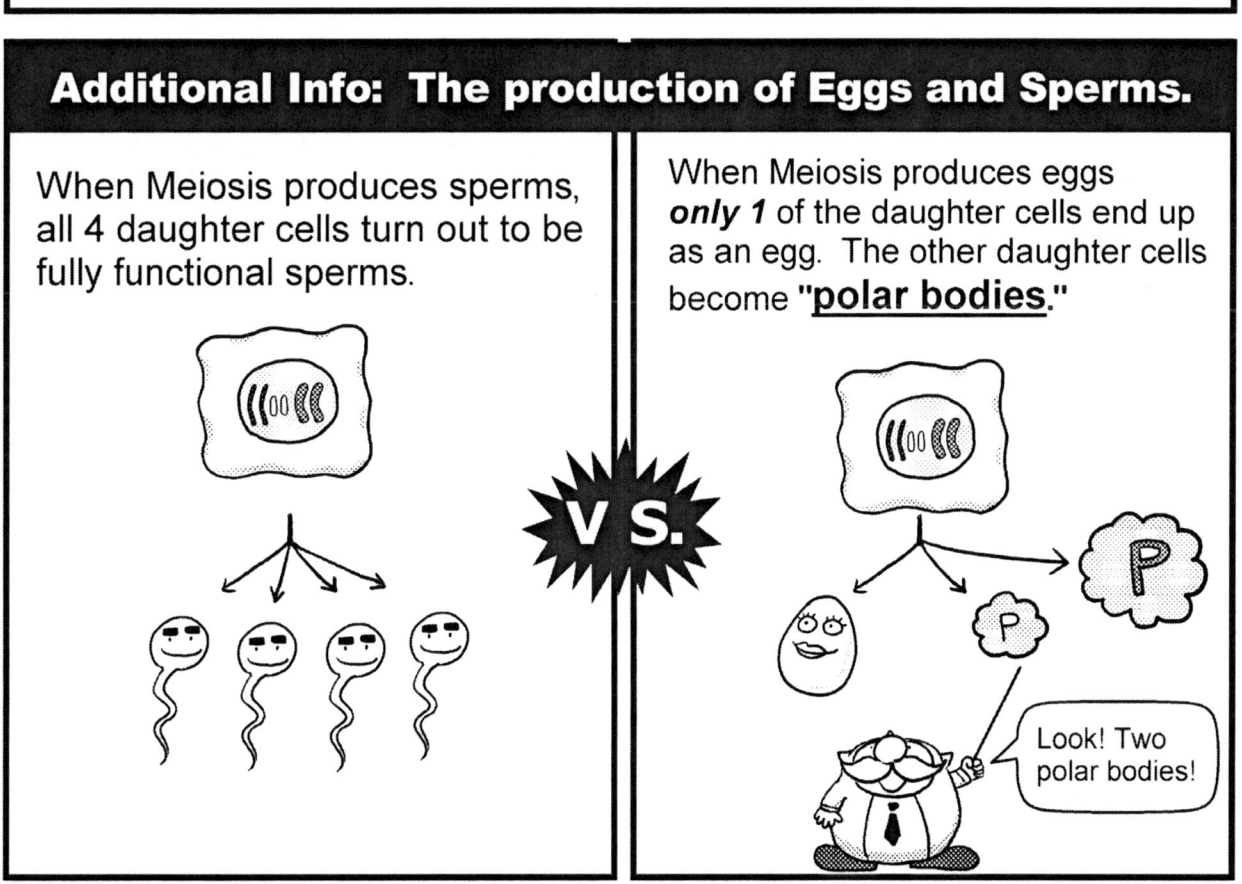

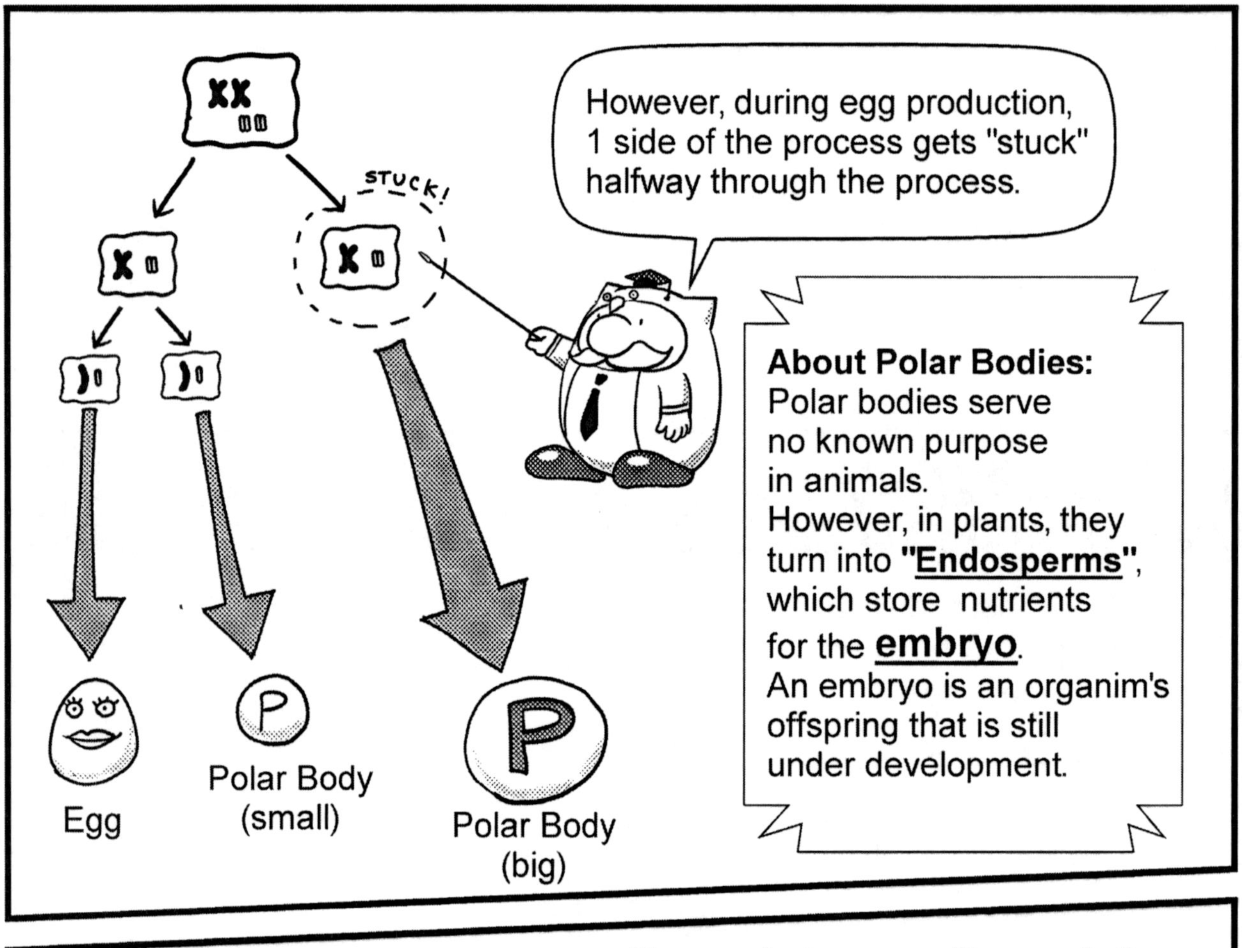

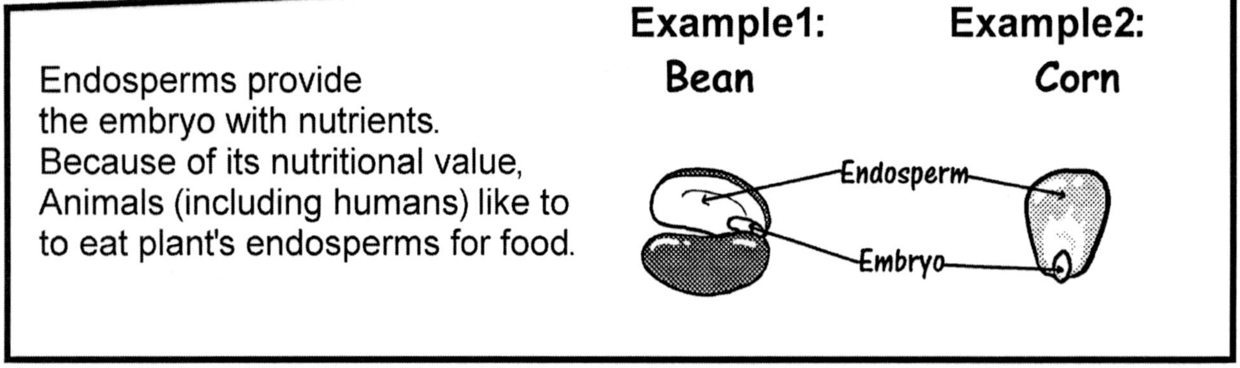

Important Fact on PROPHASE 1:

During Prophase1 of Meiosis, **"Crossover"** always happens:

Step 1:
The **Crossover** process starts with 2 *homologous* chromosomes standing next to each other.

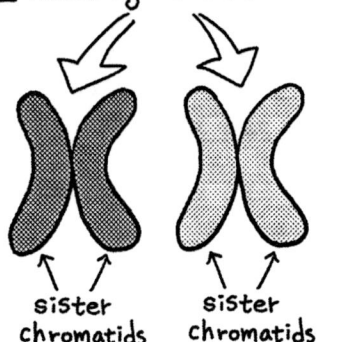

Step 2:
The homologous chromosomes touch each other with 1 of their *sister chromatids.*

Step 3:
The parts of the *chromatids* that overlapped end up being *exchanged* between the 2 homologous chromosomes.

Crossover has a shuffling/mixing effect on the genes. This causes a greater *variety* in one's offspring.

"Variety is a good thing!"

Chapter 22: Sexual Reproduction's Advantage

Science Standard Bio/LS . 2 . d :
Students know new combinations of alleles may be generated in a zygote through the fusion of male and female gametes (fertilization).

Science Standard Bio/LS . 2 . e :
Students know why approximately half of an individual's DNA sequence comes from each parent.

NEW VOCABS

* **Asexual Reproduction:** Reproduction done by mitosis. It does not involve any meiosis or gametes.

* **Reproduction:** The act of an organism producing more organisms of its own type.

* **Sexual Reproduction:** This type of reproduction involves the use of gametes produced from meiosis. Sexual reproduction is comlpeted when the male gamete (sperm) fertilizes the female gamete (egg), which produces a new organism.

When an organism's reproduction involves *gametes*, we call it "**Sexual Reproduction**"

When an organism reproduces by **mitosis** (instead of using gametes), we call that process "**Asexual Reproduction**"

Sexual Reproduction's Advantage --
It causes more variation in one's offspring...**for 2 reasons:**

(1) The offspring (zygote) has **50%** of DNA from its father... and **50%** from its mother. So it does *not* look exactly like either parent. Instead, it is an unique mixture.

(2) The Prophase1 **crossover** that happens during meiosis further increases offspring variety. (* See Chapter 21 for details)

COMING UP: Why variety in the offspring is a good thing.

We will use rabbits as an example.

page 193
Unit 3: Chapter 22

Chapter 23: Sex Chromosomes

Science Standard Bio/LS . 2 . f :
Students know the role of chromosomes in determining an individual's sex.

NEW VOCABS

* **Autosomes:** Chromosomes that contain genes for all the characteristics not directly related to an organism's gender. Basically, any chromosome that is NOT a sex-chromosome is an autosome.

* **Sex Chromosomes:** Chromosomes that determine an organism's sex (gender).

* **X Chromosome:** One of the two types of sex chromosomes. It causes female characteristics in an organism. However, the X chromosome contains more than just genes that cause the female traits. It also contains other genes that are vital to an organism's survival.

* **Y Chromosome:** One of the two types of sex chromosomes. It causes male characteristics in an organism.

Every organism with gender has *a pair* of **Sex Chromosomes.**	For Example: Humans have a total of **23 pairs (46 pieces)** of chromosomes. 
Among the 23 pairs, 1 pair are **Sex Chromosomes.** 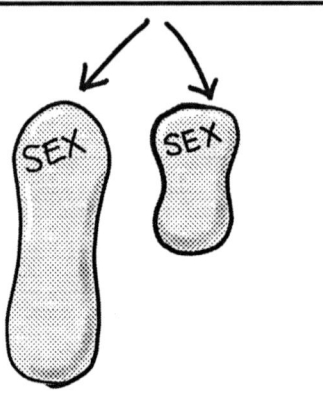	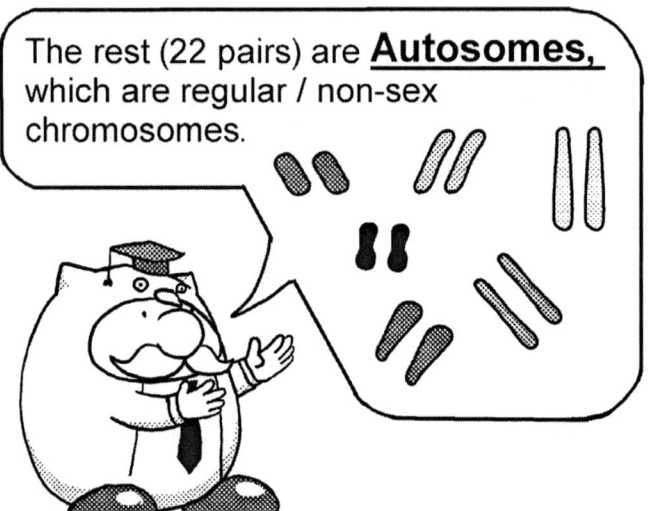 The rest (22 pairs) are **Autosomes,** which are regular / non-sex chromosomes.
There are **2 types** of sex chromosomes: X and Y 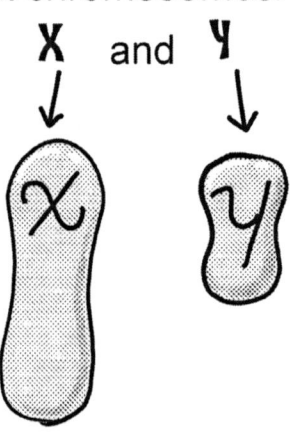	 X chromosome = **female** gender — Y chromosome = **male** gender

If Your Sex Chromosomes are . . .

1 X + 1 Y → makes you "Male"

1 X + 1 X → makes you "Female"

"What about 1Y+1Y?"

"1Y + 1Y does **not** work... The offspring will *die*!"

The Y chromosome only contains the genes that cause the male gender.

But the X chromosome contains the female genes AND other genes vital to your survival!

"No one can survive without me"

Chapter 24: Predicting One's Offspring

Science Standard Bio/LS.3.a:
Students know how to predict the probable outcome of phenotypes in a genetic cross from the genotypes of the parents and mode of inheritance (autosomal or X-linked, dominant or recessive).

NEW VOCABS

* **Autosomal gene:** A gene located on an autosome (reminder: automsome = non-sex chromosome).

* **Autosomal trait:** A trait controlled by autosomal gene(s).

* **Punnette Squares:** A method used to predict the genotypes of 2 parents' offsprings. But this method has a limitation-- it can only be used to predict autosomal genes.

* **Sex-linked gene:** A gene located on the X-chromosome. (Not the Y-chromosome, which is too small to contain sex-linked genes)

* **Sex-linked trait:** A trait controlled by sex-linked gene(s).

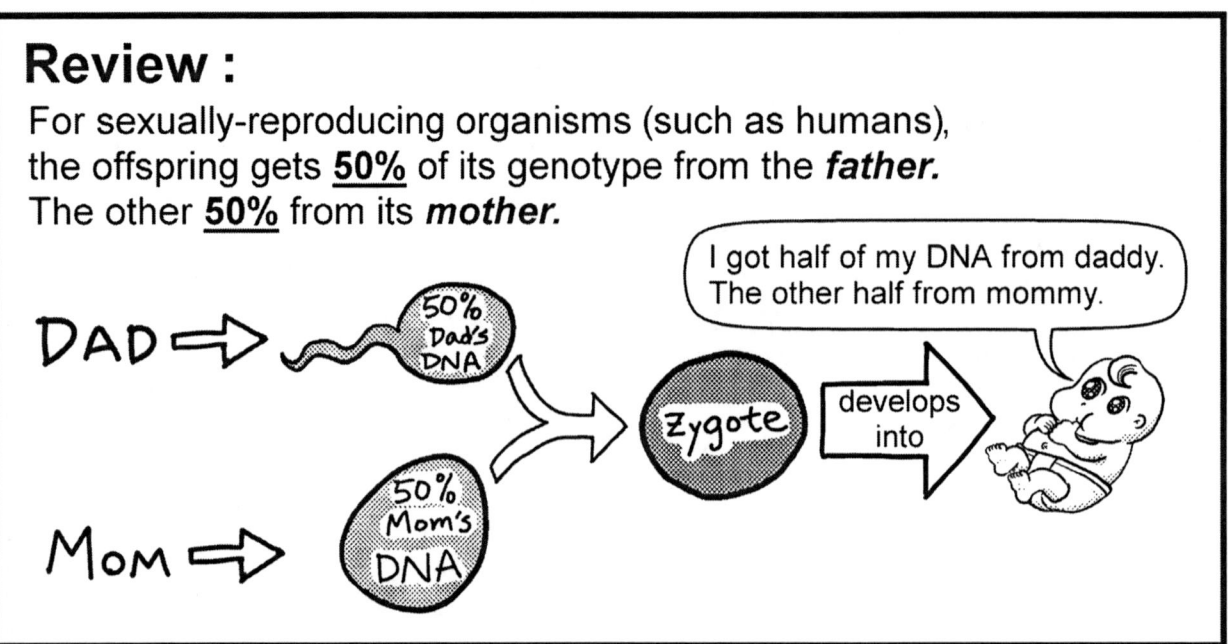
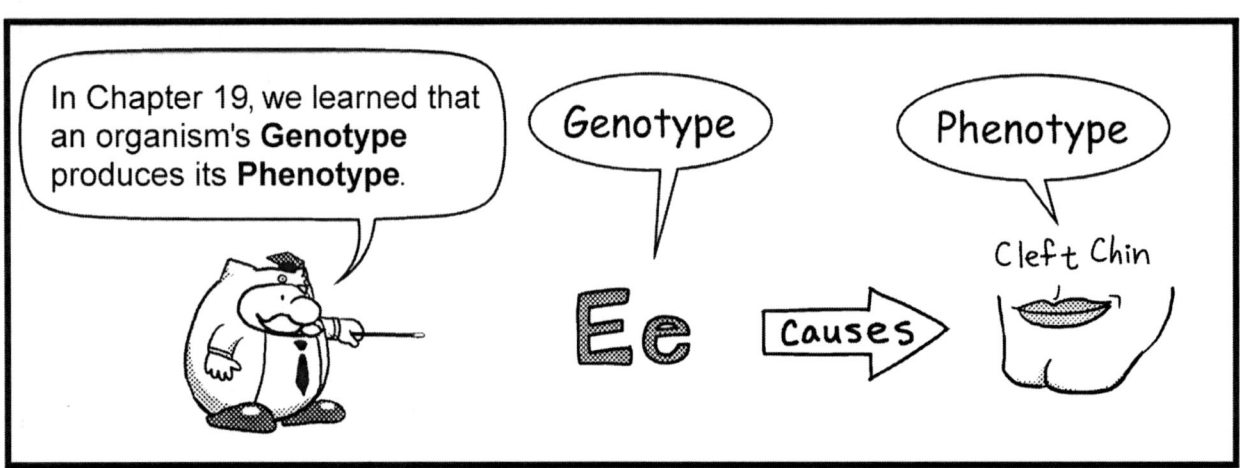

How Punnette Squares Work:

Step 1 :

First, decide on what trait you want to examine :

"Let's look into the chance of our child having a clefted chin!"

"Good idea !"

Step 2 :

Write down the 2 parents' genotype for that particular trait.

e e   E e

Step 3 :

Draw a 2 x 2 table. Label the father's genotype on the upper side, and label the mother's on the left side. (Or you can switch their places, the outcome will be the same)

"My sperms can either carry "E" allele **or** "e" allele!"

"My genotype is ee. So I can **only** produce eggs that have the "e" allele."

page 201

Unit 3: Chapter 24

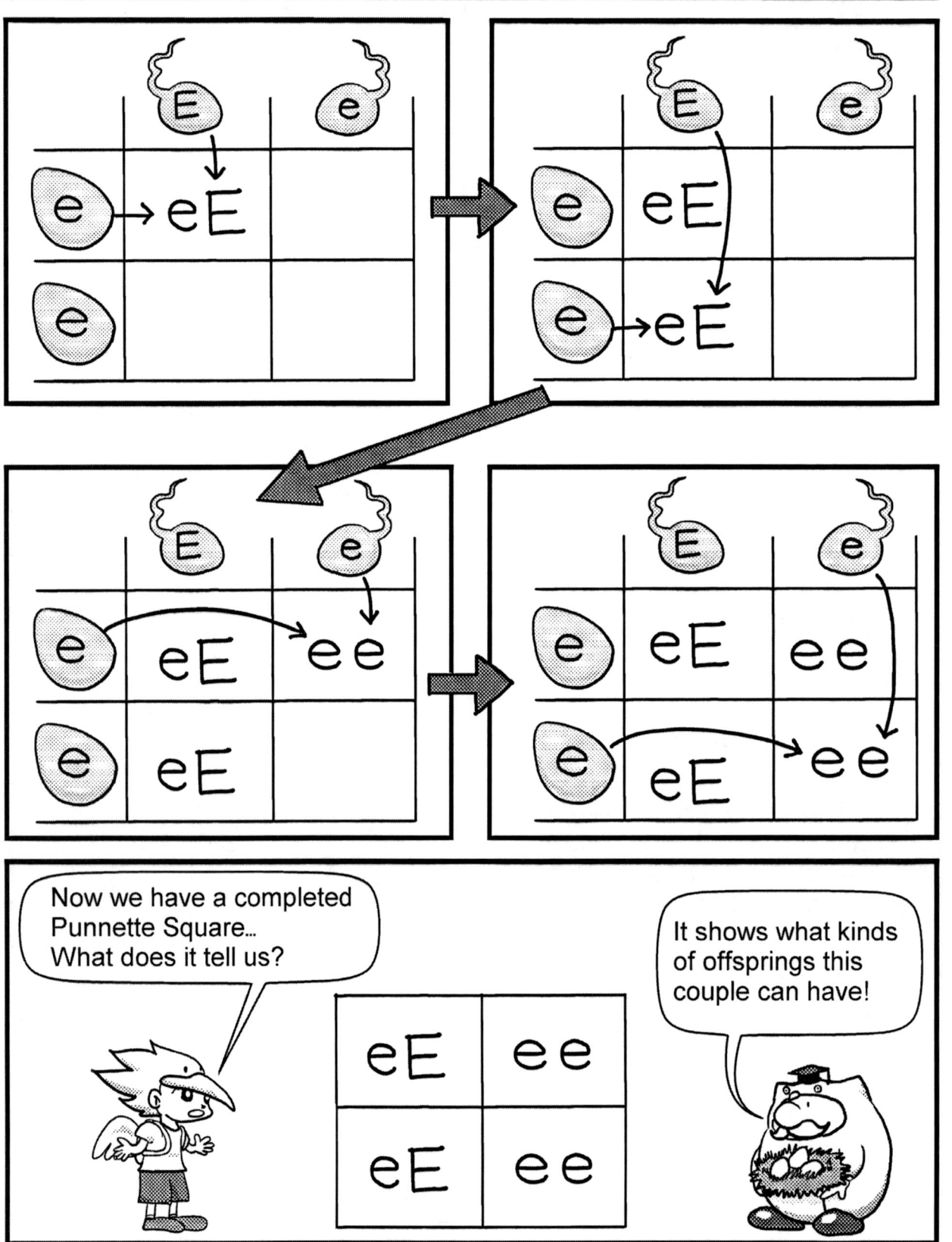

EXAMPLE: Widow's Peak.

Here we have 2 parents who are both "Aa":

 Aa  Aa

First, we set up a Punette Square with parents' alleles.

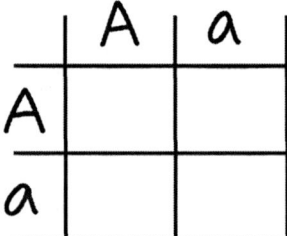

2nd, we fill out the Punnette Square.

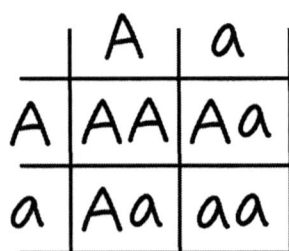

Then, we analyze the results, each tile has a 25% chance of being in the offspring.

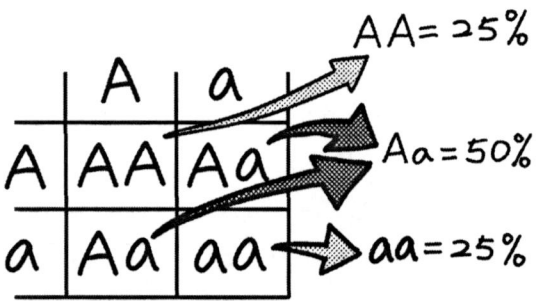

AA = 25%
Aa = 50%
aa = 25%

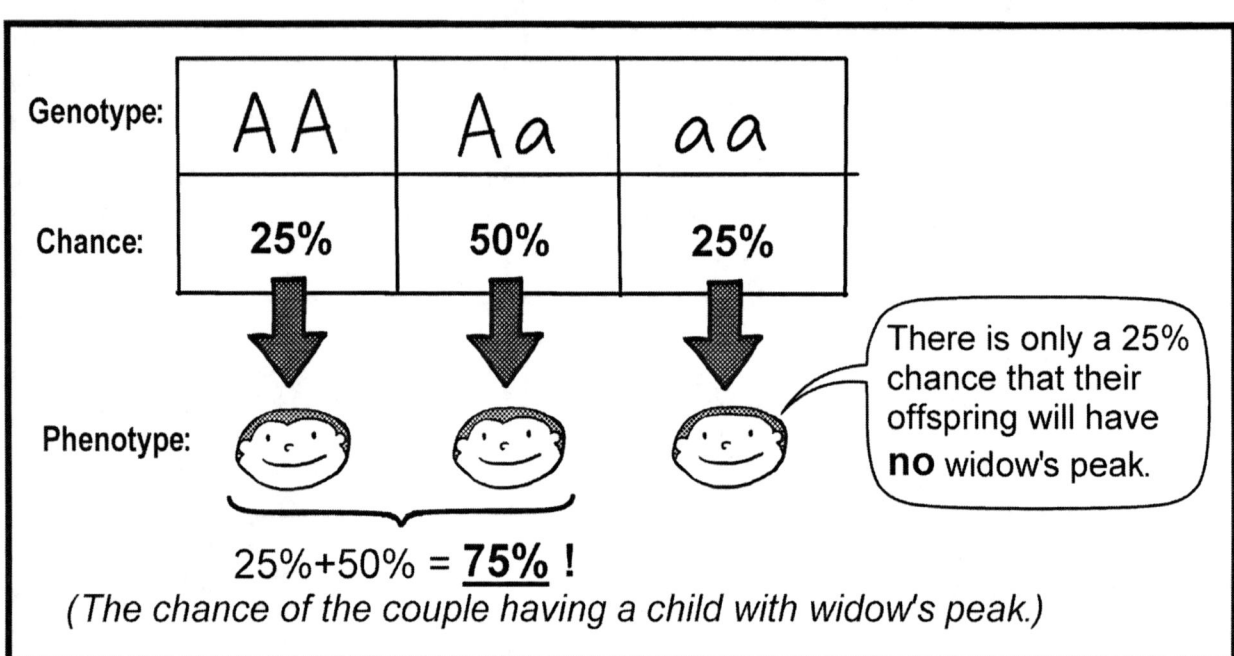

There is only a 25% chance that their offspring will have **no** widow's peak.

25% + 50% = **75%** !
(The chance of the couple having a child with widow's peak.)

Punnette Squares is only used on genes of **autosomes**. Autosomes are *non-sex* chromosomes..

Punnette Squares *cannot* be used to predict genes located on the **sex chromosomes!**

There are 2 types of sex chromosomes:

X (female) & Y (male)

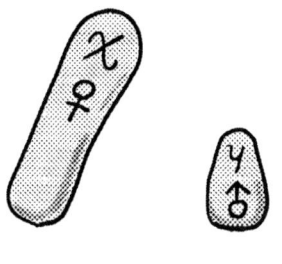

Genes on the **X-chromosomes** are called **"sex-linked genes"**.

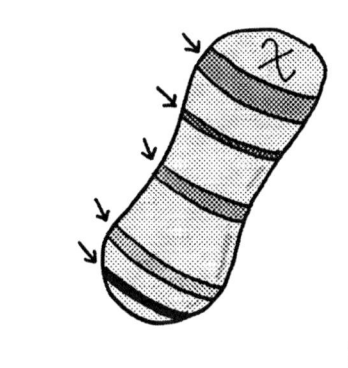

The Y chromosome does **not** carry any sex-lnked gene.

I'm too small to carry any sex-linked gene...

Traits caused by sex-linked genes are called **"Sex-Linked Traits"**
Below are 2 examples:

Color-Blindness
(Trouble in seeing certain colors)

Hemophilia
(Blood's inability to clot)

Chapter 25: DNA and RNA

Science Standard Bio/LS . 5 . a :
Students know the general structures and functions of DNA, RNA, and protein.

NEW VOCABS

* **Adenine:** 1 of the 5 types of nucleobases. Its symbol is "A".

* **Cytosine:** 1 of the 5 types of nucleobases. Its symbol is "C".

* **Guanine:** 1 of the 5 types of nucleobases. Its symbol is "G".

* **Thymine:** 1 of the 5 types of nucleobases. Its symbol is "T". It is only found in DNA, not RNA.

* **mRNA:** mRNA stands for messenger ribonucleic acid. It copies down DNA's instructions and send it to rRNA to tell rRNA how to assemble a protein.

* **Nucleobases:** They are an important part of DNA and help connect the 2 strands of DNA together. There are 5 types of nucleobases: adenine (A), guanine (G), thymine (T), cytosine (C), and uracil (U).

* **rRNA:** rRNA stands for ribosomal ribonucleic acid. It helps put amino acids together into proteins.

* **tRNA:** tRNA stands for transfer ribonucleic acid. Each tRNA carries an amino acid and it will transfer the amino acid to rRNA, which assembles them into a protein.

* **Uracil:** 1 of the 5 types of nucleobases. Its symbol is "U". It is only found in RNA, not DNA.

Unit 3: Chapter 25

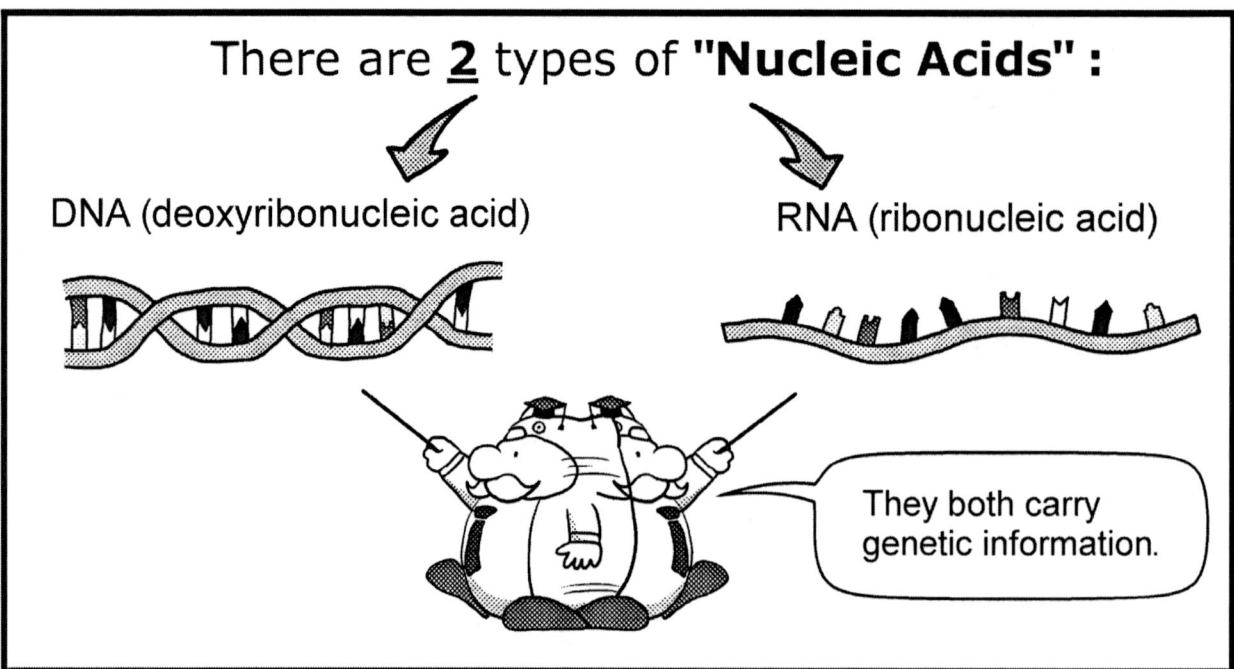
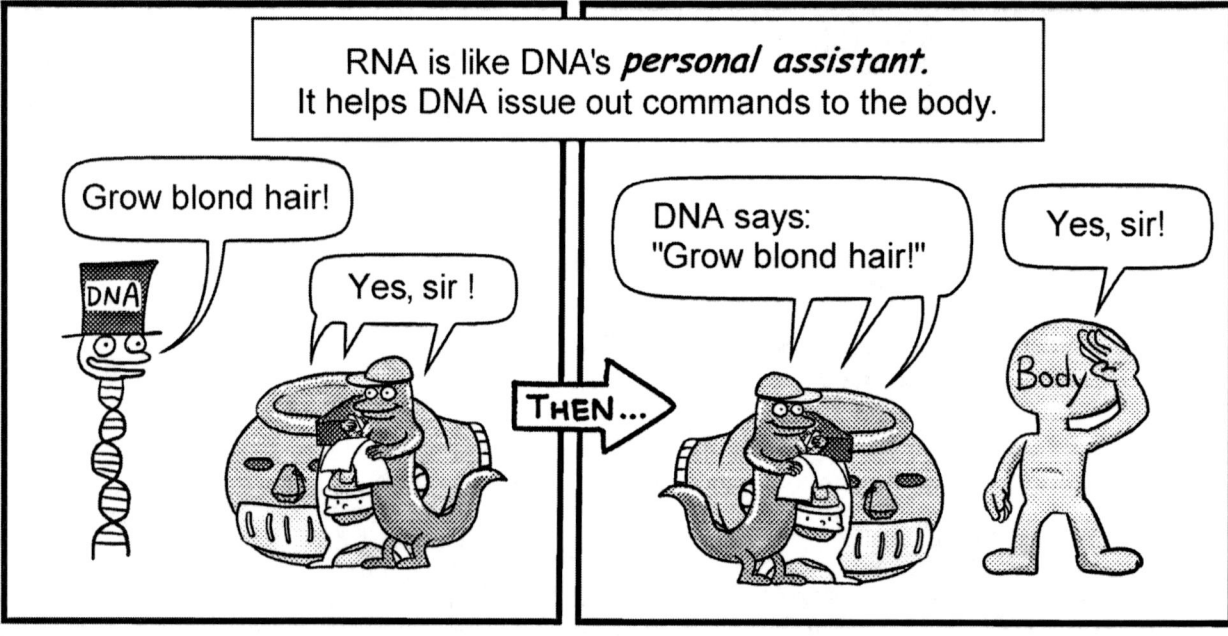
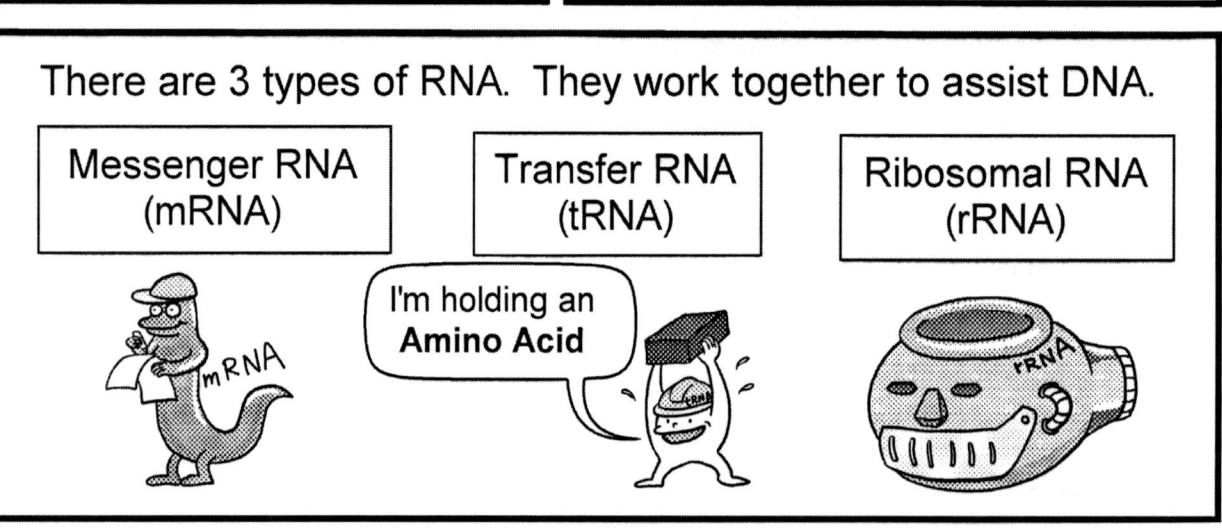

Here is how the 3 kinds of RNA work together to **assist the DNA:**

Step 1: mRNA copies down the command issued by the DNA.

"Tell the body to make **curly hair**!"

"OK, let me write it down..."

Step 2: mRNA tells this message to the rRNA.

"DNA wants you to make **curly hair**."

"O K"

Step 3: tRNA comes and throw the proper *amino acids* into rRNA to form a **protein** that will create the effect the DNA wants.

Special Protein that causes "curly hair"

Summary of how DNA controls you:

DNA controls RNA's.
Then, RNA controls **proteins**.
And protein controls the **body**.

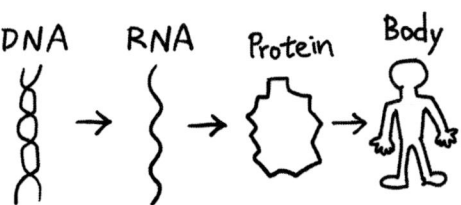

Protein:
(1) controls an organism's body.
(2) builds most of its tissues.

Next topic:
An in-depth look at DNA's **structure.**

DNA's double-helix shape contains many "rungs".
Each "rung" is made of **2** different **nucleobases**, as shown below:

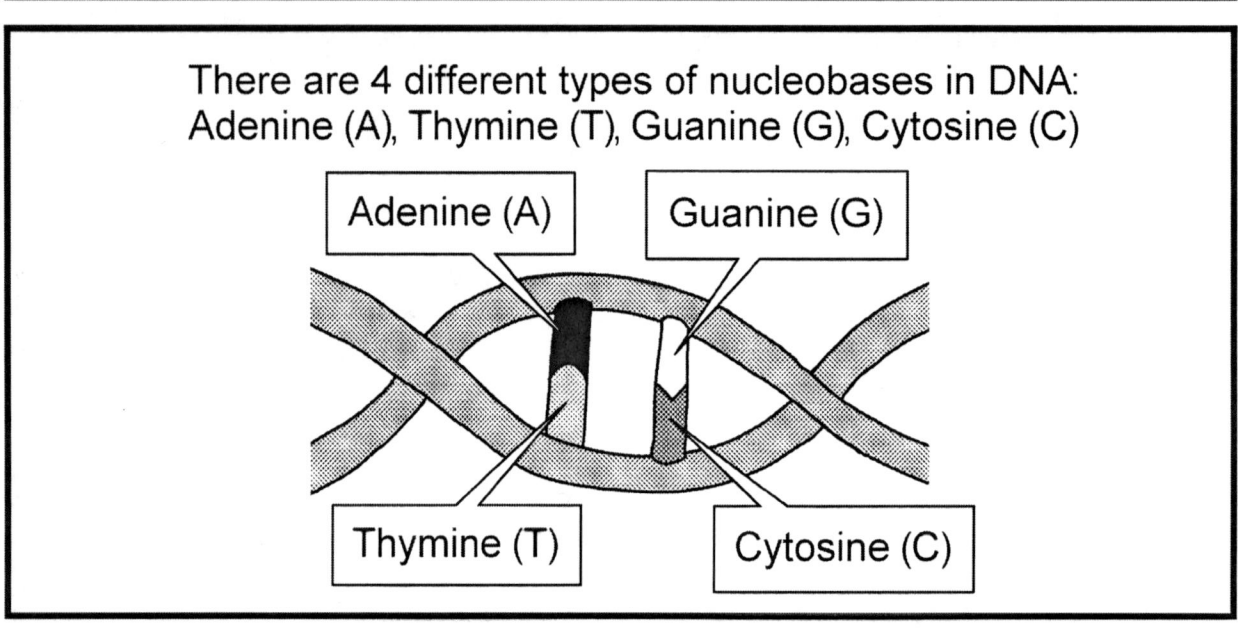

There are 4 different types of nucleobases in DNA:
Adenine (A), Thymine (T), Guanine (G), Cytosine (C)

Adenine (A) Guanine (G) Thymine (T) Cytosine (C)

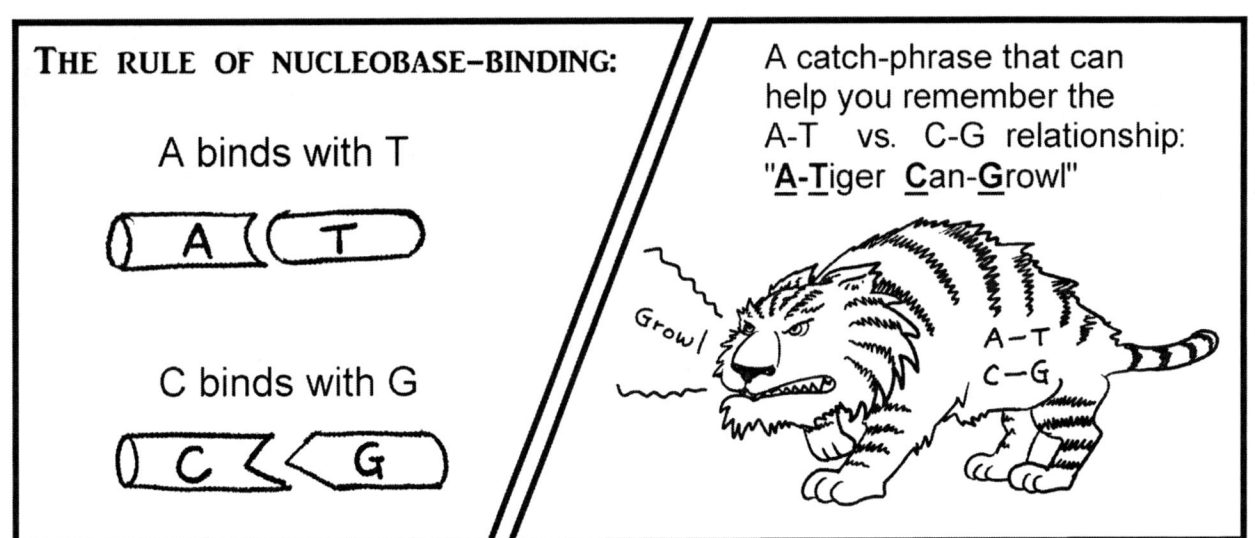
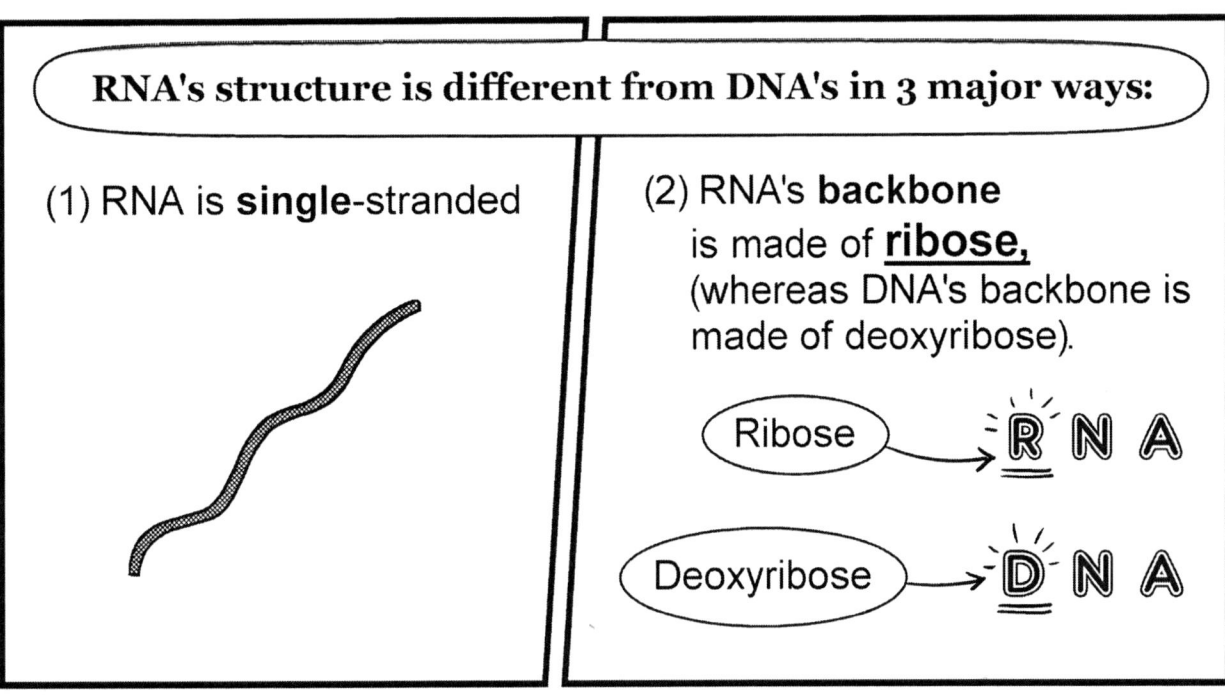

page 214
Unit 3: Chapter 25

Unit 4: Evolution

Chapter 26: Genetics and Evolution

Science Standard: 7.3.a
Students know both genetic variation and environmental factors are causes of evolution and diversity of organisms.
Science Standard: Bio/LS.7.c
Students know new mutations are constantly being generated in a gene pool.
Science Standard: Bio/LS.7.d
Students know variation within a species increases the likelihood that at least some members of a species will survive under changed environmental conditions.

NEW VOCABS

* **Adaptation:** A feature that allows an organism to have a better chance to survive an environment.

* **Artificial Selection:** Also known as "selective breeding." It is a human-guided process in which humans breed only plants or animals with certain trait(s) in order to cause change to a population. (It works like natural selection—except in this case, the humans are doing the "selection".

* **Evolution:** The transformation of a population, caused by natural selection over numerous generations and a long period of time.

* **Genetic Diversity:** The amount of variety of genetic characteristics in a population.

* **Genetic Mutation:** A random change in an organism's genes

* **Macro-Evolution:** A large-scale evolution that changes the population from 1 type of organisms to a different type of organisms.

* **Micro-Evolution:** A small-scale evolution that involves small changes to a population of organisms.
  This process requires adequate amount of genetic diversity and natural (or artificial) selection.

* **Natural Selection:** A process that causes only the fit organisms to survive, the unfit ones will eventually die out.
  This process is often described as: "Survival of the fittest."

 **Adaptation:** A feature that allows an organism to have a better chance to survive in an environment.

EXAMPLES OF Adaptation:

A crab's **pincers** help it catch food.

A polar bear's **fur** helps it stay warm.

A cheetah's **speed** helps it catch food.

A spider's **knowledge** on web-making helps it catch food.

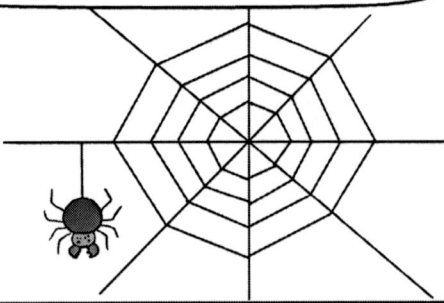

DID YOU KNOW…?

Spiders are **born with** the **knowledge** of how to make webs.

A knowledge known at birth (which is called an "instinct") counts as an adaptation.

An organism's adaptations act like its survival equipments.

"Why do organisms need adaptations?"

"Because it is hard to survive in nature!"

In nature, organisms face many challenges:

Predators
Limited space
Heat
Cold
Catching/Finding Food

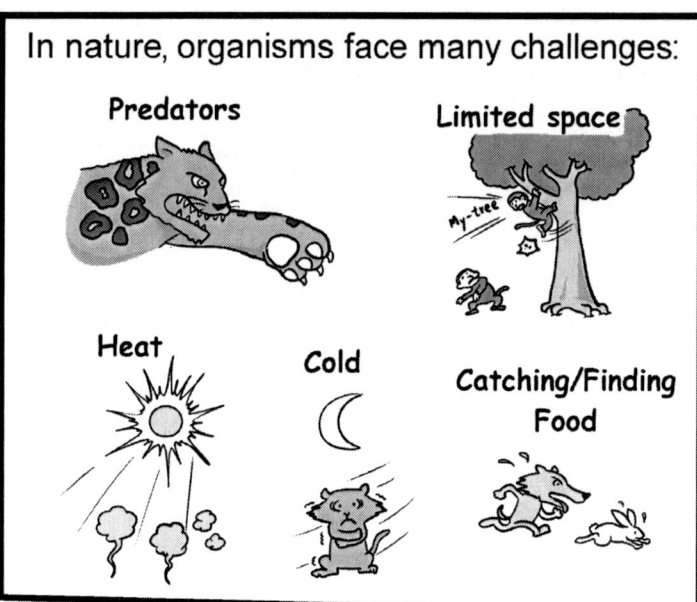

Nature's challenges result in **"Natural Selection"**, A process that causes only the fit ones to survive, the unfit will eventually die out.
In other words: "Survival of the fittest."

In some cases, to be "fit" is to be bigger and stronger.

FIT   UNFIT

But in other cases, bigger is NOT always better!

EXAMPLE:
The ancestor of modern bears were a type of bears called "Ursus Minimus." Just as people comes in different shapes and sizes, these bears also varied in shapes and sizes.

At first, life was easier because the climate suited the bears well.

But one day, the climate changed.

Tall trees started to grow.

Before, there was more food on the ground.

But now, there is little food on the ground.

In this new climate, you have to climb trees to get to food!

Therefore, the smaller and more agile bears have more access to food.

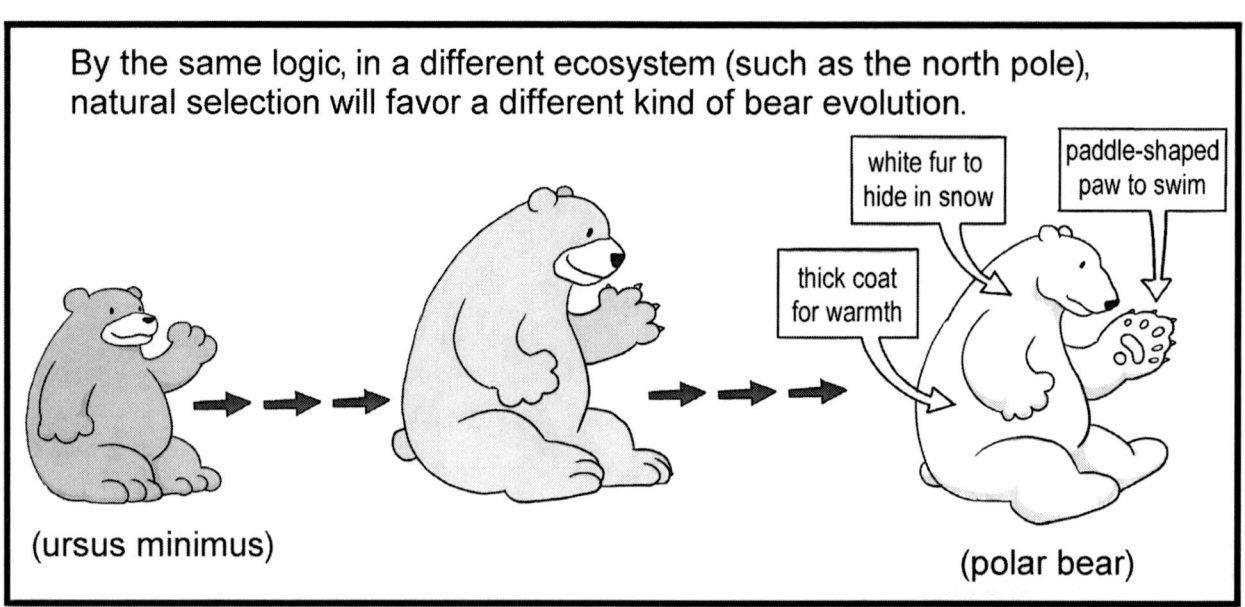

Humans can also cause evolution!

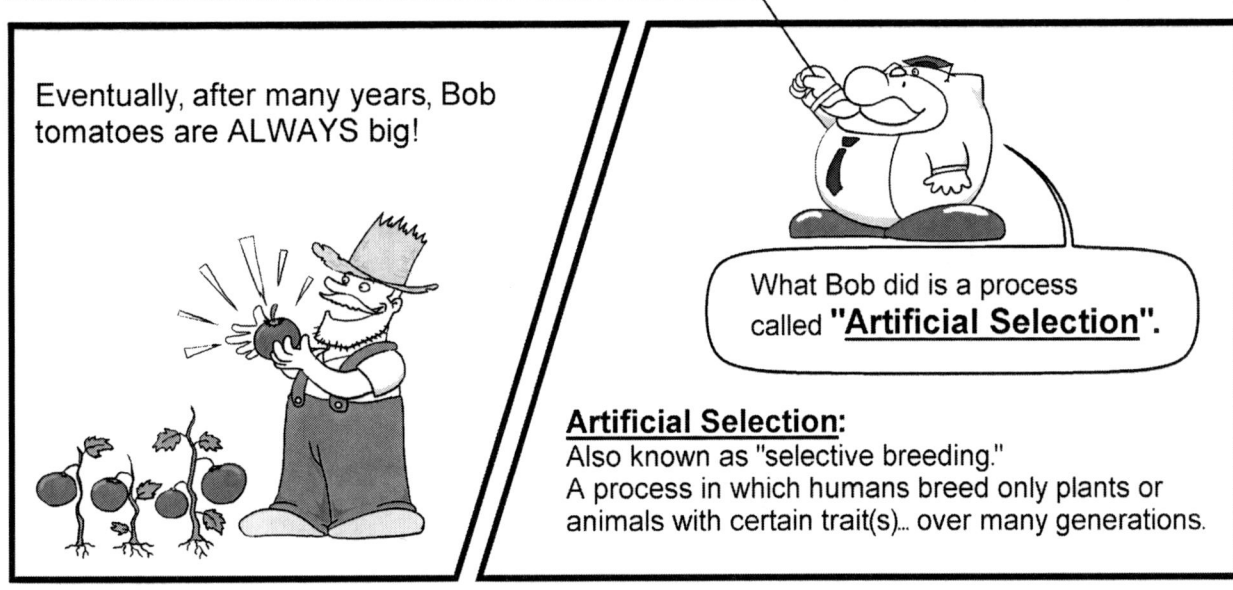

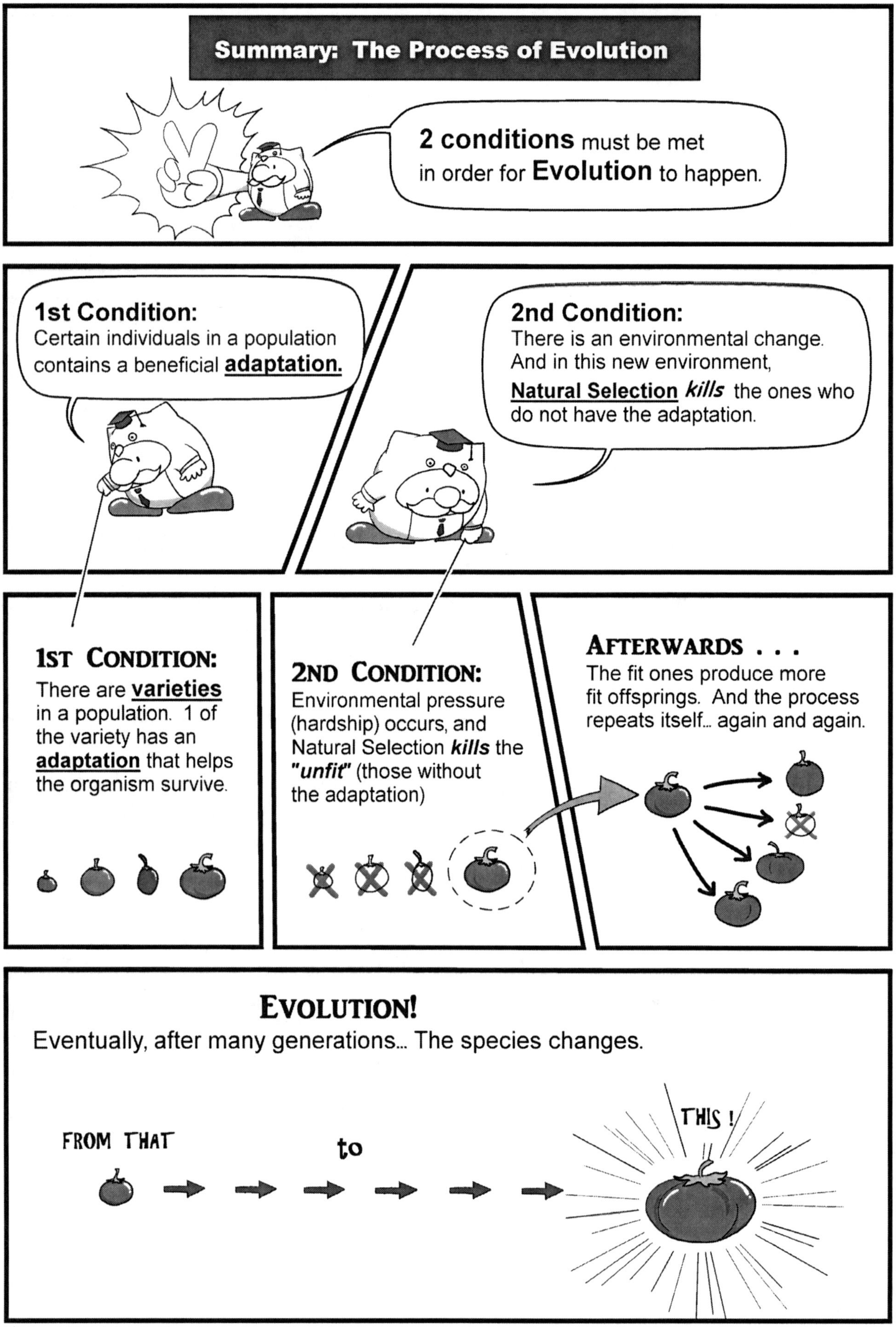

Summary: The Process of Evolution

2 conditions must be met in order for **Evolution** to happen.

1st Condition: Certain individuals in a population contains a beneficial **adaptation.**

2nd Condition: There is an environmental change. And in this new environment, **Natural Selection** *kills* the ones who do not have the adaptation.

1ST CONDITION:
There are **varieties** in a population. 1 of the variety has an **adaptation** that helps the organism survive.

2ND CONDITION:
Environmental pressure (hardship) occurs, and Natural Selection *kills* the **"unfit"** (those without the adaptation)

AFTERWARDS . . .
The fit ones produce more fit offsprings. And the process repeats itself... again and again.

EVOLUTION!
Eventually, after many generations... The species changes.

FROM THAT to THIS!

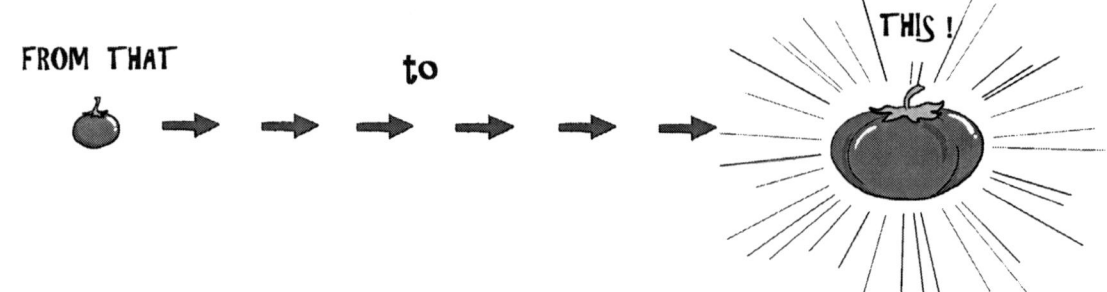

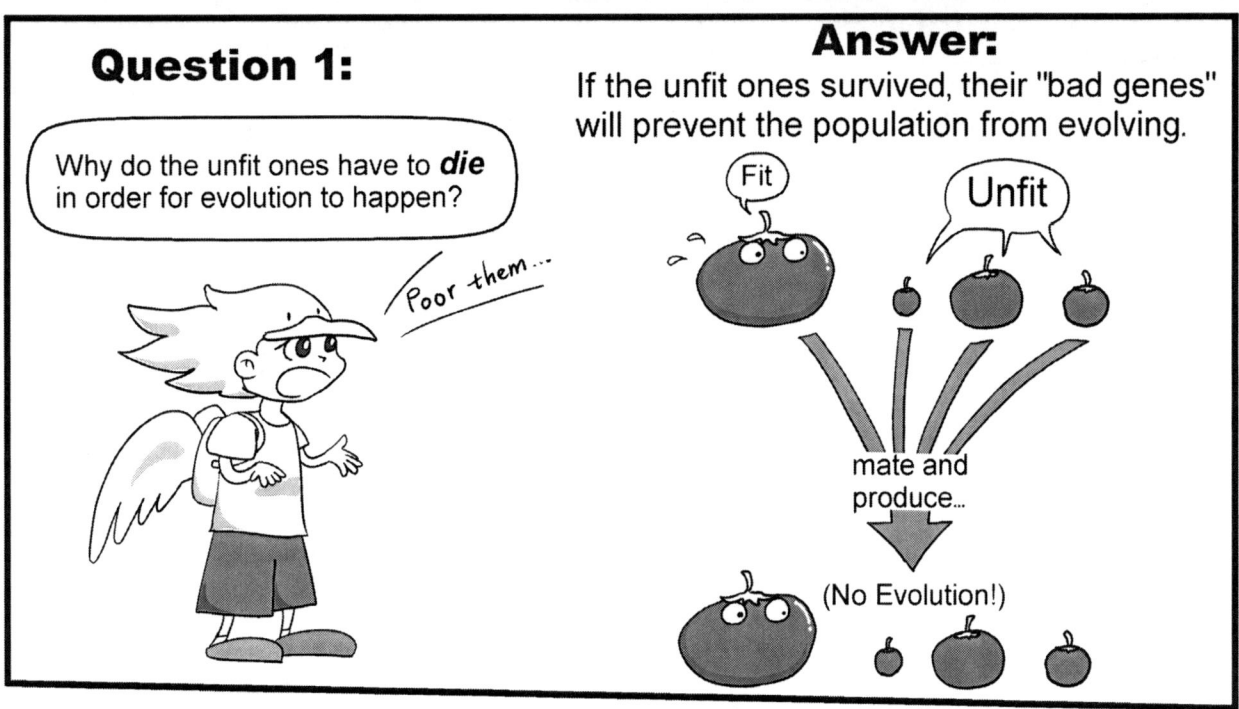

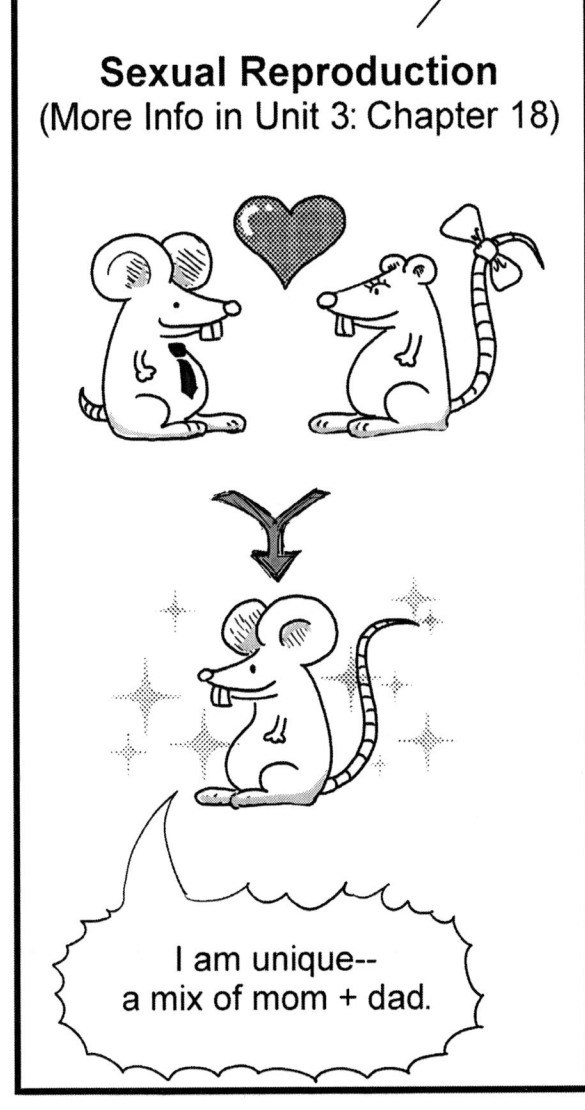

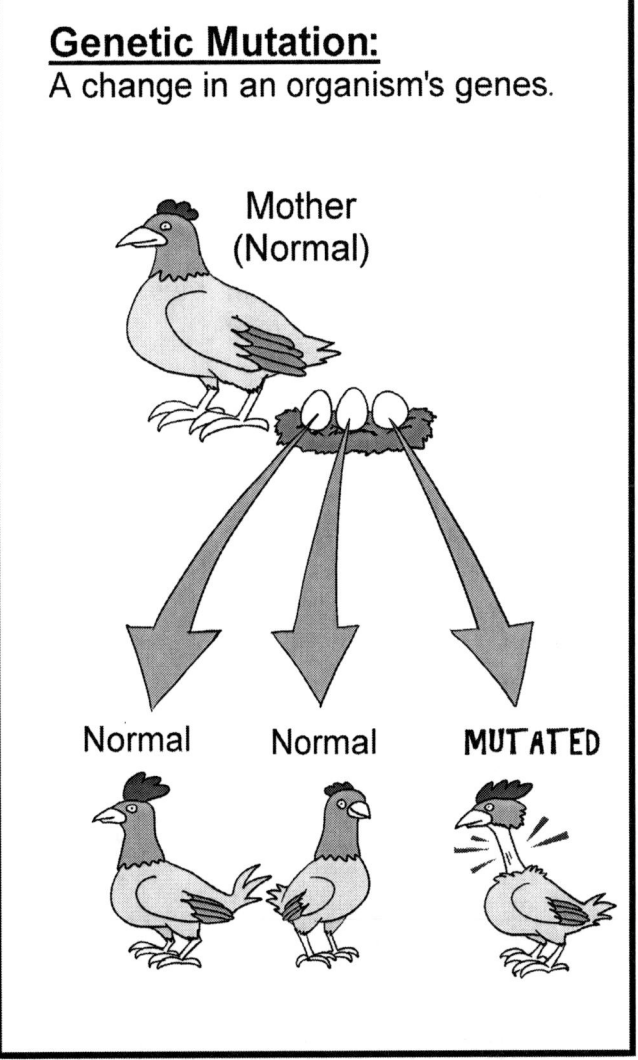

page 227

Unit 4: Chapter 26

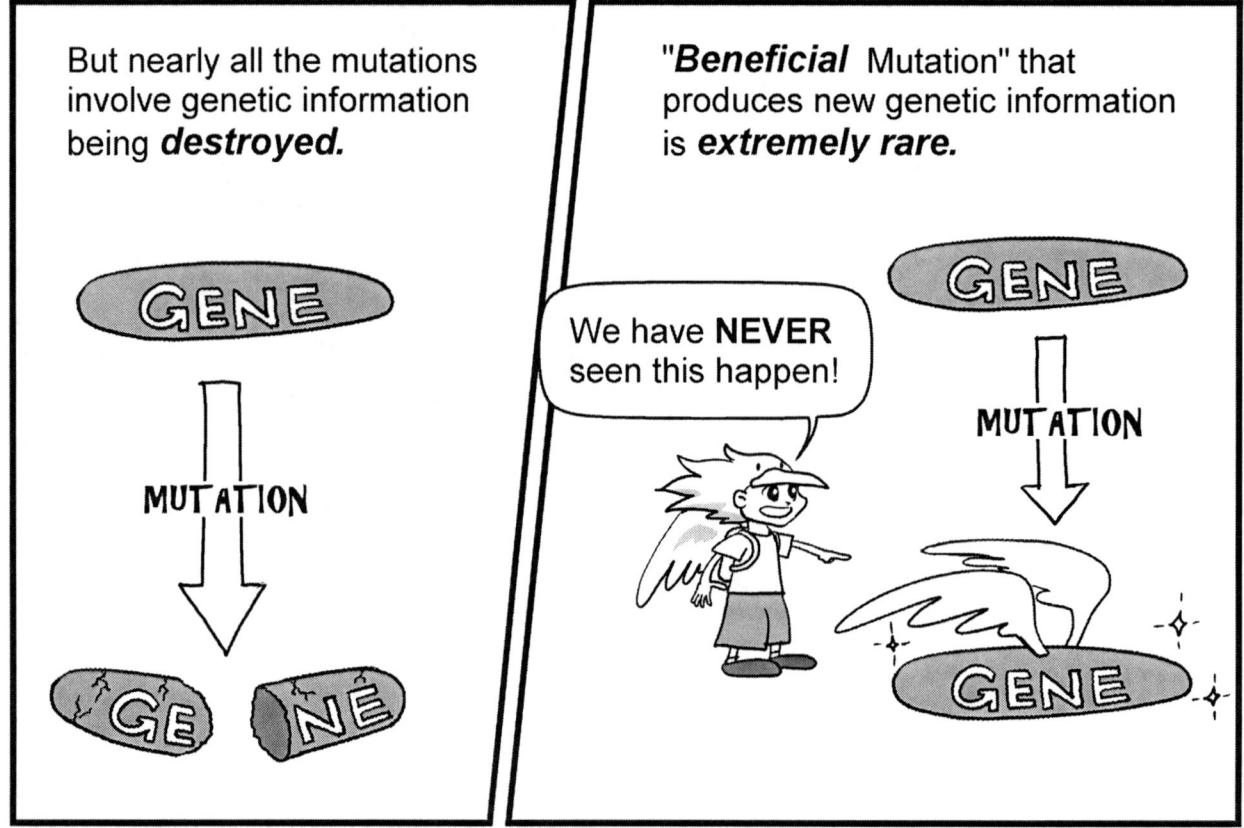

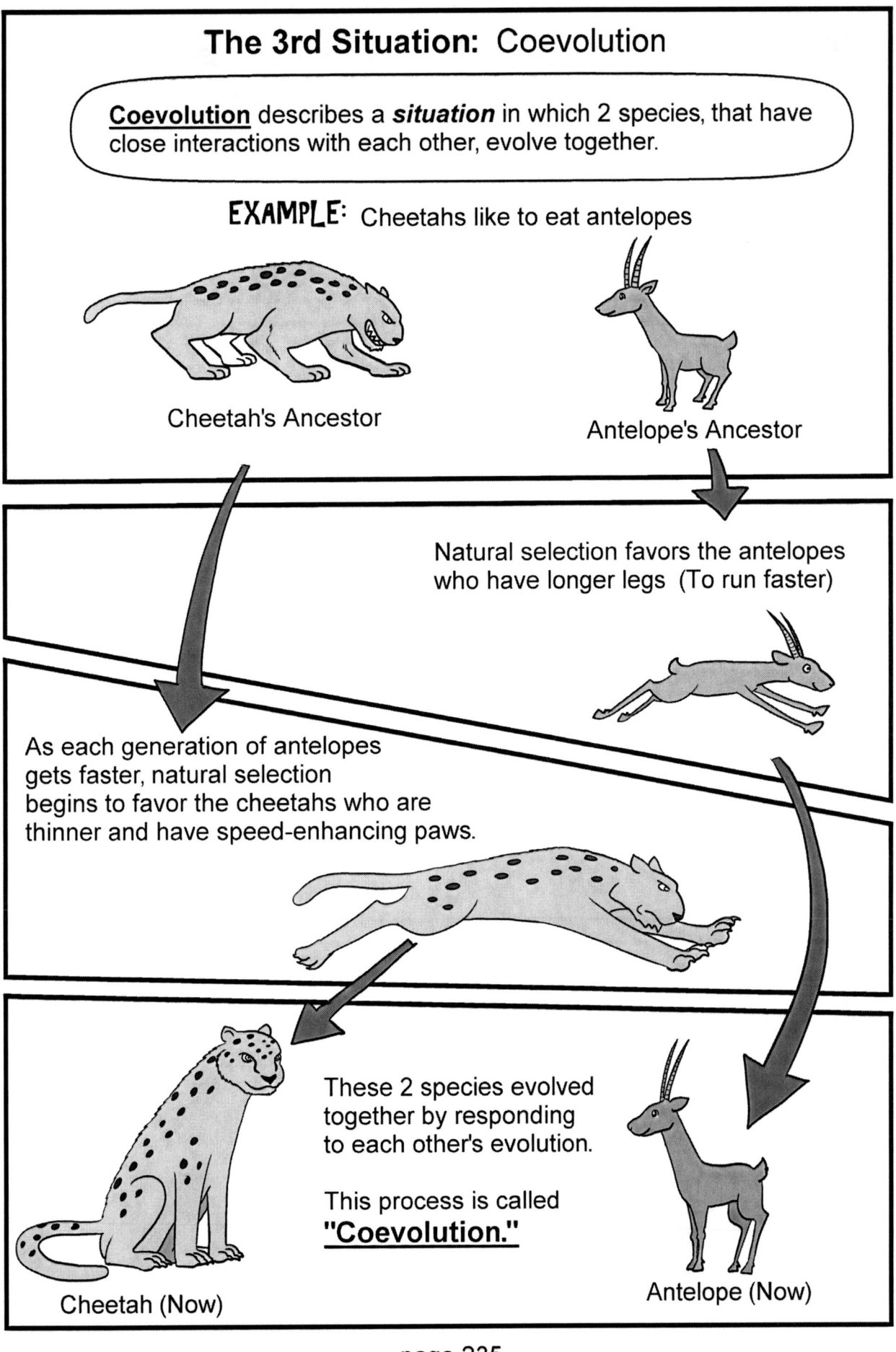

Unresolved Issues in the Theory of Macro-Evolution

"Let's take a look!"

Problem 1: New Genes Needed.
In order for macro-evolutions to happen, you will need to to gain many new genes that are too complex to have come from random mutations.
Where do these new gene come from?)

For Example:
If you want a dinosaur to evolve into a bird...

...You will need a lot of **new genetic** information.

Such As:

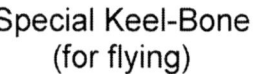

Special Keel-Bone (for flying)   Feather

A feather is extremely complex.

"Use a microscope and you will see!"

HOOKS

BARBULES (tiny barbs)

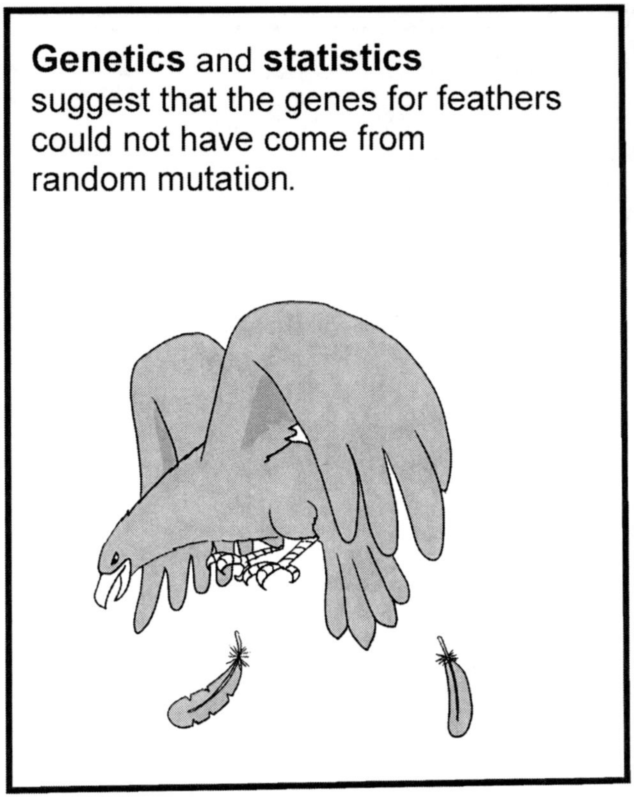

Genetics and **statistics** suggest that the genes for feathers could not have come from random mutation.

So how do birds get their feathers? (no one knows)

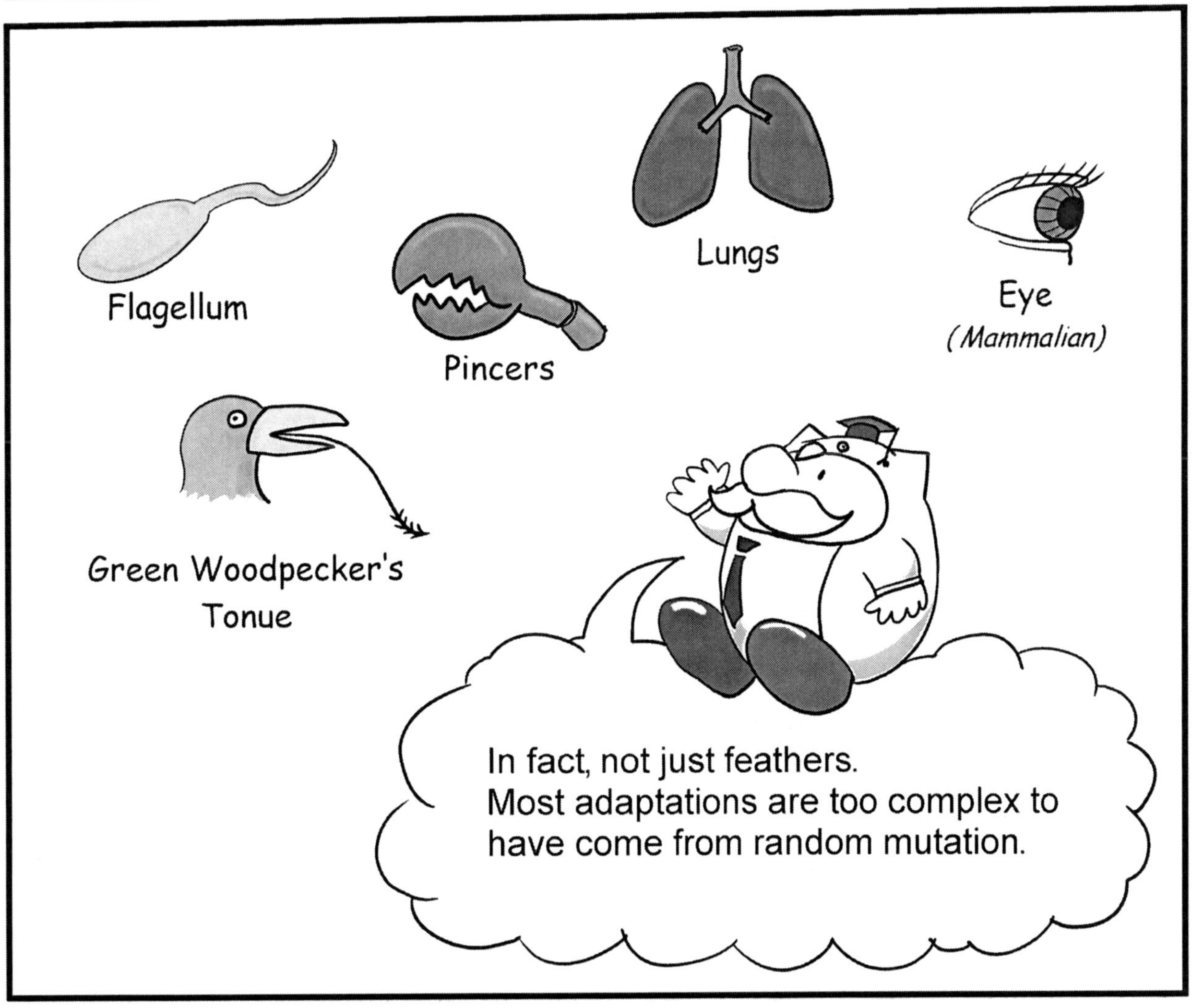

Flagellum

Pincers

Lungs

Eye (Mammalian)

Green Woodpecker's Tonue

In fact, not just feathers. Most adaptations are too complex to have come from random mutation.

page 237

Unit 4: Chapter 26

Problem 2: Adaptations with Multiple Parts.
An adaptation usually consists of multiple parts. And all parts must simultaneously exist in order for it to work!

EXAMPLE:

Giraff's long neck allows it to eat tall leaves.

But to make the blood reach its tall brain, it needs a powerful heart to pump it up!

There is one more problem.

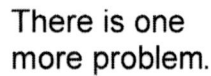

When a griaff lowers its head to drink water, the downward blood-flow will be too powerful...

The powerful blood-flow can damage its brain!

So how does a giraffe *avoid* brain damage?

Giraffe's brain has a special tissue that acts like a "sponge" to absorb this force of this blood-flow.

Here's Another Problem:

When we suddenly stand up... because our brain suddenly changes position, we may feel a little dizzy.

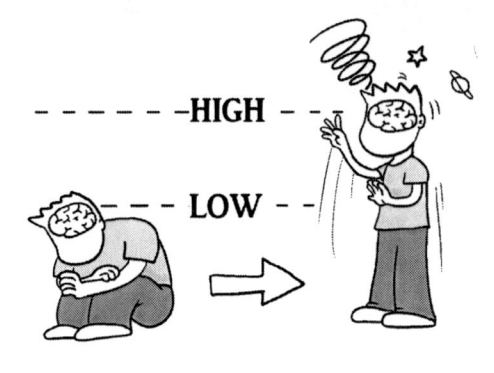

Because a giraffe is so tall, this effect is severe enough to cause it to lose its balance and fall. (even pass out!)

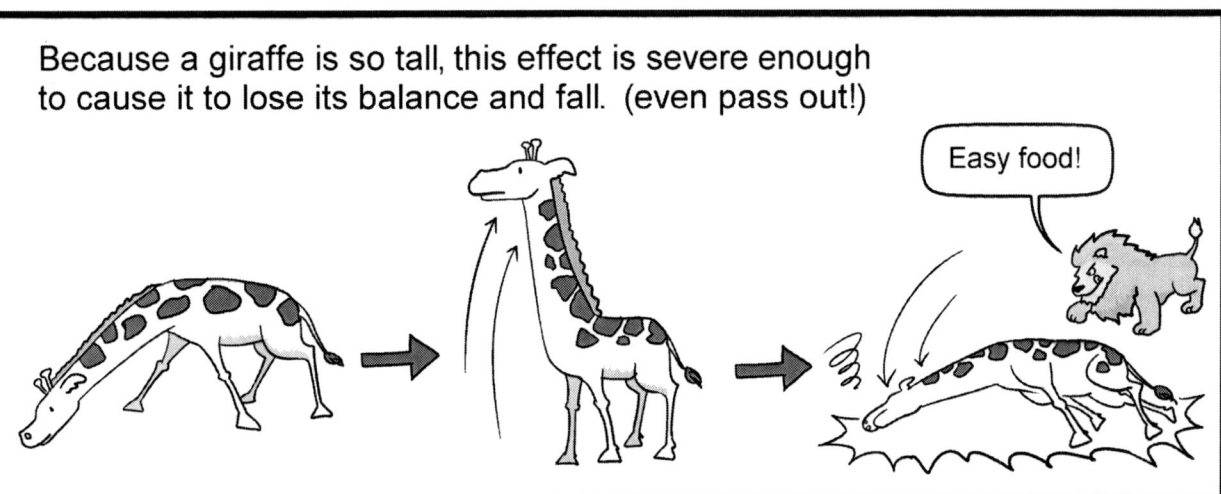

But because of this sponge-like tissue at its brain, a giraffe always feels fine when it stands up.

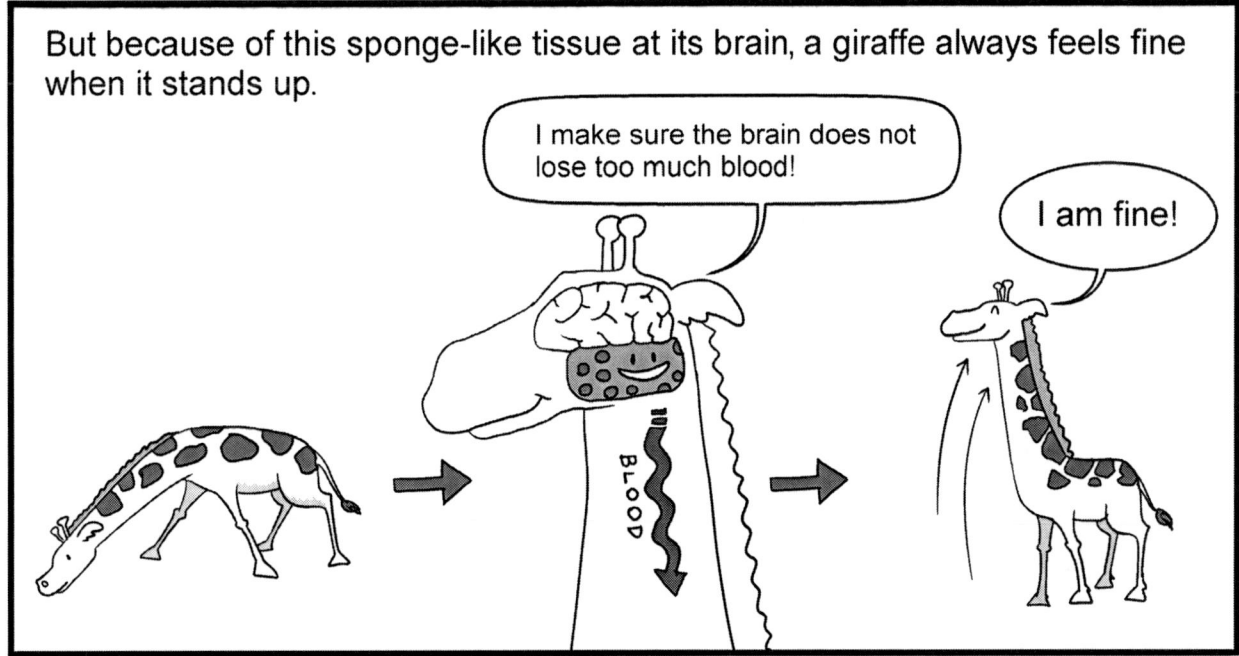

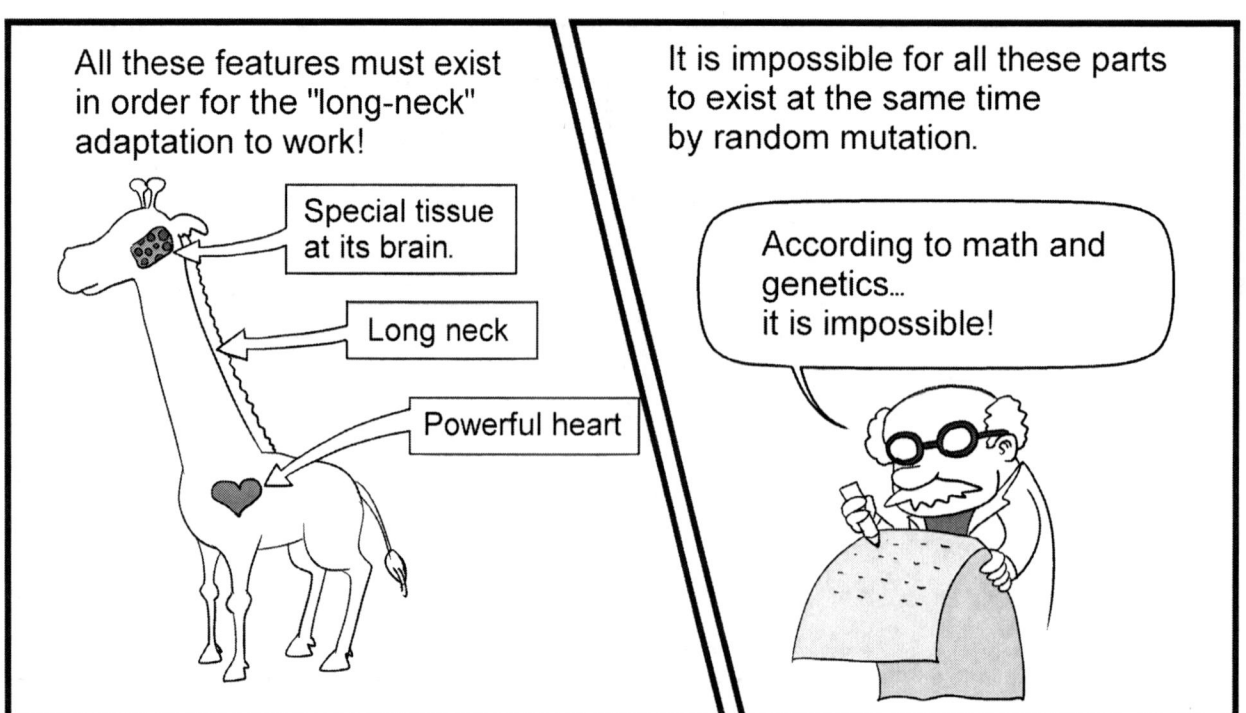

Problem 3: The Intermediate Stages
It is hard to survive when you are at an awkward "intermediate" (in-between) stage!

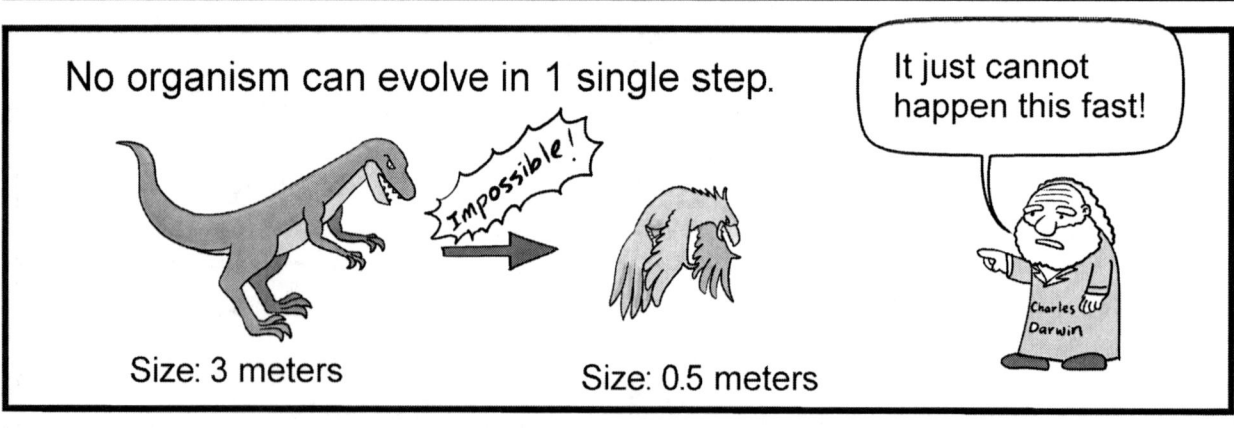

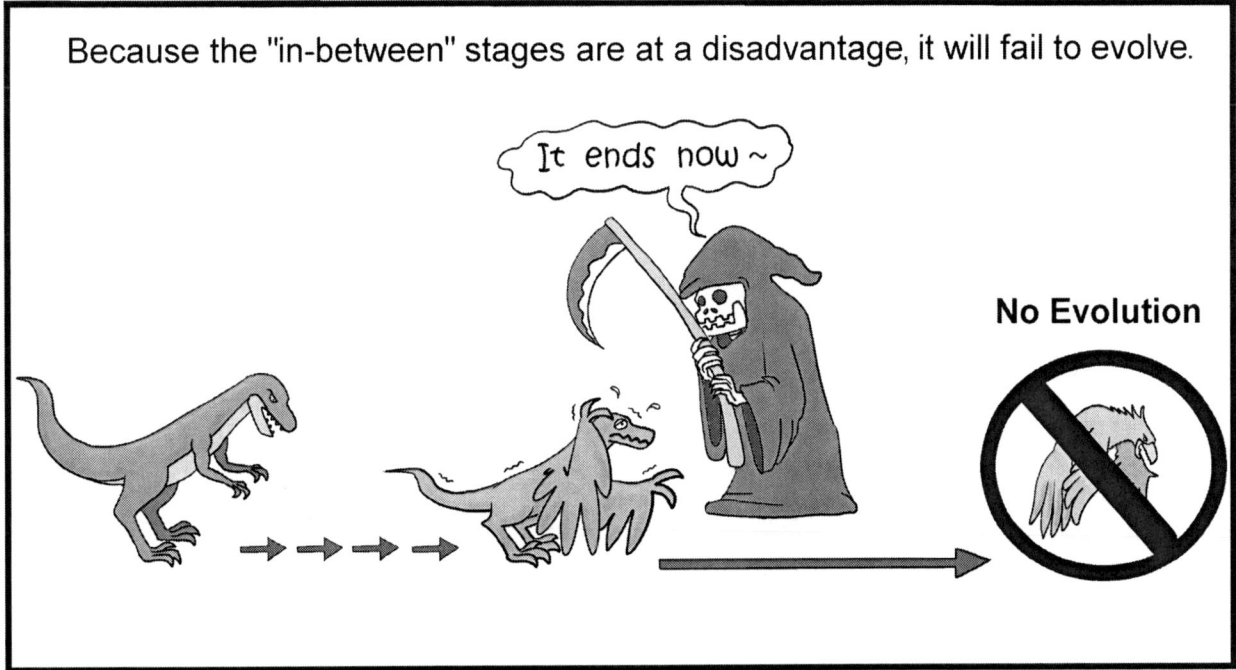

Chapter 27: Charles Darwin

Science Standard: 7.3.b
Students know the reasoning used by Charles Darwin in reaching his conclusion that natural selection is the mechanism of evolution.

NEW VOCABS

* **Adaptive Radiation:** A "common-ancestor" species evolving into different species in different environments, which feature different methods of natural selections, over many generations.

* **Common Ancestor:** A type of ancient organism who is the ancestor of 2 (or more) varieties of modern-day organisms.

* **Fossil:** the preserved remains of an organism from the past.

* **Uniformitarianism:** An idea which states: "The way that natural processes work today is the same way they have been working in the ancient past... In other words, all natural processes have always been working in the same manner (uniform manner) since the beginning of time."

A **"Common Ancestor"**, as the term suggests, is an ancient organism who is the ancestor of 2 (or more) varieties of modern-day organisms.

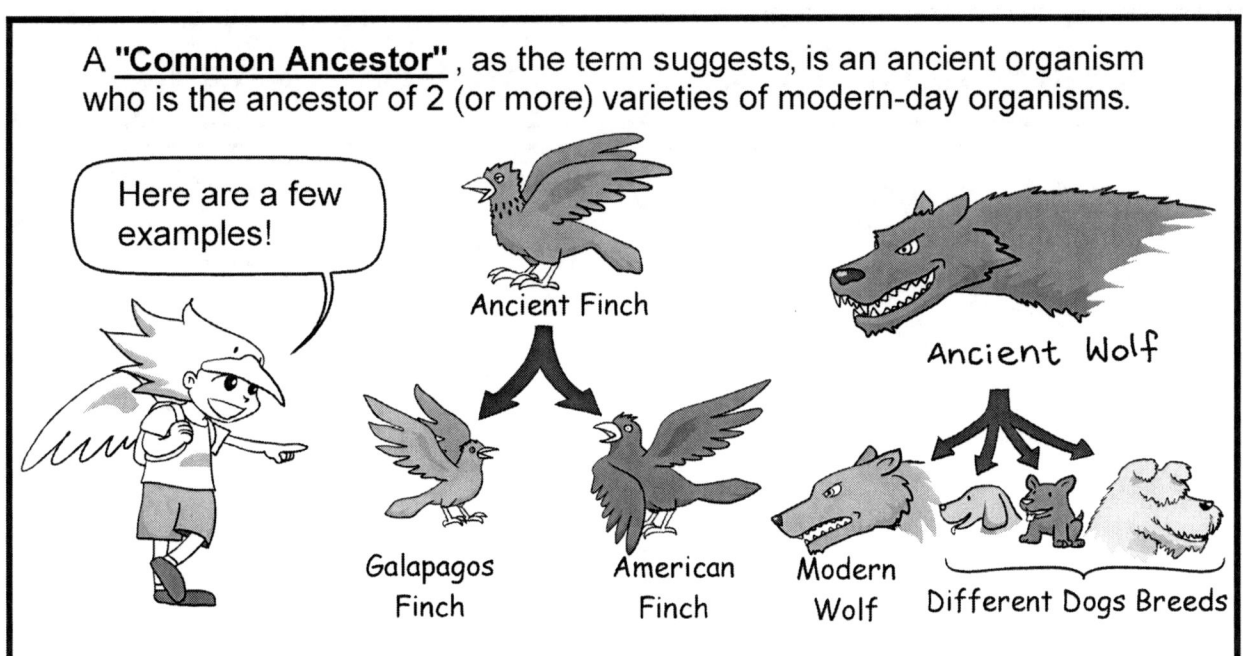

Here are a few examples!

The concept of "Common Ancestor" gave rise to the concept of **"Adaptive Radiation"**

Adaptive Radiation means a "common-ancestor" species evolving into different species in different environments.

Common Ancestor

Different Species

Unit 4: Chapter 27

Example of Adaptive Radiation : Finches

At first, there was only one population of finch

One day, they left each other to live in different places.

Good luck~

Bye~

One group lives in the Galapagos Islands

Bug

GALAPAGOS ISLAND

The other group lives in South America

Nut

SOUTH AMERICA

After many generations of Natural Selection

After many generations of Natural Selection

After many generations, We are now 2 different species!

(Galapagos Finch)

(American Finch)

page 246
Unit 4: Chapter 27

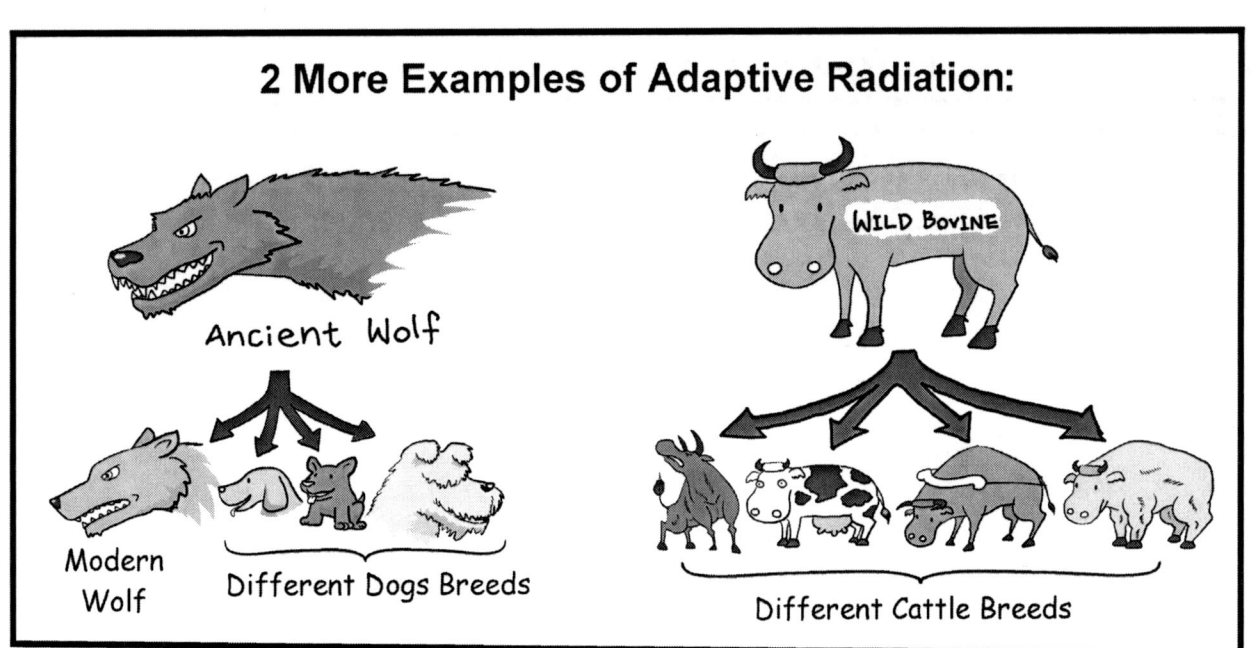

New Topic: Fossils on the Mountains

Darwin also found fossils of ancient clams up in the mountains.

This suggests that the mountains used to be under water.

Darwin suggests:

Since I have never seen a mountain forming before my eyes, it must be a very *slow* process.

Millions of years later

From this, Darwin Concludes:

The earth must be very old! (billions of years)

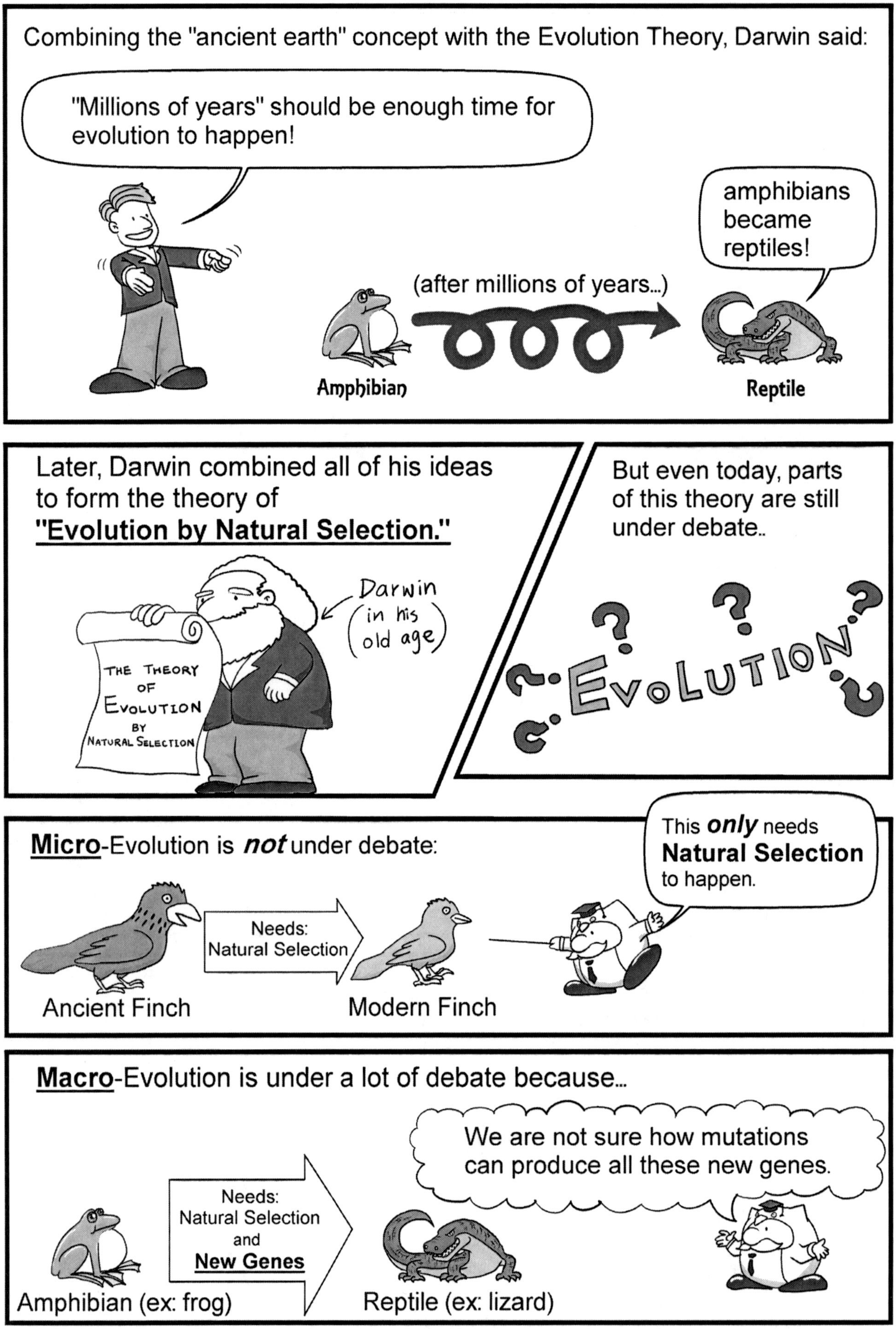

Chapter 28: Fossils and the Evidence for Evolution

Science Standard: 7.3.c
Students know how independent lines of evidence from geology, fossils, and comparative anatomy provide the bases for the theory of evolution.
Science Standard: Bio/LS.8.e
Students know how to analyze fossil evidence with regard to biological diversity, episodic speciation, and mass extinction.

NEW VOCABS

* **Background Extinctions:** Extinctions that happens at a slow rate, a few organisms at a time. This is constantly happening as part of the natural process..

* **Cambrian Explosion:** The mysterious, sudden increase of species variety during the Cambrian Period

* **Comparative Anatomy:** It is the study of structural similarities between 2 different types of organisms. For example, both humans and cats have 2 bones in their forearm (which allows them to twist/rotate their forearms).

* **Fossils:** The preserved remains (example: bones) or impressions (example: footprints) of prehistoric organisms.

* **Geologic Column:** A concept that suggests the earth's land comes in layers. Each layer is laid at different times in earth's history—with the older layers at the bottom, and the younger layers at the top.

* **Index Fossil:** A fossil that is believed to be unique to a certain period of time in earth's history. Therefore, it can be used to identify the rock layer it is found in.

* **Mass Extinction:** A sharp decrease in biodiversity, involving the loss of numerous species over a short period of time..

* **Punctuated Equilibrium:** (Also known as episodic speciation) This theory says that evolution can happen very quickly in a brief period of time, followed by long period of no change at all.

* **Vestigial Structure:** An organism's physical structure that has lost all (or most) of its original function.

Unit 4: Chapter 28

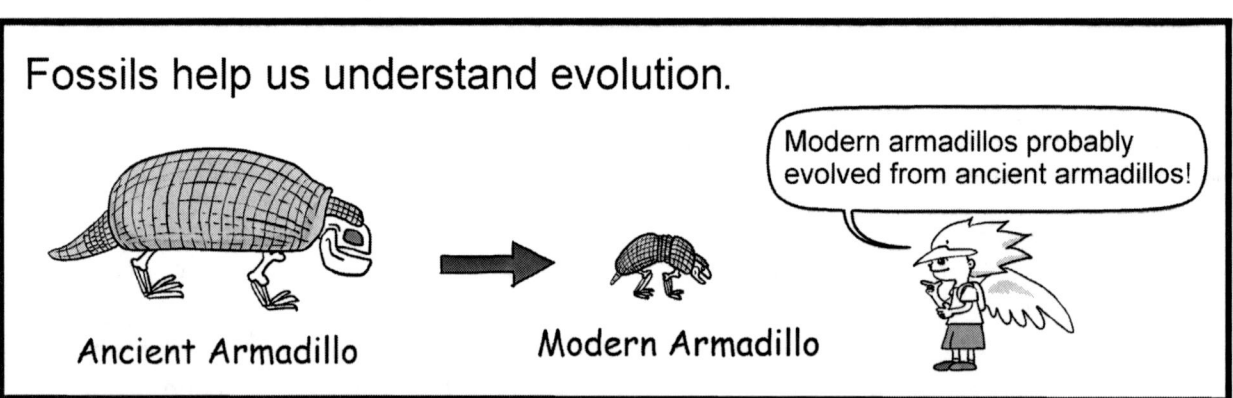

However, the lack of certain fossils may also raise doubts on the theory of evolution itself.

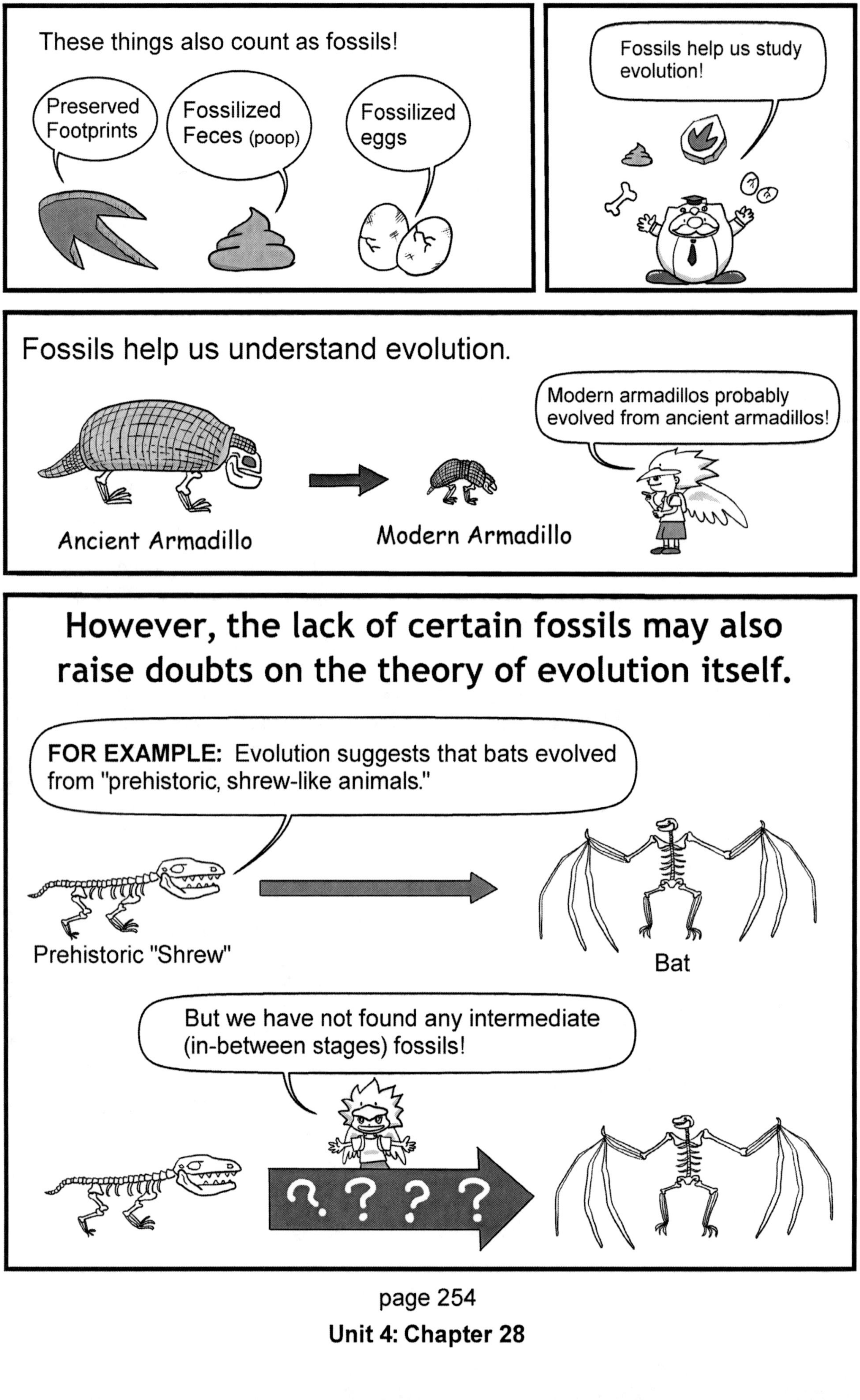

FOR EXAMPLE: Evolution suggests that bats evolved from "prehistoric, shrew-like animals."

Prehistoric "Shrew" → Bat

But we have not found any intermediate (in-between stages) fossils!

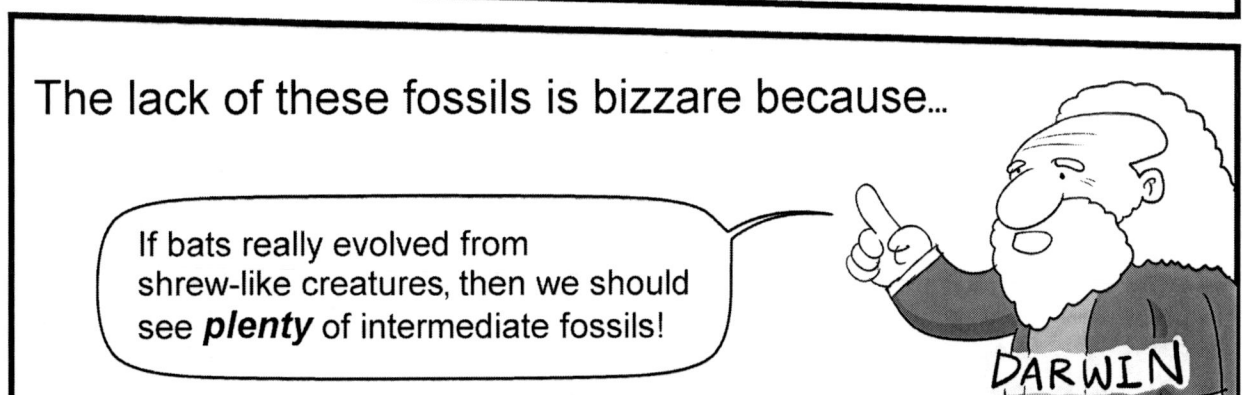

Not just bats, most of the organisms do not have enough intermediate (in-between) fossils to explain where they came from.

So, where did we come from?

To explain this mystery, the following idea was proposed in 1972.:

Punctuated Equilibrium: (Also known as Episodic Speciation)
This theory says that evolution can happen **very quickly** in a **brief** period of time, followed by long period of no change at all.

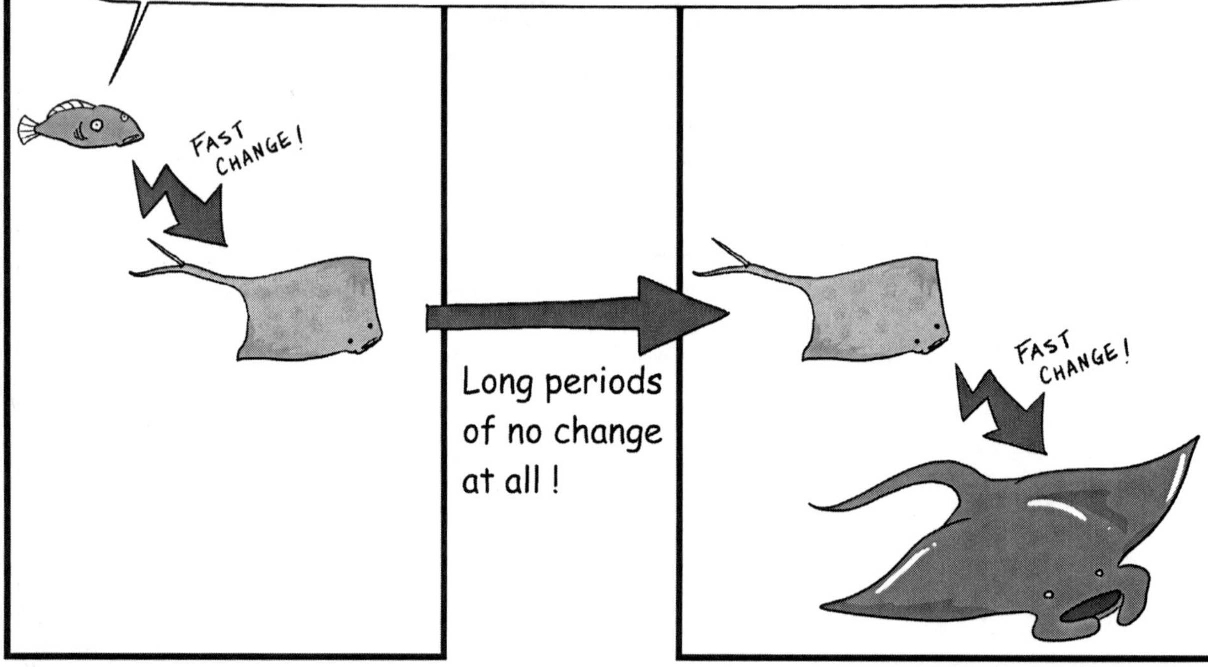

FAST CHANGE!

Long periods of no change at all !

FAST CHANGE!

page 256
Unit 4: Chapter 28

Today, the lack of intermediate fossils is still a mystery.

Coming Up: we will look at how scientists determine the age of a rock.

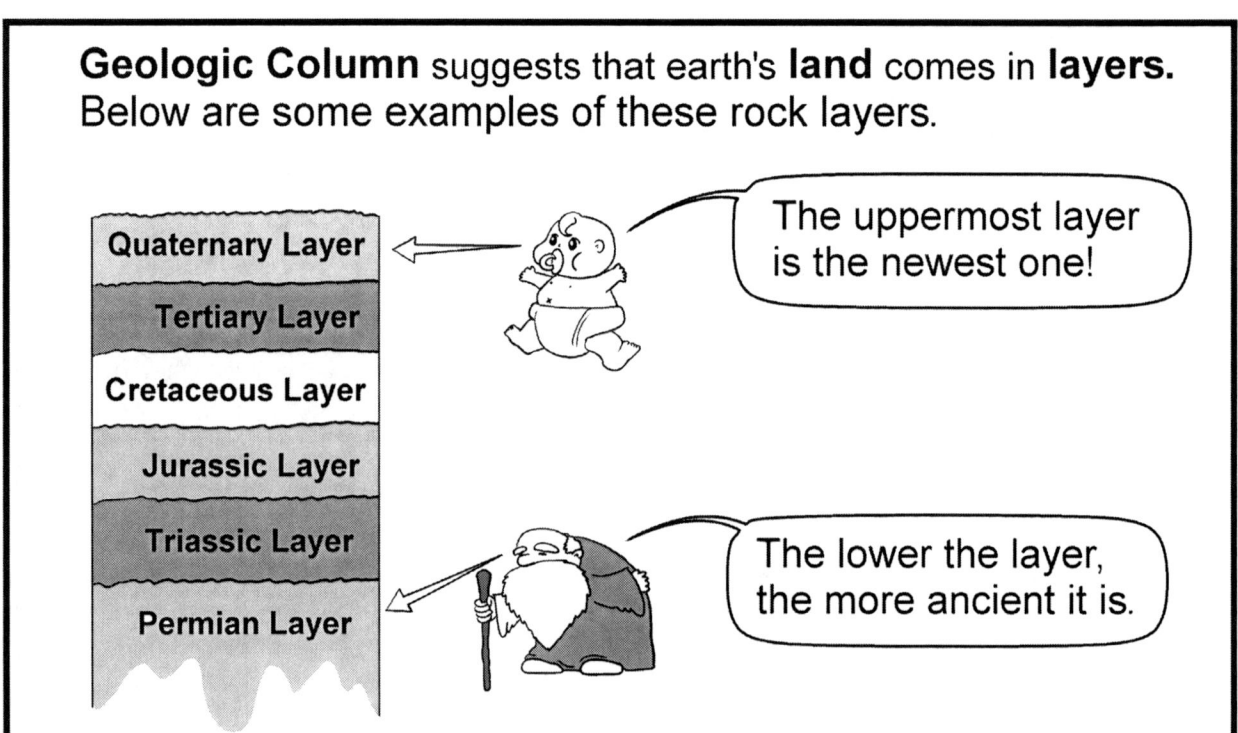

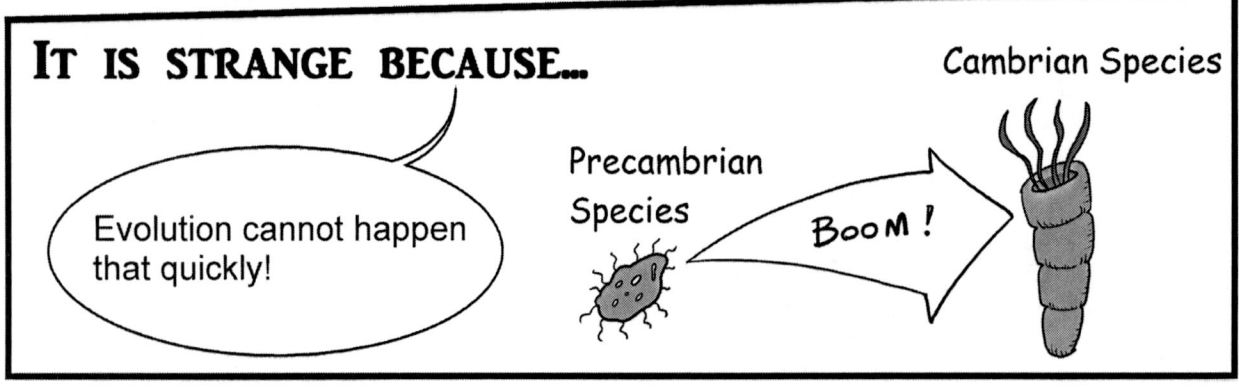

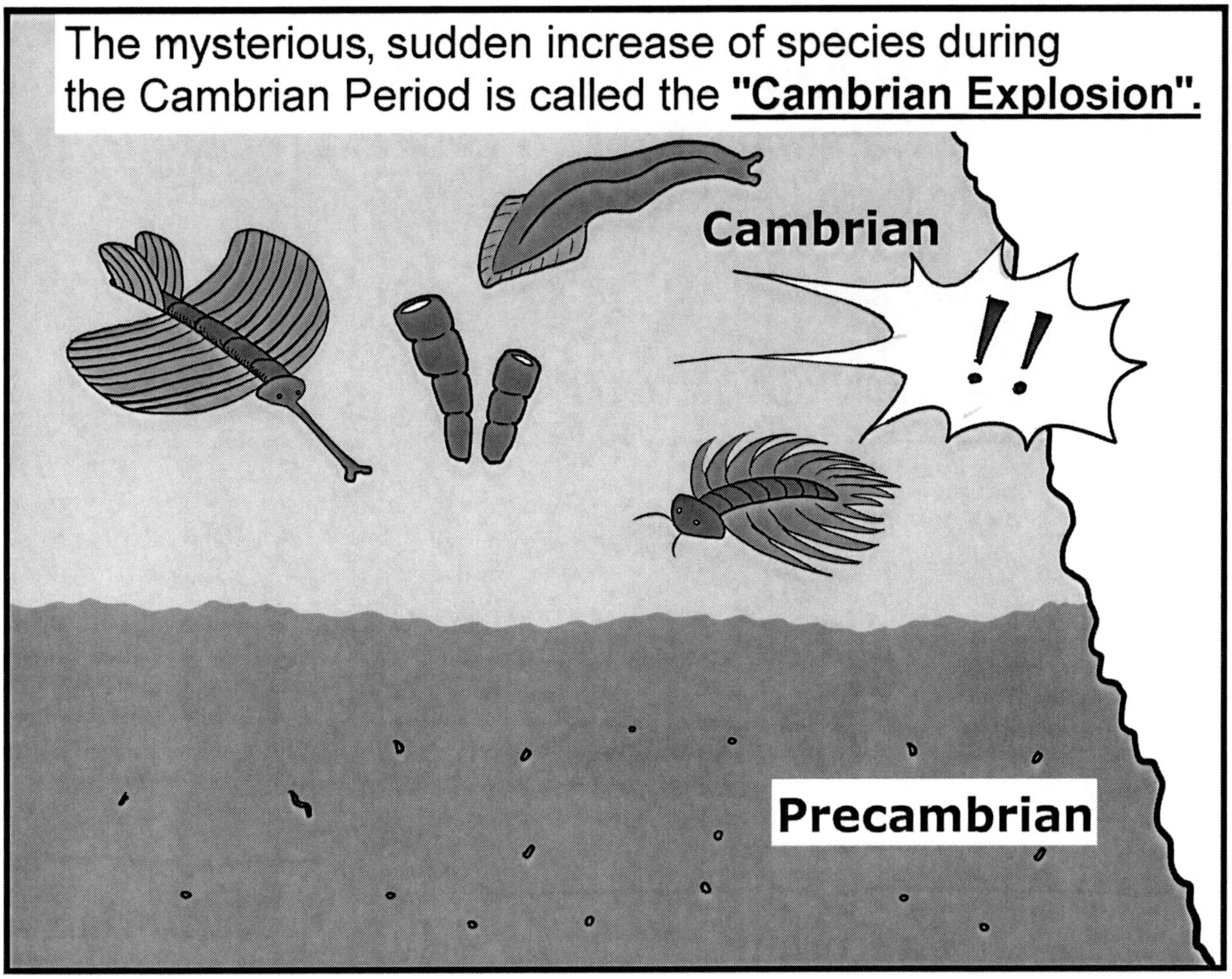

The mysterious, sudden increase of species during the Cambrian Period is called the **"Cambrian Explosion"**.

Cambrian

Precambrian

Some people try to explain this with "Punctuated Equilibrium".

But Punctuated Equilibrium itself is not a proven theory, so it cannot be used as an explanation.

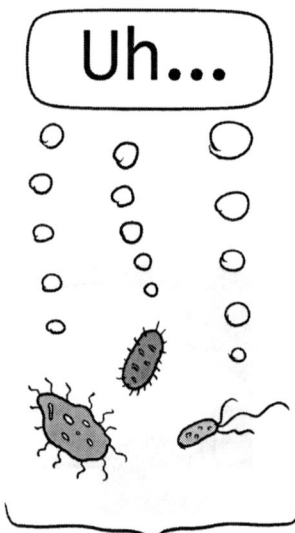

Therefore, scientists believe that there have been several "Mass Extinctions" in the past.

Example: The mass extinction of dinosaurs was one of them.

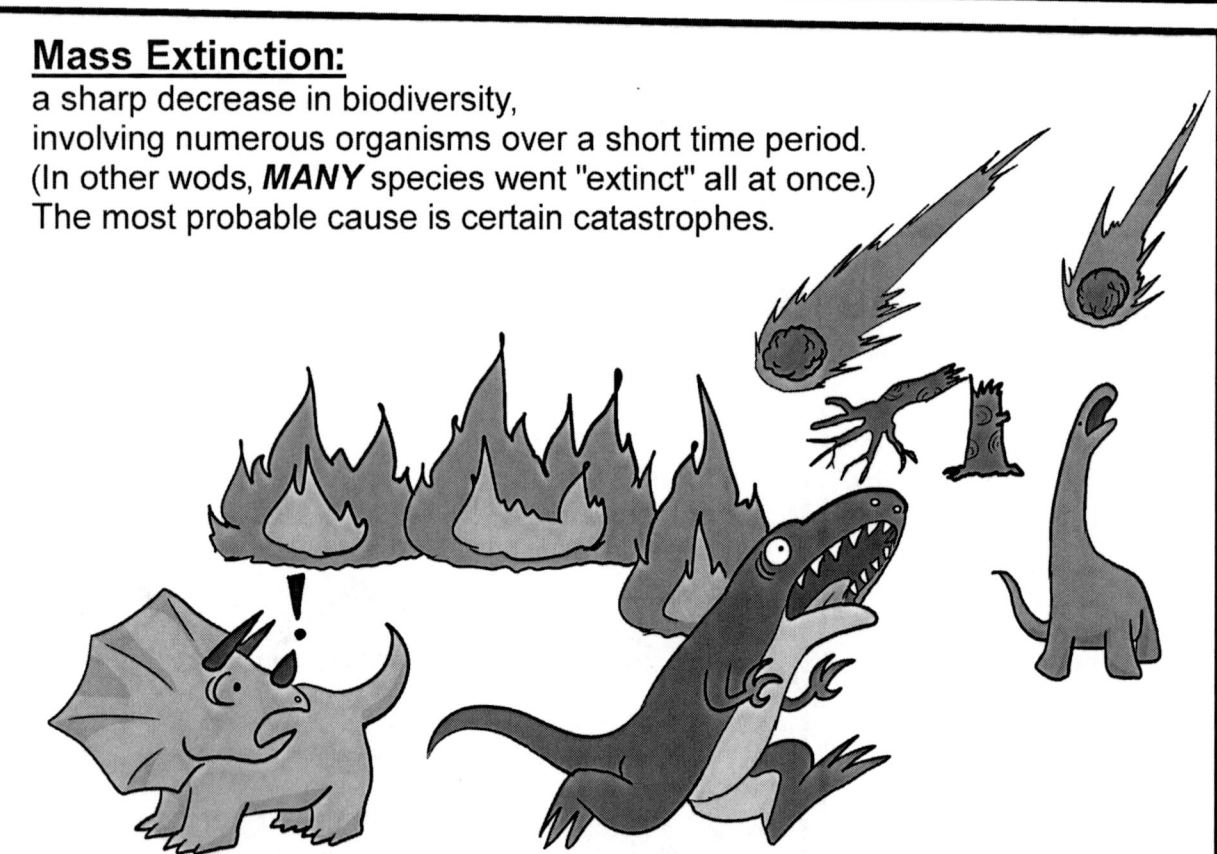

Mass Extinction:
a sharp decrease in biodiversity,
involving numerous organisms over a short time period.
(In other wods, **MANY** species went "extinct" all at once.)
The most probable cause is certain catastrophes.

Most scientists believe that there were several mass extinctions in the past....
One of them was the mass extinction of the dinosaurs.

Mass Extinctions are *rare.*
But the second type of extinction, called **"Background Extinctions,"** are constantly happening in nature.

Background Extinctions:
Extinctions that happens at a slow rate, a few species at a time. This is constantly happening as part of the natural process..

Background Extinction Example: between 1990~2000AD, 4 species have slowly gone extinct.

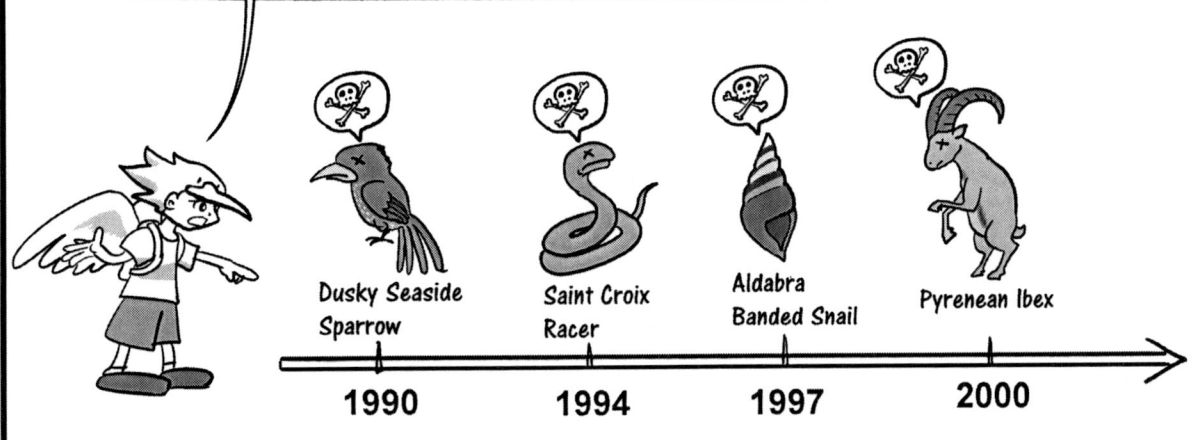

It is sad that Background Extinctions are happening all the time!

Come on... background extinction is a **normal, natural** process!

page 267

Unit 4: Chapter 28

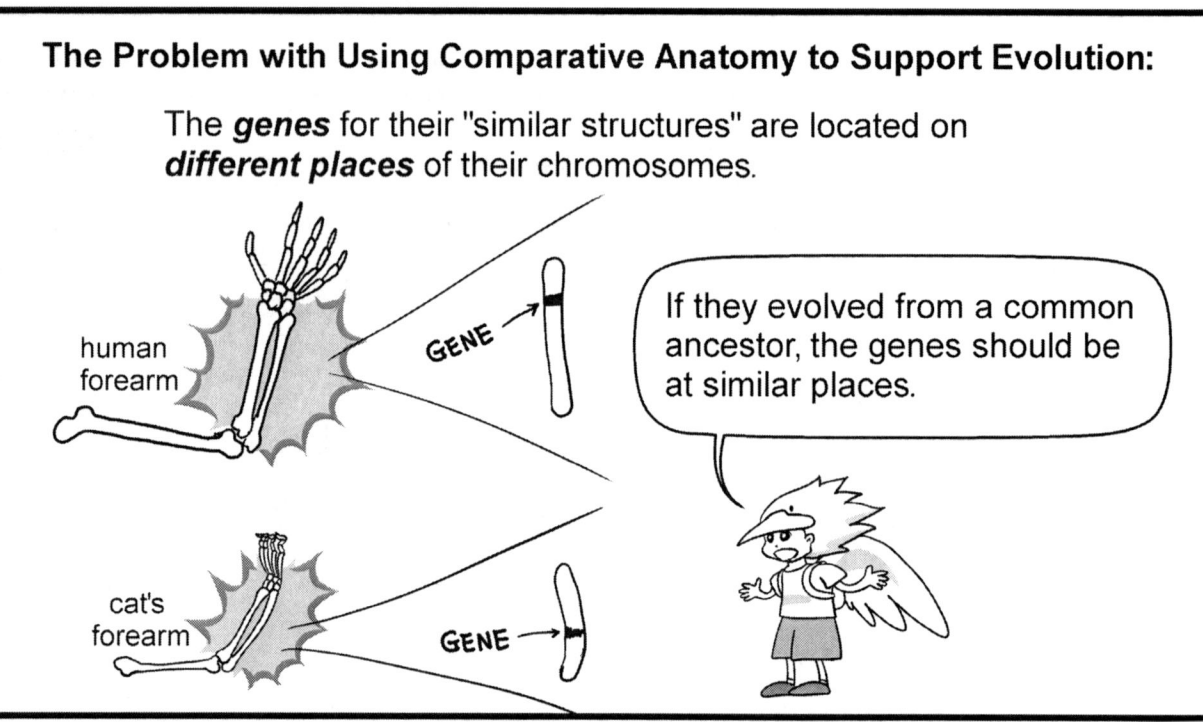

New Topic: Vestigial Structures

Vestigial Structure:
A structure that has lost all (or most) of its original function.

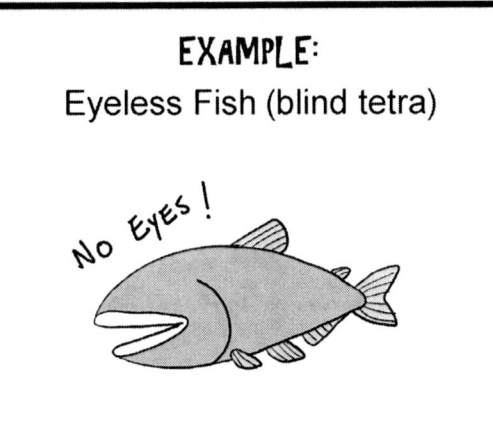

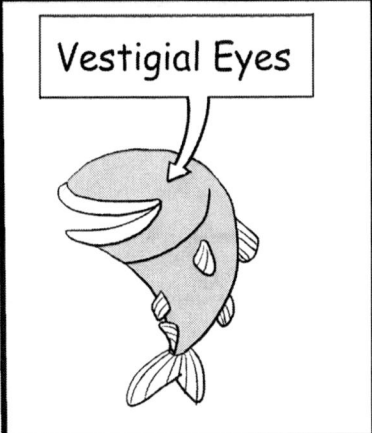

page 269

Unit 4: Chapter 28

The fish with mutated eyes is actually less likely to get hurt!

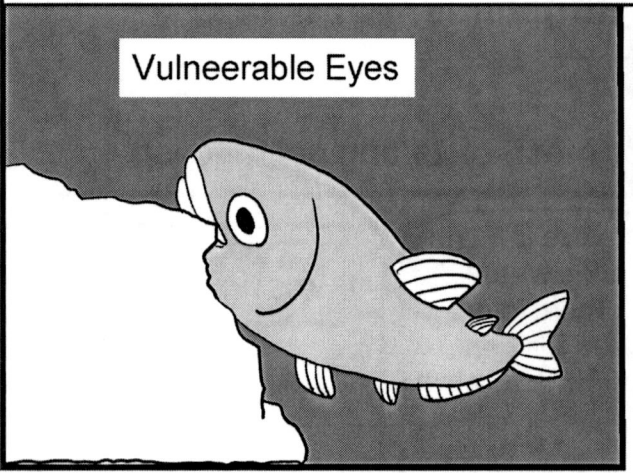
Vulneerable Eyes

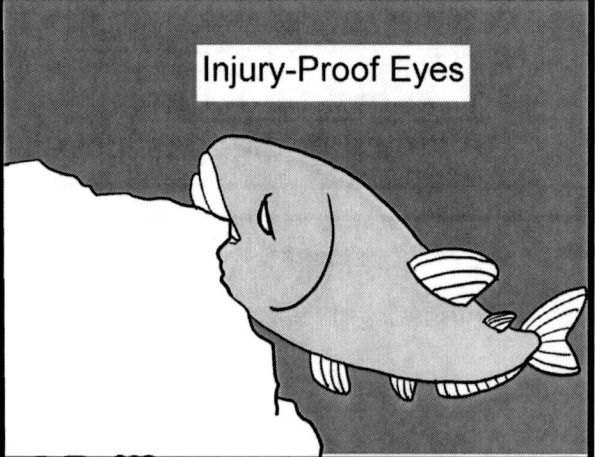
Injury-Proof Eyes

Because natural selection favors the fish with this birth defect... eventually, every fish in this cave has this birth defect!

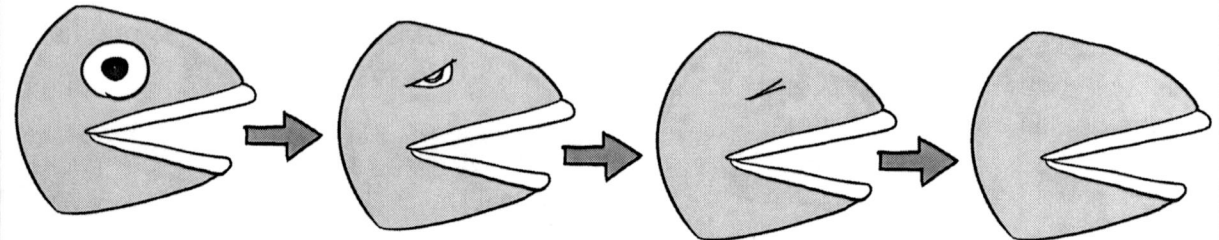

Chapter 29: Natural Selection and Phenotype

Science Standard Bio/LS . 7 . a
Students know why natural selection acts on the phenotype rather than the genotype of an organism.
Science Standard Bio/LS . 7 . b
Students know why alleles that are lethal in a homozygous individual may be carried in a heterozygote and thus maintained in a gene pool.

NEW VOCABS

* **Allele Frequency:** a number that describes how common an allele is in a population's gene pool. In other words, it shows the % (or proportion) of the gene pool that consists of that allele.

* **Carrier:** Someone whose recessive allele is masked (hidden) by the dominant allele and not expressed (shown) in its phenotype.

* **Gene Pool:** The sum of all the variety of genes in a population.

* **Genotype:** The type of alleles an organism has.

* **Phenotype:** The outward phsyical characteristics of an organism, as determiend by its genotype.

An organism's phenotype comes from its genotype. (Review: Chapter 19 of Unit 2). Phenotype: Albino fur — Genotype: aa	Your genotype consists of 2 alleles, which can be either dominant or recessive. **Dominant** — written in capital letters — **A** **Recessive** — written in lower-case — a

If **both** of your alleles are recessive, you will have the recessive phenotype.

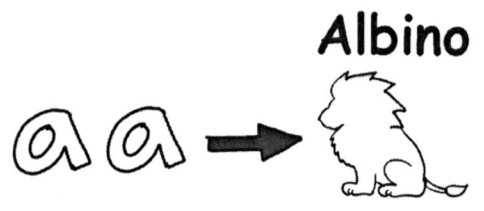

Albino

However, If you only have 1 recessive allele, you will not have the recessive phenotype.

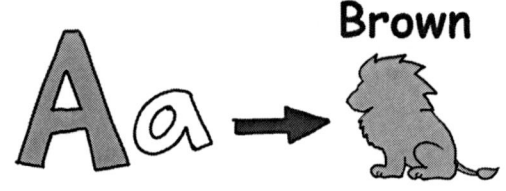

Brown

Review Vocabulary: Carrier.

A carrier is someone who is secretly hiding a recessive allele while its phenotype (outer appearance) is dominant.

This individual is a **"carrier of the albino allele."**

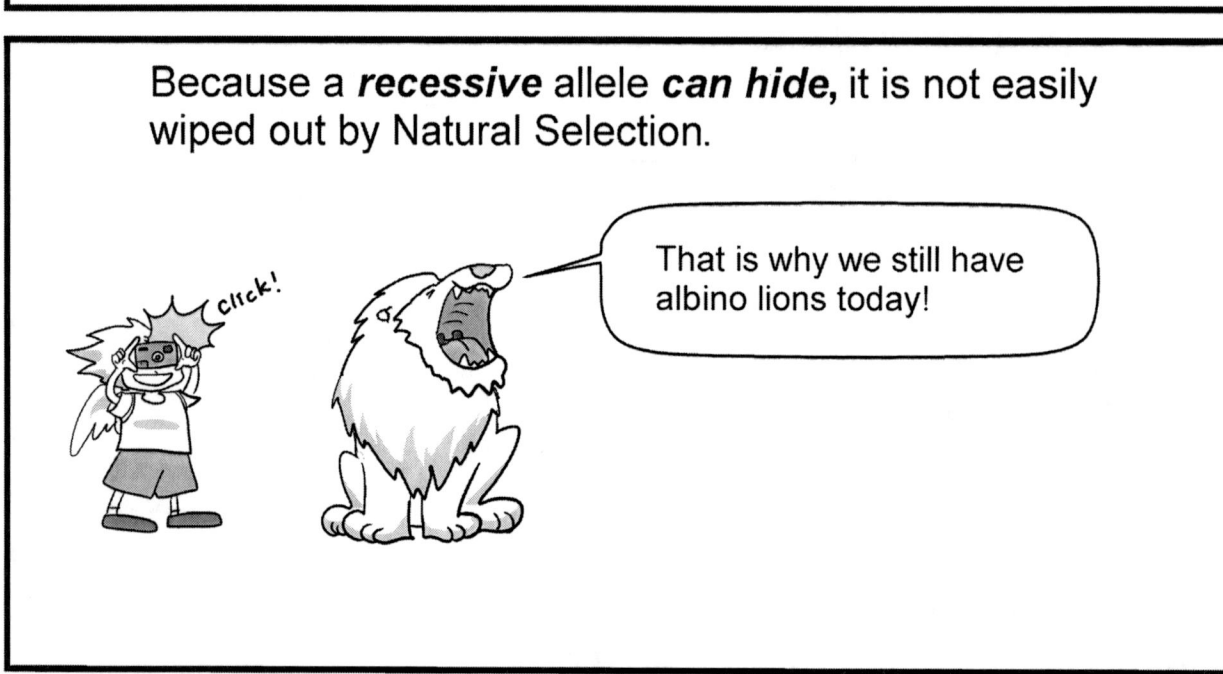

ANOTHER EXAMPLE:

Cystic Fibrosis is a recessive allele that has always remained within the human population.

"Cystic Fibrosis is recessive, so we will use a lower-case letter as its symbol!"

"Your genotype must be "f f" to actualy get the disease!"

Homozygous FF (healthy)	Heterozygous Ff (healthy)	Homozygous ff (Cystic Fibrosis)

Cystic Fibrosis's main symptom: The various parts of your body start making too much mucus.

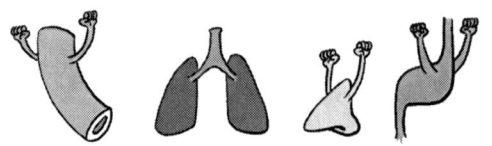

Cystic Fibrosis causes a person to be frequently sick...

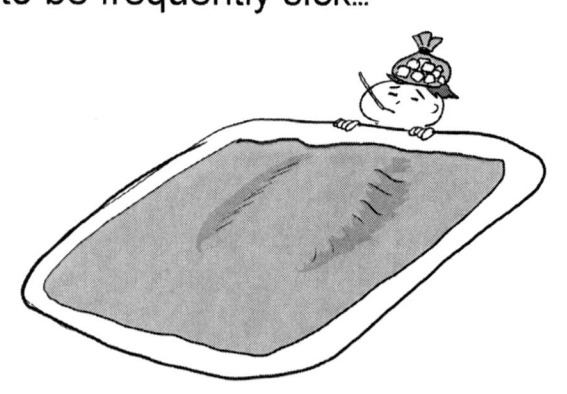

In fact, cystic fibrosis is a lethal disease that has killed many people.

However, because it comes from a recessive allele, it has remained within the human **"gene pool"**.

"Gene Pool" is a term used to describe the sum of all the variety of genes in a population.

EXAMPLE: The human gene pool is the sum of all the variety of genes in the human population.

New Vocabulary: Allele Frequency

Allele Frequency is a number that describes how common an allele is in a population's gene pool. In other words, it shows the % (or proportion) of the gene pool that is made of that allele.

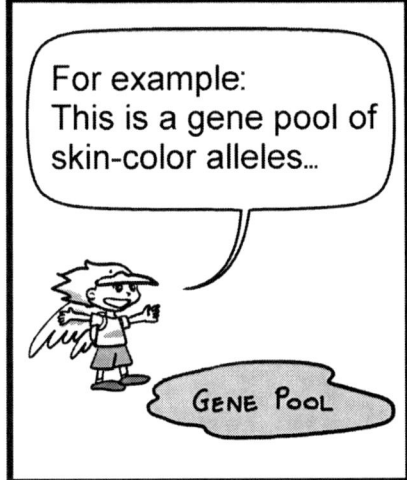

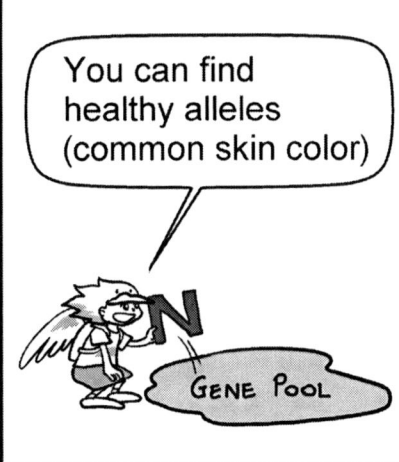

Chapter 30: Three Types of Natural Selections

Science Standard Bio/LS . 8 . a :
Students know how natural selection determines the differential survival of groups of organisms.

NEW VOCABS

* **Directional Selection:** A type of Natural Selection in which 1 particular phenotype is favored. Therefore, 1 extreme is considered very fit while the other extreme is considered as unfit. The medium (median) ones are neither strongly favored nor disfavored.

* **Disruptive Selection:** A type of Natural Selection in which the medium (median) ones are considered as unfit, while the 2 extremes are considered as fit.

* **Stabilizing Selection:** A type of Natural Selection in which the medium (median) ones are considered as fit, while the 2 extremes are considered as unfit.

1st Type: Stabilizing Selection

The medium (median) ones are fit, while the 2 extremes are unfit.

EXAMPLE: Body size of groundhogs.

Stablizing Selection decreases the number of individuals in the 2 extremes while increasing the number of the medium ones.

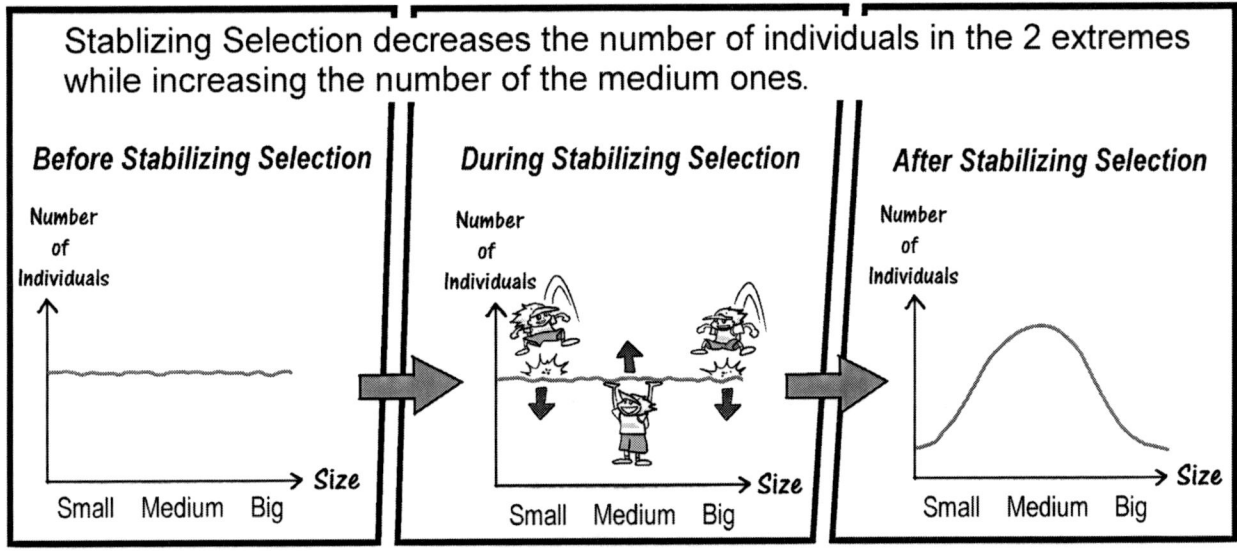

2nd Type: Disruptive Selection

(The opposite of stabilizing Selection.)
The 2 extremes are fit, while the medium ones are unfit.

EXAMPLE: Moths' colors.

Black moths are **"fit"**

...because we can hide in dark trees.

Gray moths are **"unfit"**.

Our color cannot hide in rocks or trees!

White moths are **"fit"**

...because we can hide on white rocks.

Disruptive Selection increases the number of individuals in the 2 extremes while decreasing the medium-range individuals.

Before Disruptive Selection — *During Disruptive Selection* — *After Disruptive Selection*

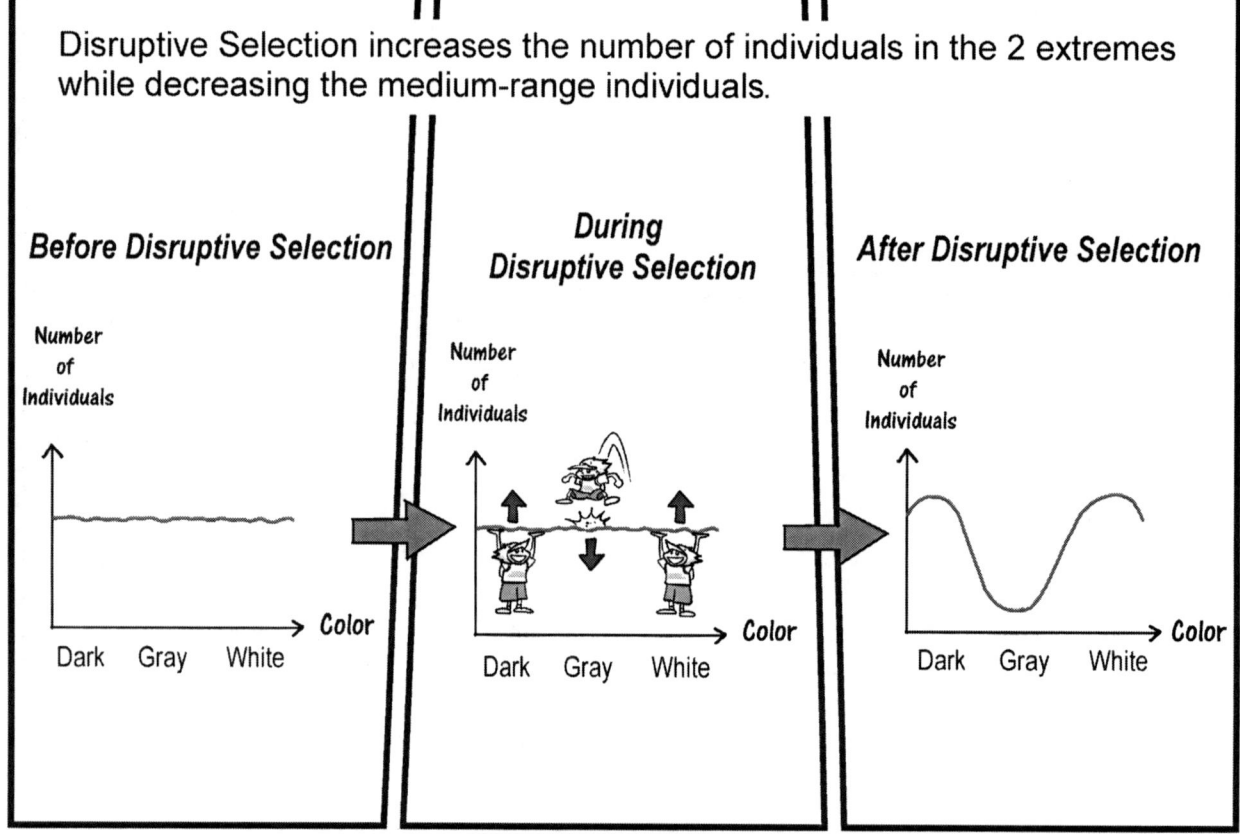

3rd Type: Directional Selection

1 phenotype is favored. Therefore, 1 extreme is considered very fit. The medium ones are OK, and the other extreme is unfit.

EXAMPLE: Cheetah's speed

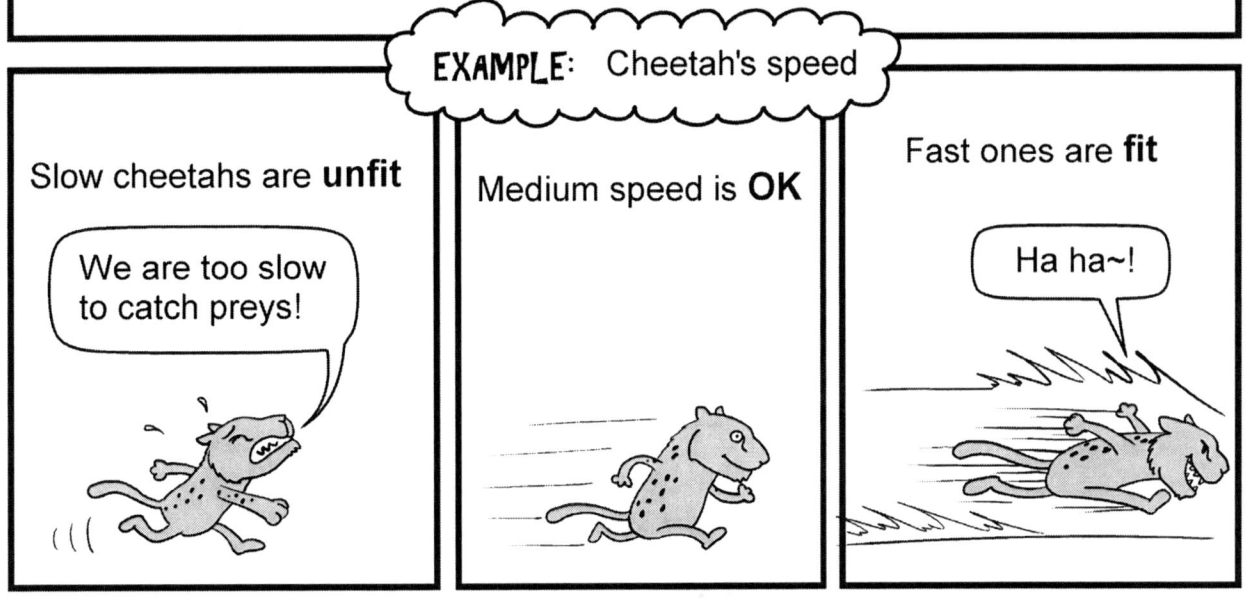

Directional Selection causes 1 extreme to have a high number while causing the other extreme to have a very low number.

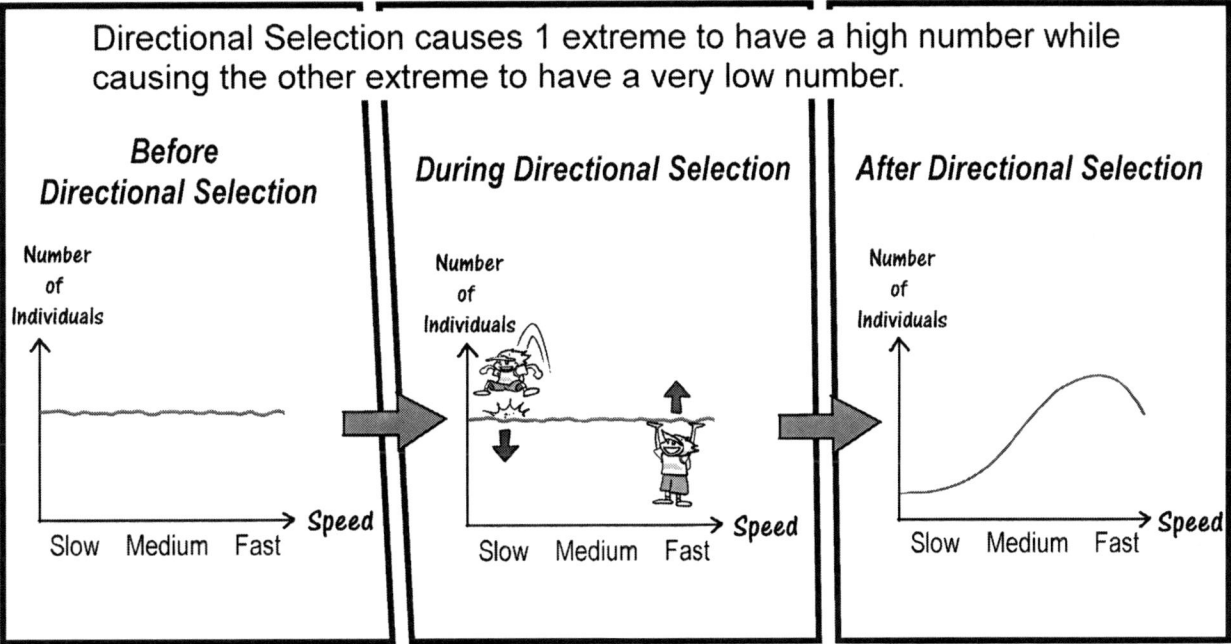

Long-Term Effect:

If directional selection goes on for a long time, the **whole graph** may **shift** toward the one side that is **fit**!

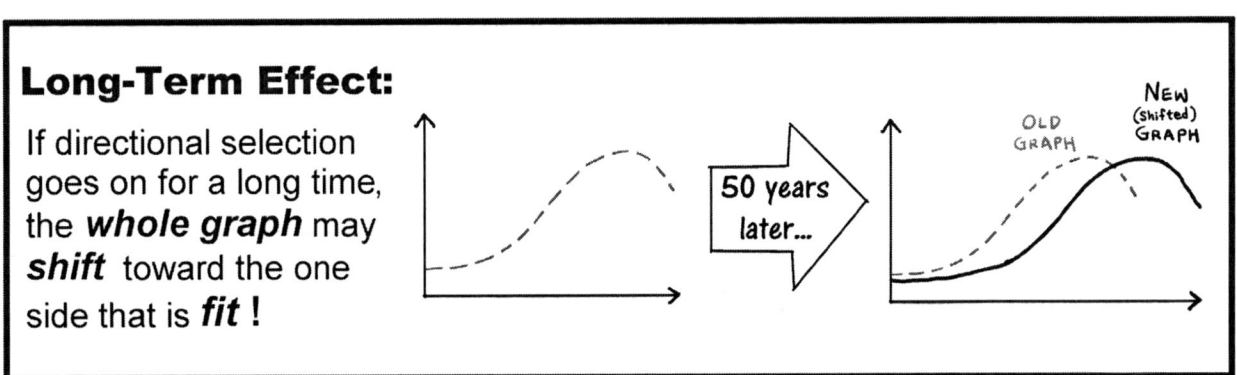

Chapter 31: Diversity and Survival

Science Standard Bio/LS . 8 . b :
Students know a great diversity of species increases the chance that at least some organisms survive major changes in the environment.

NEW VOCABS

* **Genetic Diversity:** The amount of variety of genetic characteristics in a population.

* **Species Diversity:** The amount of variety of different types of species in an environment.

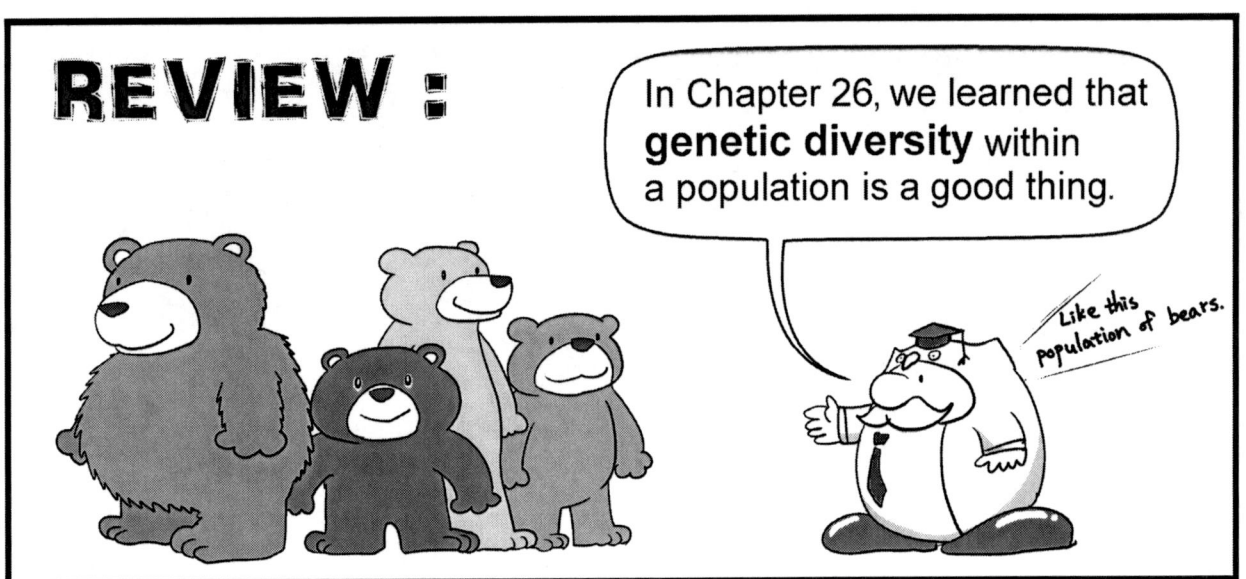

Genetic Diversity usually means variation within a population (of the same species).

Species Diversity means the variety of different types of species in an environment.

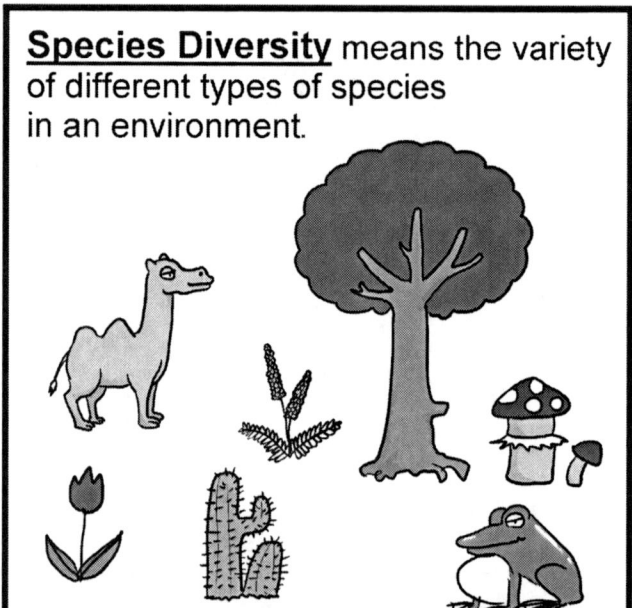

Like Genetic Diversity, Species Diversity is also a good thing because it increases the chance that at least some species will survive an environmental change.

EXAMPLE:

We start with a community that has a high species diversity.

If the environment becomes too dry:

 will survive.

If the climate becomes too wet:

 will survive.

If the climate becomes too cold:

 will survive.

In conclusion, diversity is always a good thing, whether it is:

Species Diversity OR Genetic Diversity (within a population)

Think about it. In action movies, isn't there usually a team of people with a *diverse* range of abilities?

Diversity helps with survival!

Unit 4: Chapter 31

Unit 5: Anatomy and Physiology

Chapter 32: Cells, Tissues, Organs, and Systems

Science Standard: 7.5.a
Students know plants and animals have levels of organization for structure and function, including cells, tissues, organs, organ systems, and the whole organism.

NEW VOCABS

* **Cardiac Muscle Tissue:** A type of muscle that makes up the heart.

* **Connective Tissue:** A type of tissue that supports and connects different parts of the body.

* **Differentiated Cell:** A cell that has been specialized to perform a specific function. Examples are muscle cells, epithelial cells, nervous cells...etc.

* **Differentiation:** The process in which an undifferentiated cell turns into a differentiated cell that has a specific function.

* **Epithelial Tissue:** The tissue that covers the surface of many body structures. Common examples are skin and membrane tissues.

* **Muscle Tissue:** The tissue that is responsible for causing movements. There are 3 main types of muscle tissues: skeletal muscle, smooth muscle, and cardiac muscle.

* **Nervous Tissue:** A type of tissue composed of nervous cells (neurons) and is responsible for sending/receiving of information throughout the body.

* **Organ:** 2 or more different types of tissues grouped together to perform a specific function.

* **Organ System:** Often simply called "system." Different organs working together to achieve a common goal.

* **Organism:** An individual living thing. A more complex type of organisms (such as animals) is made of more than one organ systems.

* **Skeletal Muscle Tissue:** Muscles that are attached to bones via tendons. They are voluntarily controlled by our conscious mind.

* **Smooth Muscle Tissue:** Muscles that are used mostly to construct some of our internal organs. It cannot be voluntarily controlled by our conscious mind.

* **Tissue:** Many of the same type of cells grouped together to perform a specific function.

* **Undifferentiated Cell:** (Example: Stem cell) A cell that has not yet decided on what type of cell it wants to become. Therefore, it has not yet turned on any specific cell-type-related gene.

A cell is the basic unit of life.

There are many types of cells.

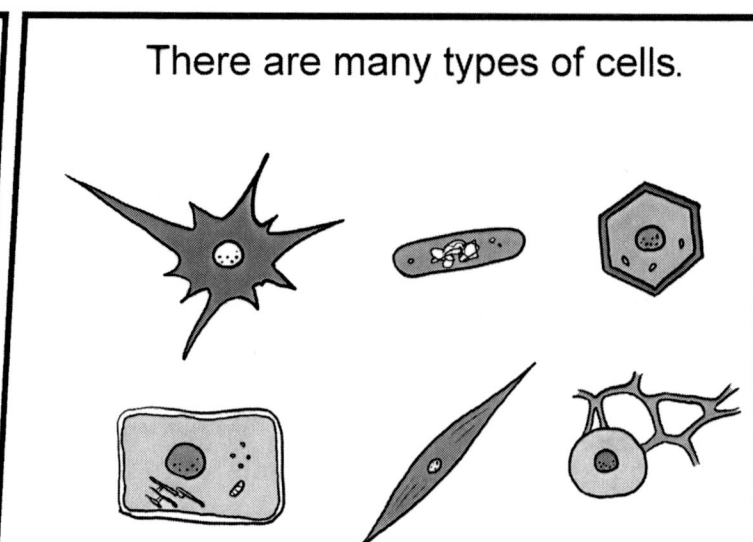

When you have many of the same type of cells grouped together, we call this a **"tissue."**

EXAMPLES:

In order to be called a "tissue", it must be made of the same kind of cells.

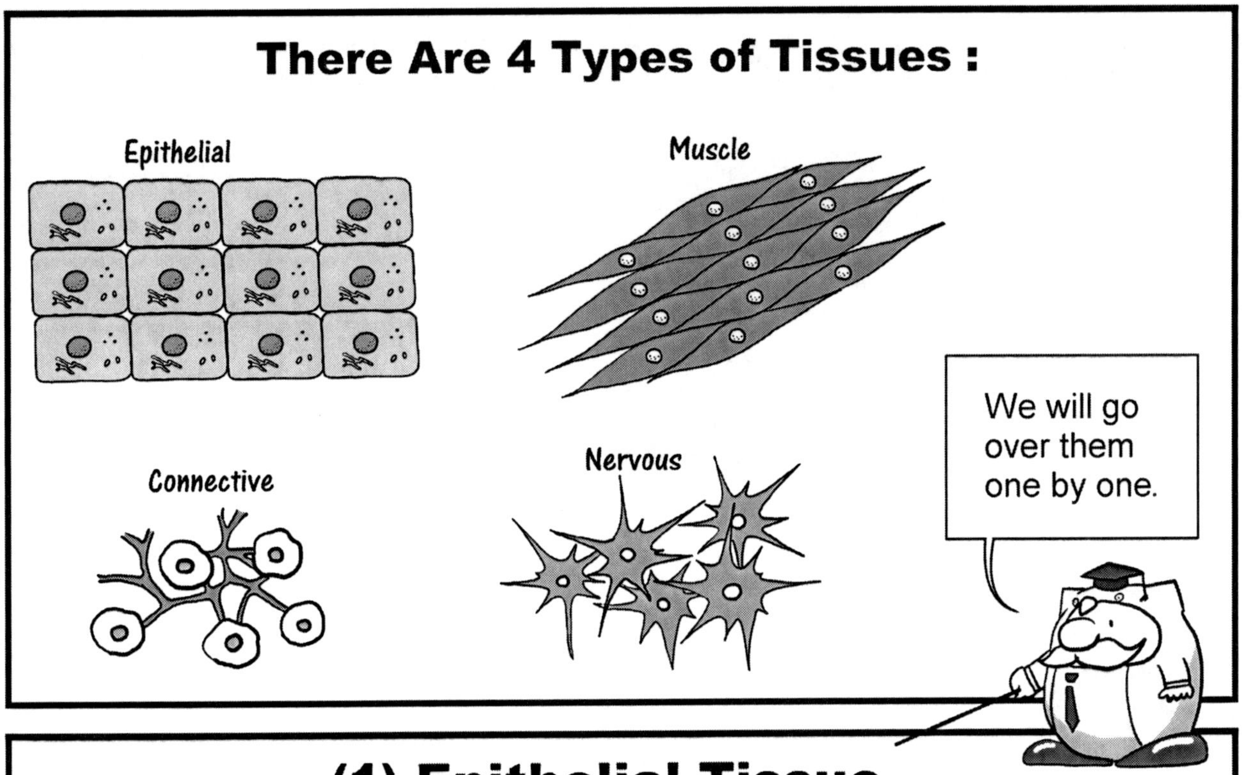

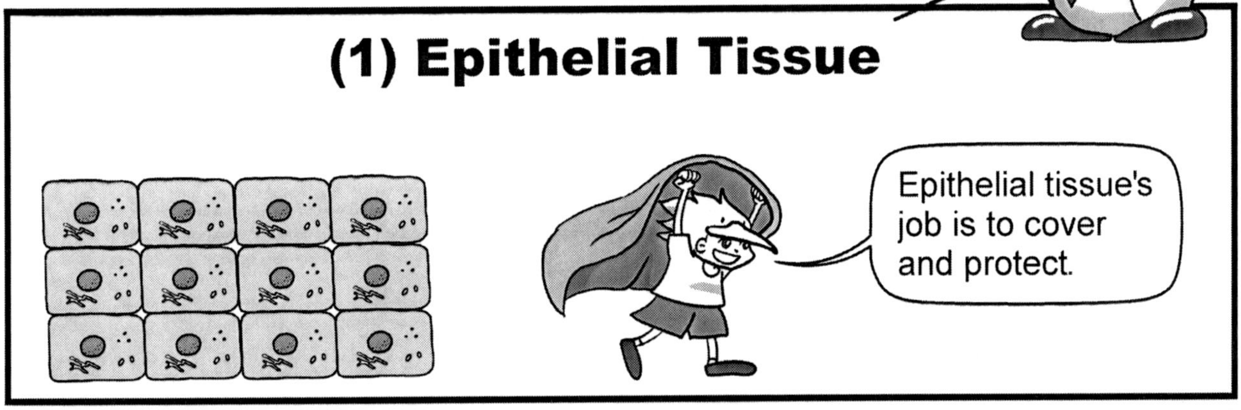

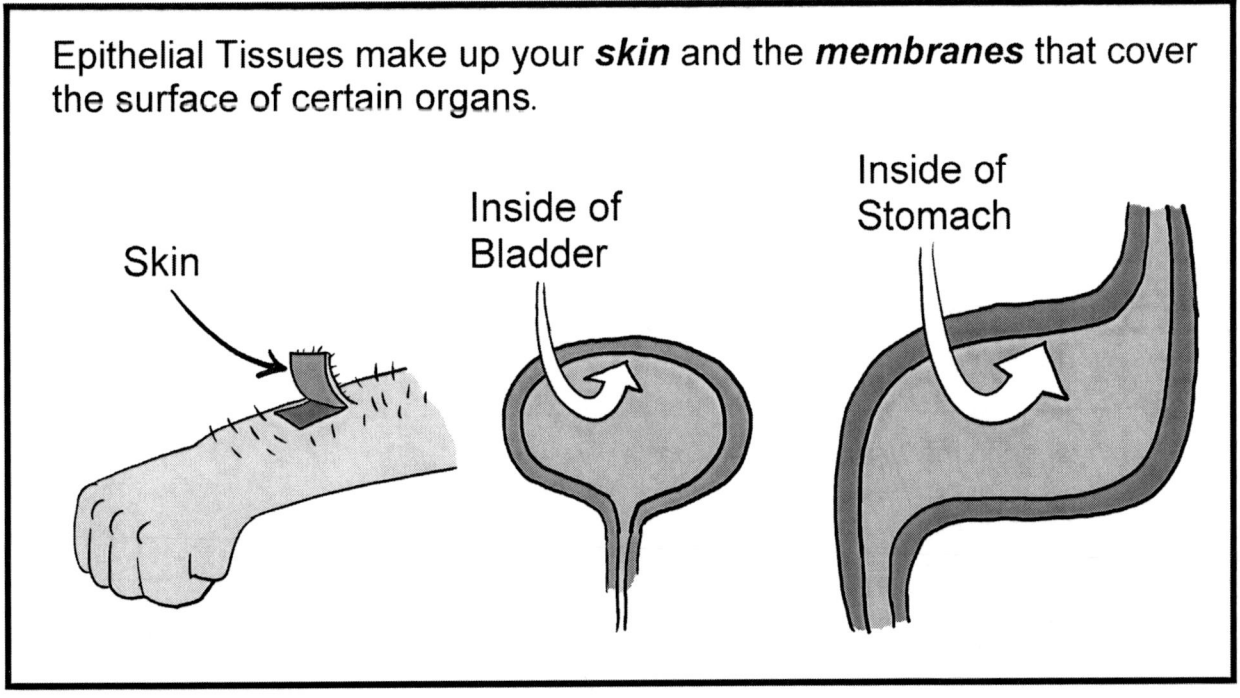

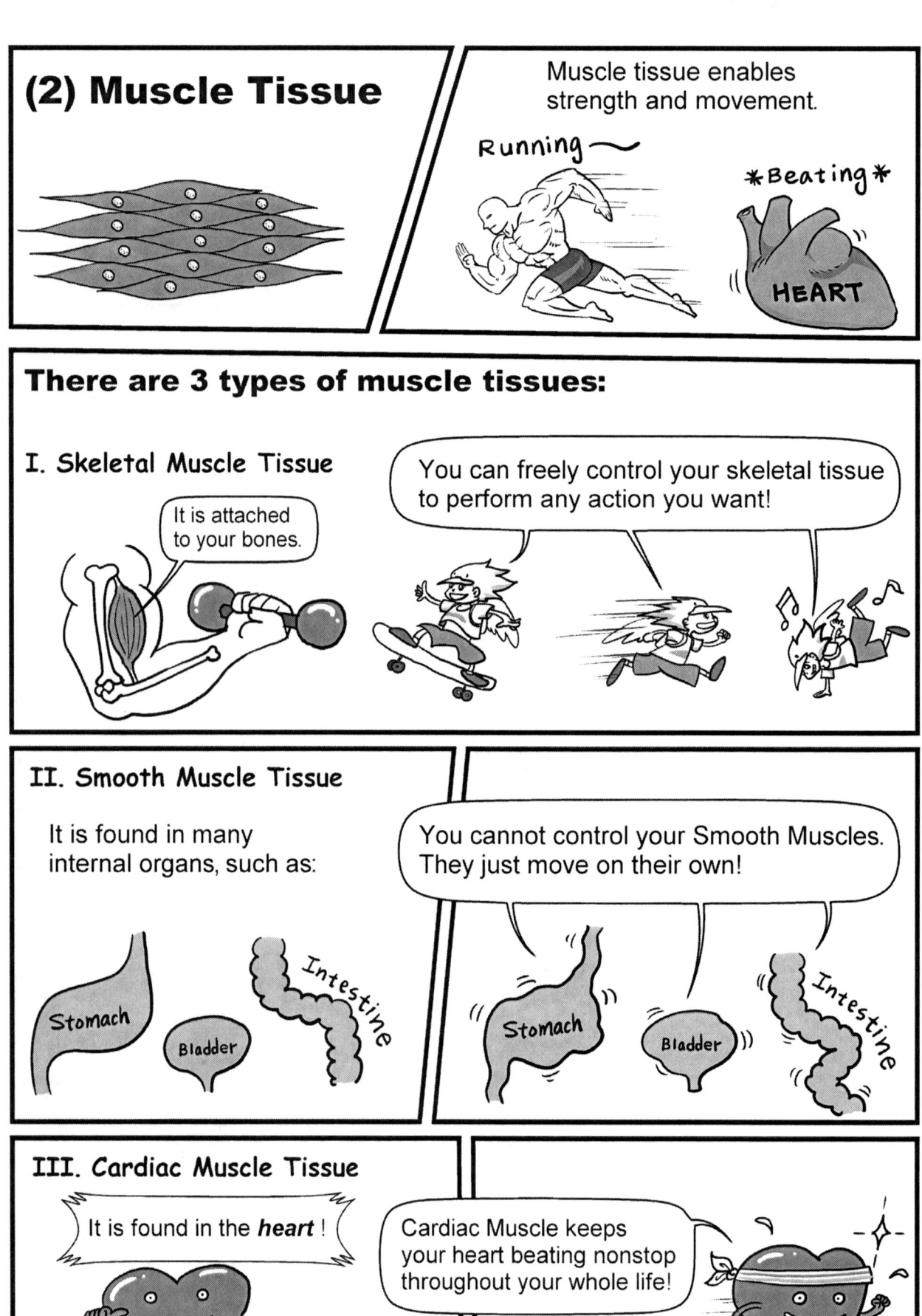

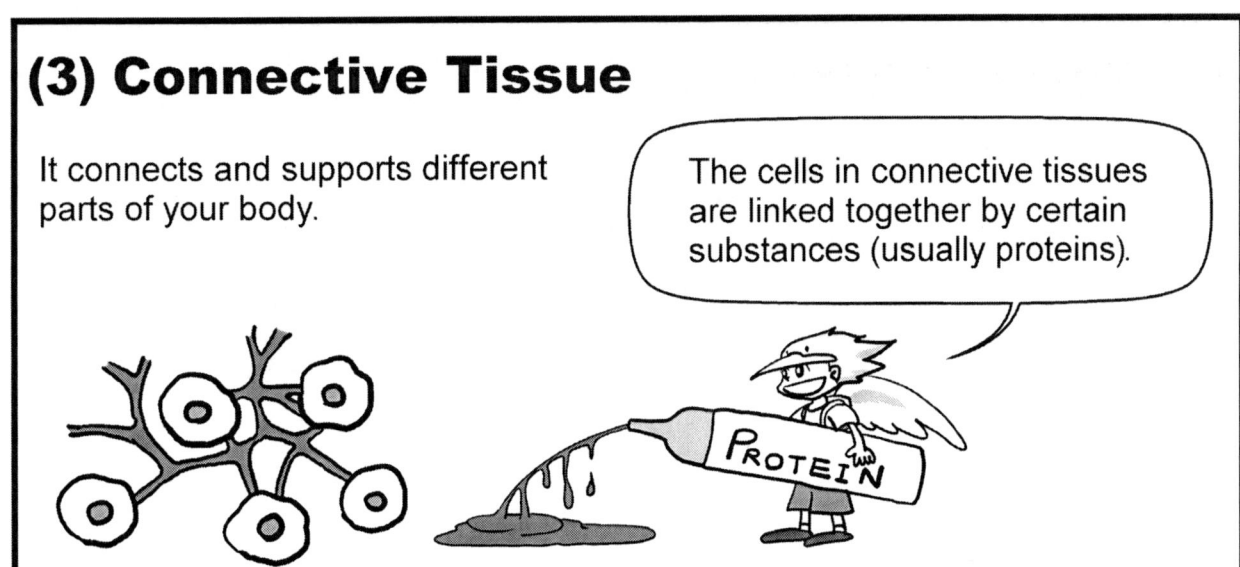

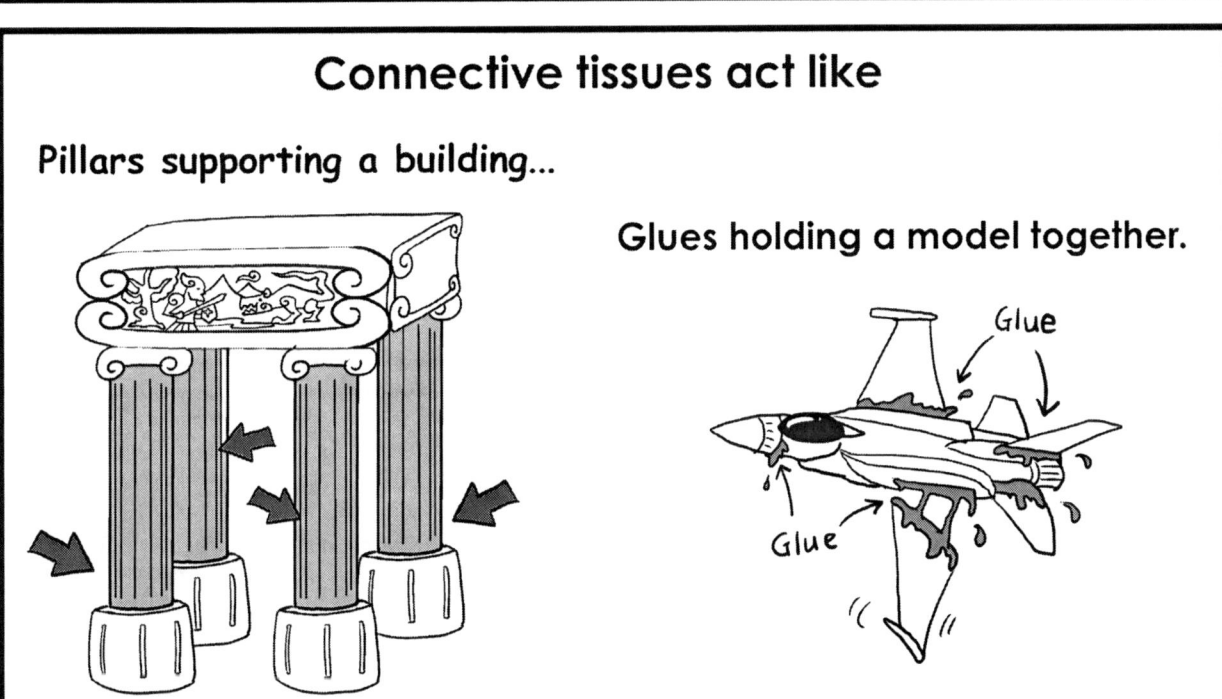

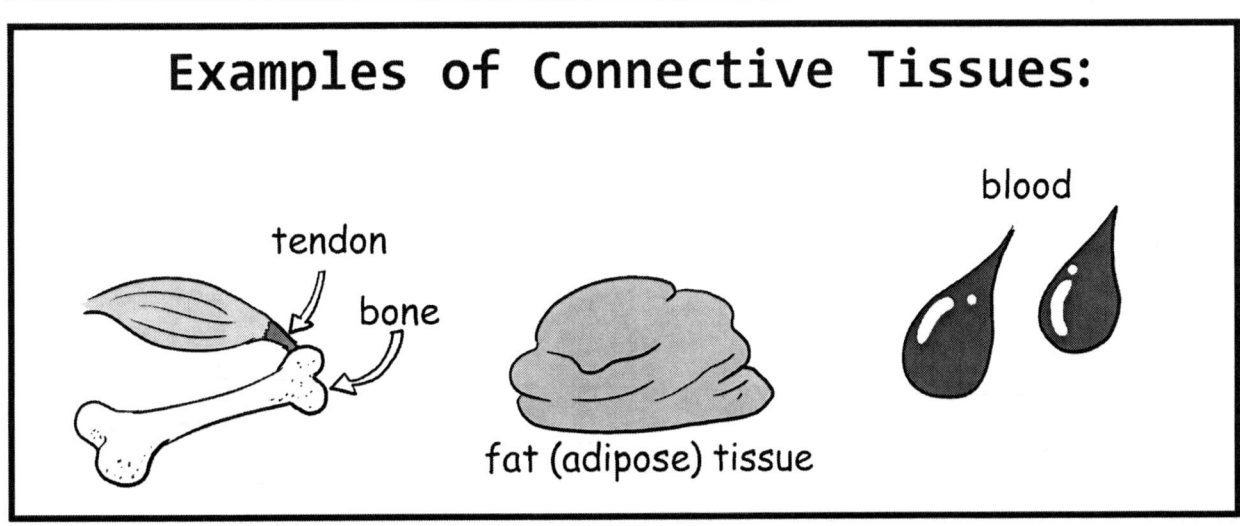

page 293
Unit 5: Chapter 32

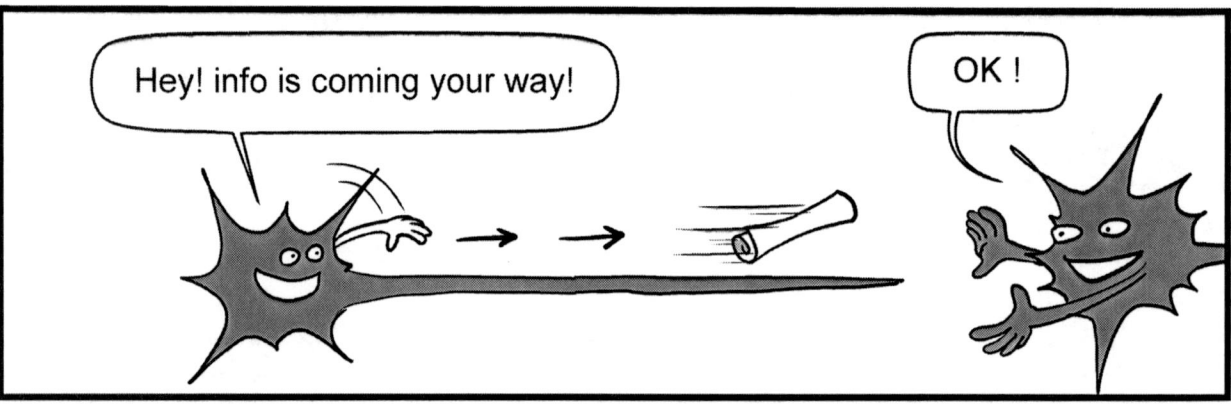

When many neurons work together, information can travel over long distances!

Here is an example of a Nervous Tissue. You can see multiple neurons working together (total 3).

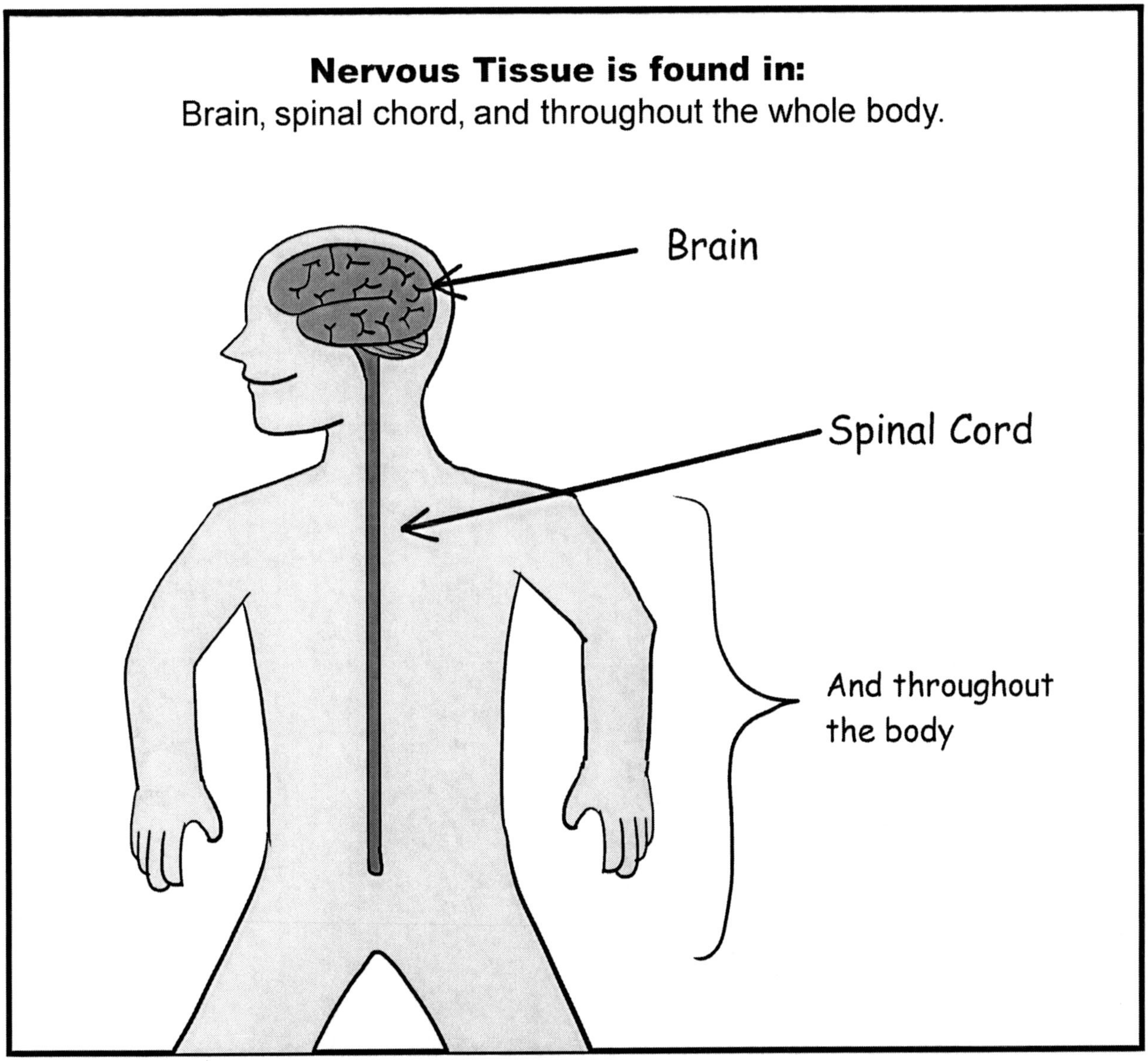

Nervous Tissue is found in:
Brain, spinal chord, and throughout the whole body.

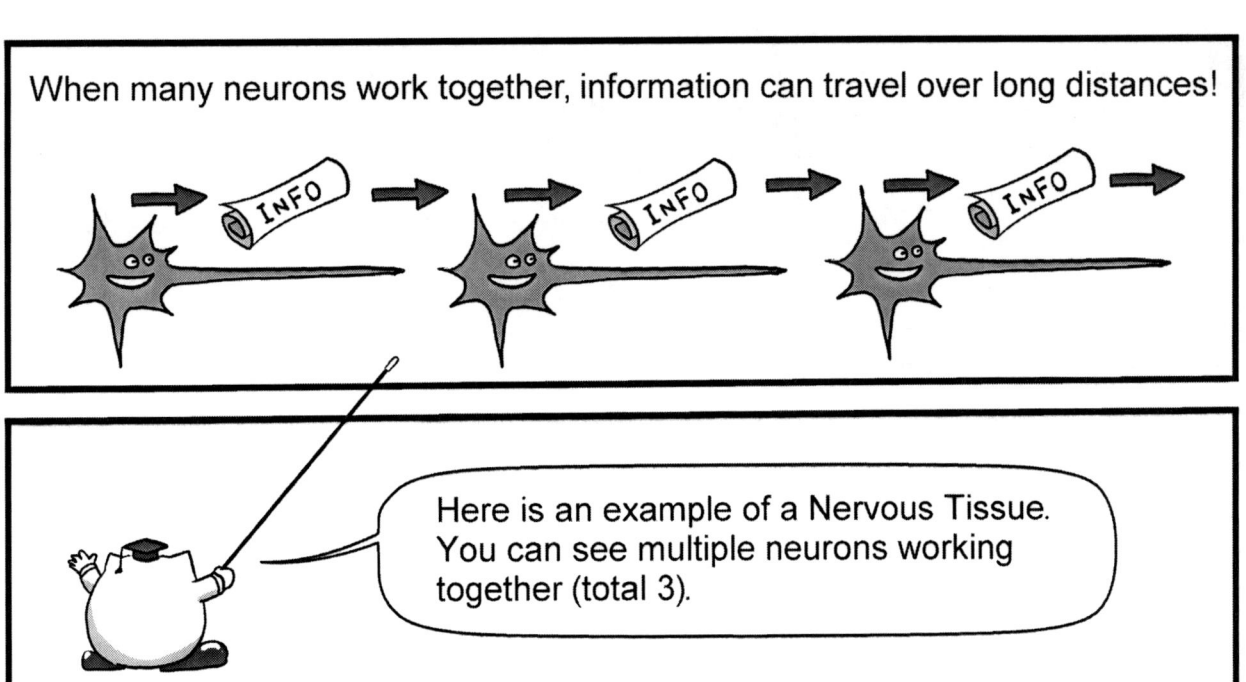

- Brain
- Spinal Cord
- And throughout the body

Organ:

2 or more tissues grouped together and working together to perform a function.

For Example: Stomach is an organ made of different kinds of tissues working together to perform a function: **Digesting food.**

Epithelial tissue

Nervous Tissue

Muscle Tissue

Connective Tissue

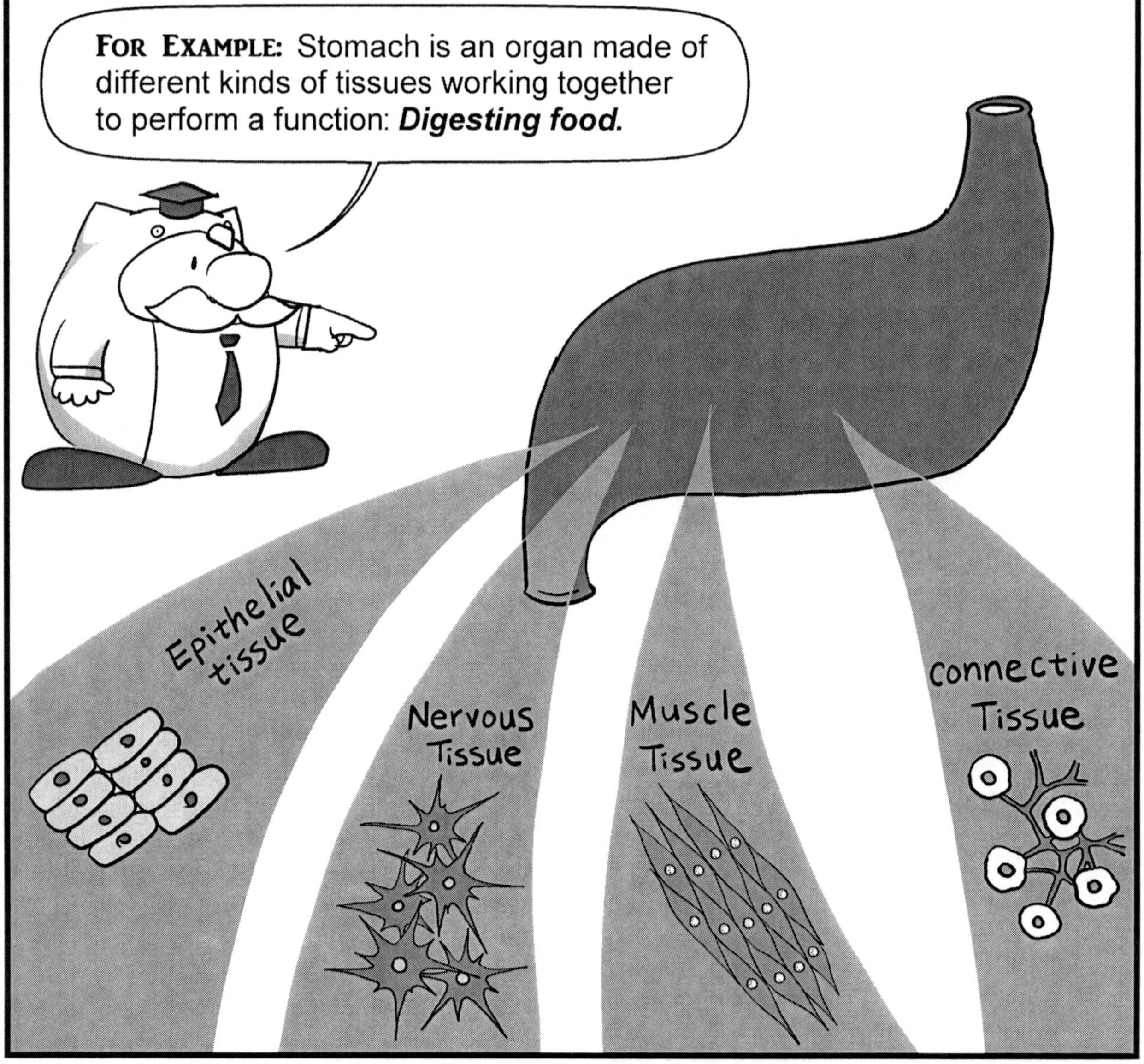

ORGAN SYSTEM:
Different organs working together to achieve a common goal.

EXAMPLE:

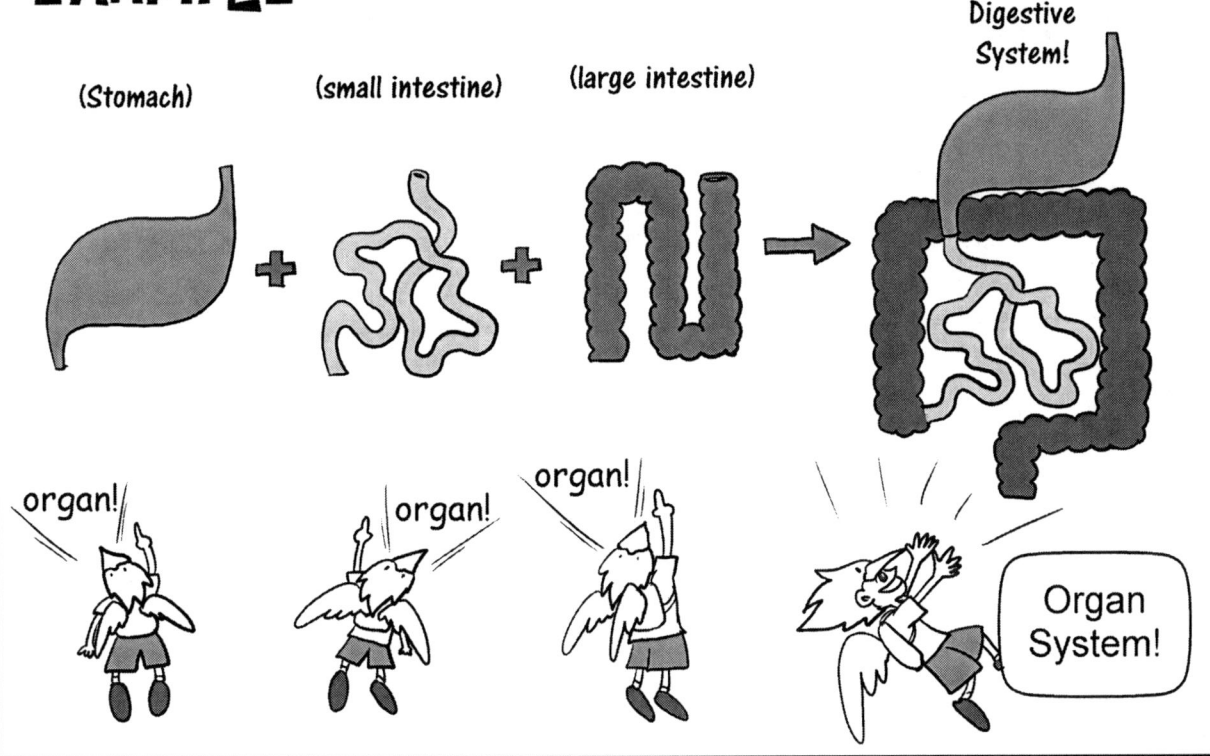

There are several types of organ systems. Below are a few examples:

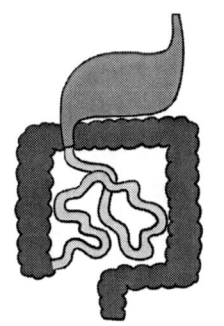

Digestive System

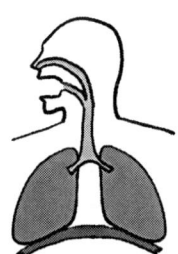

Respiratory System

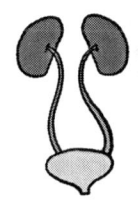

Urinary System

And many more...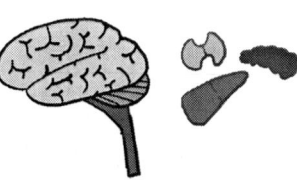

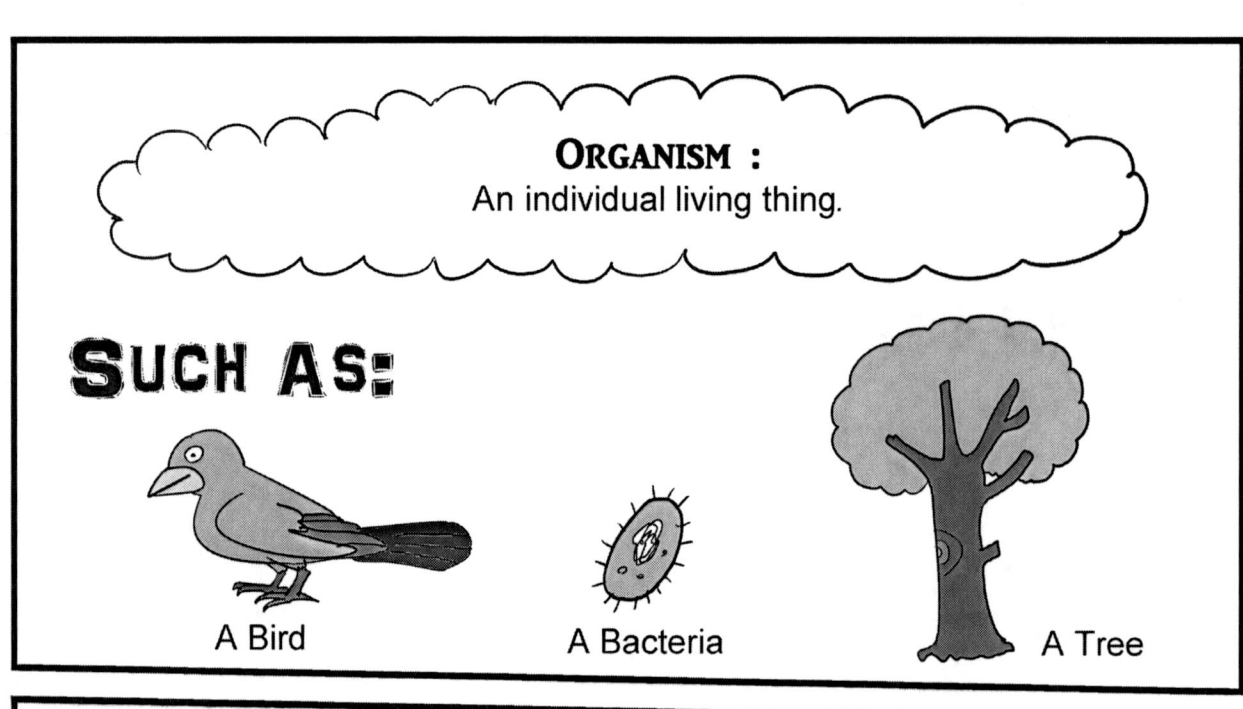

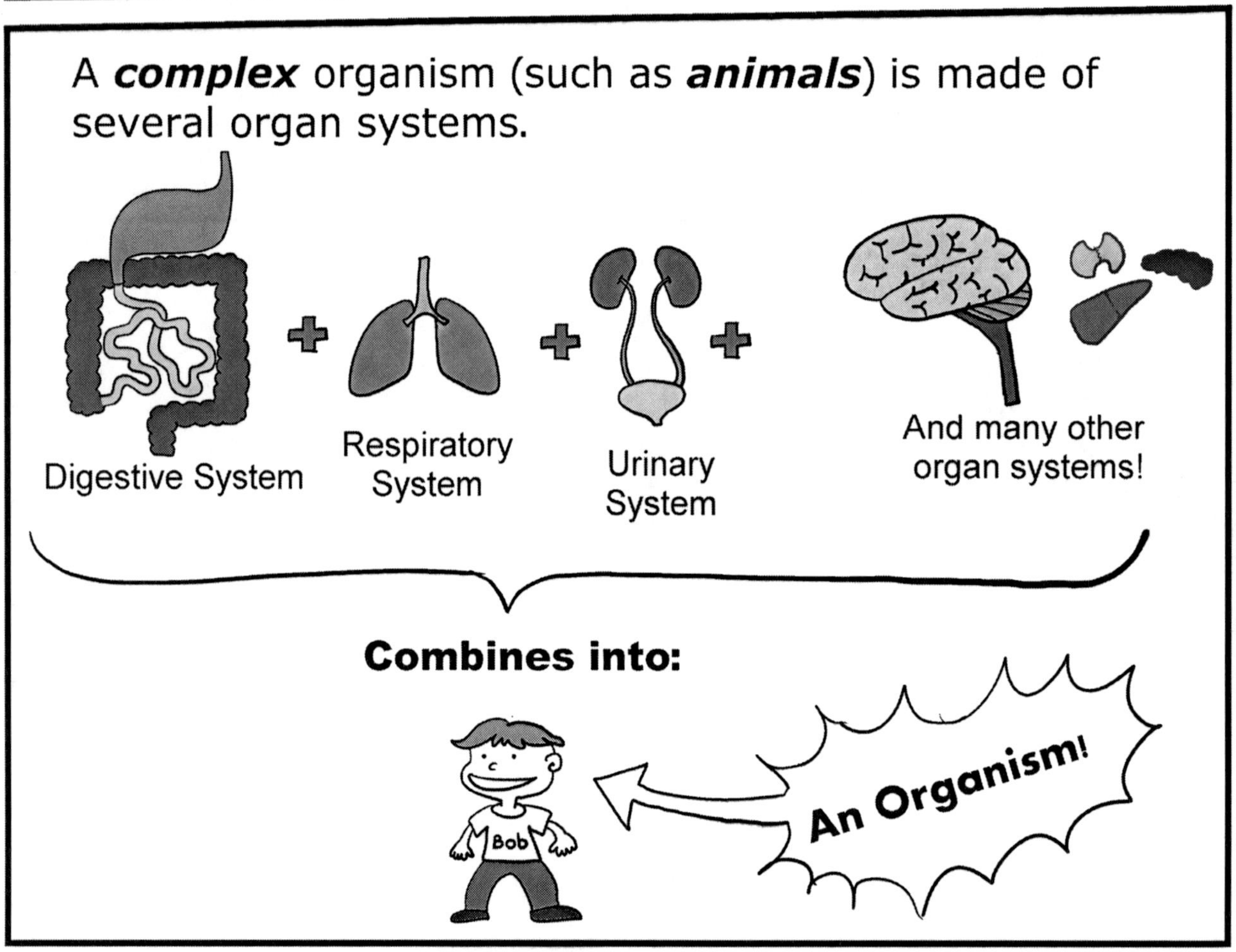

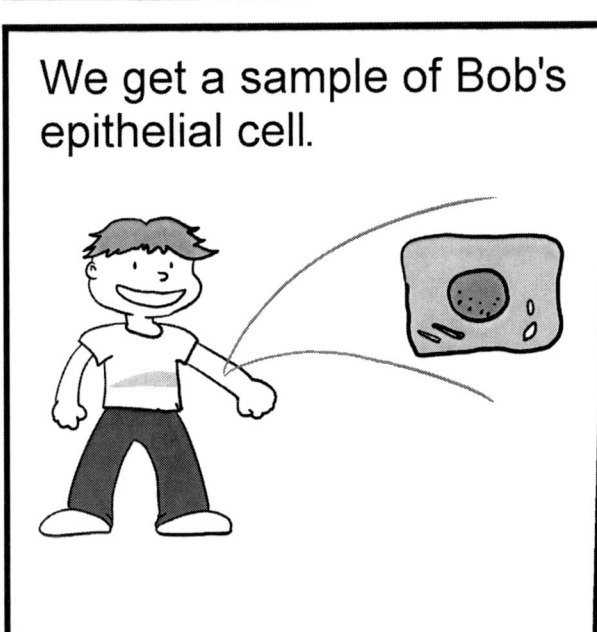
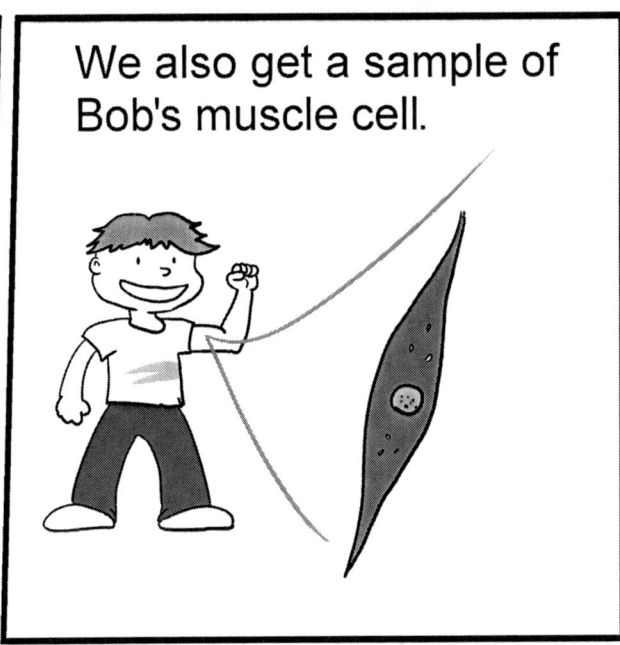
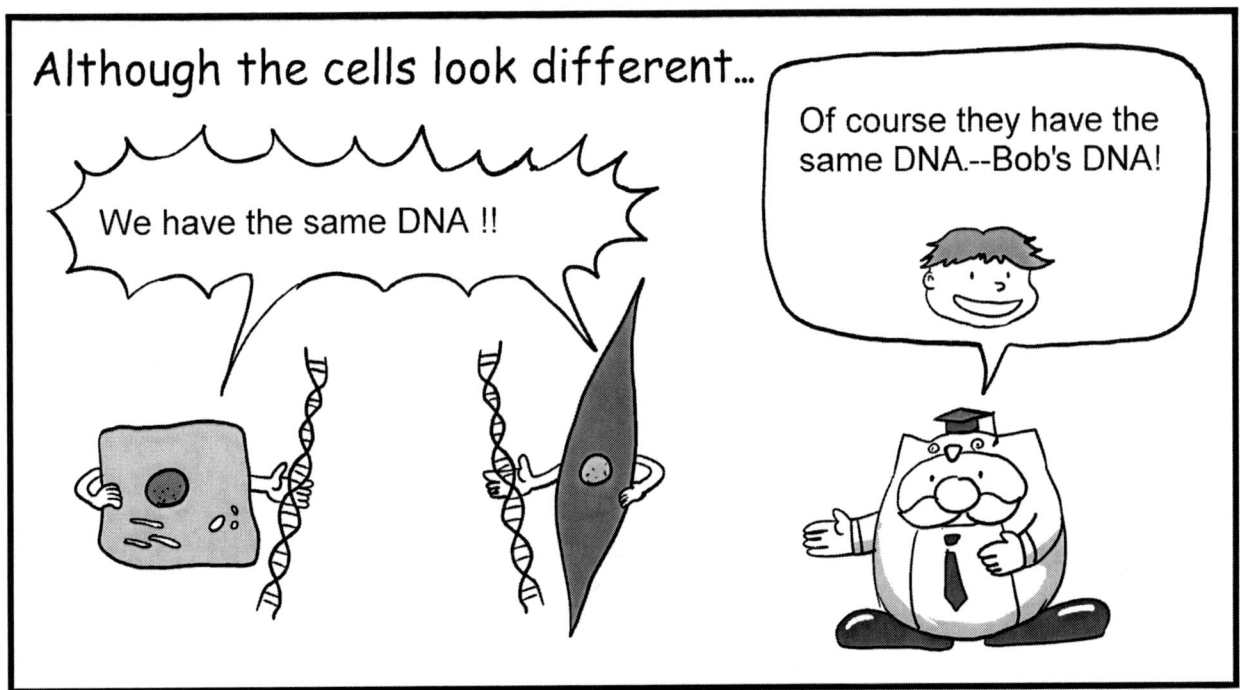

page 299

Unit 5: Chapter 32

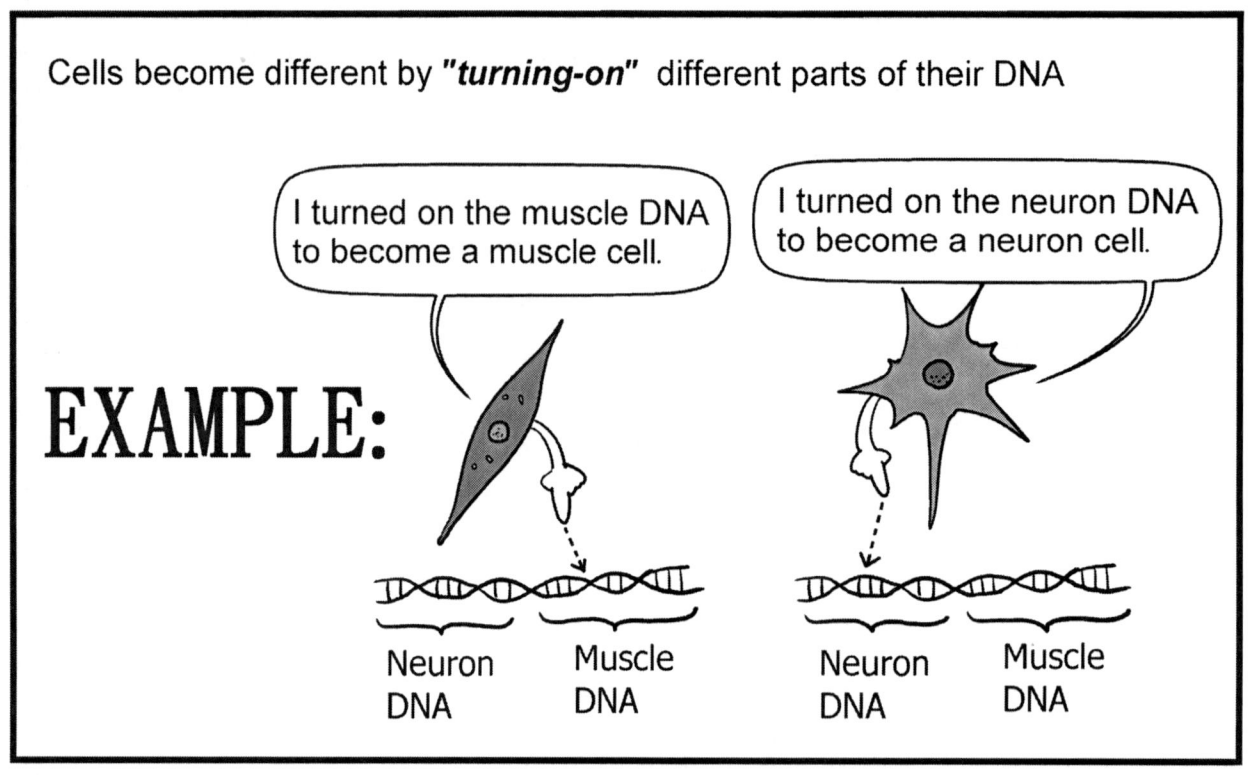

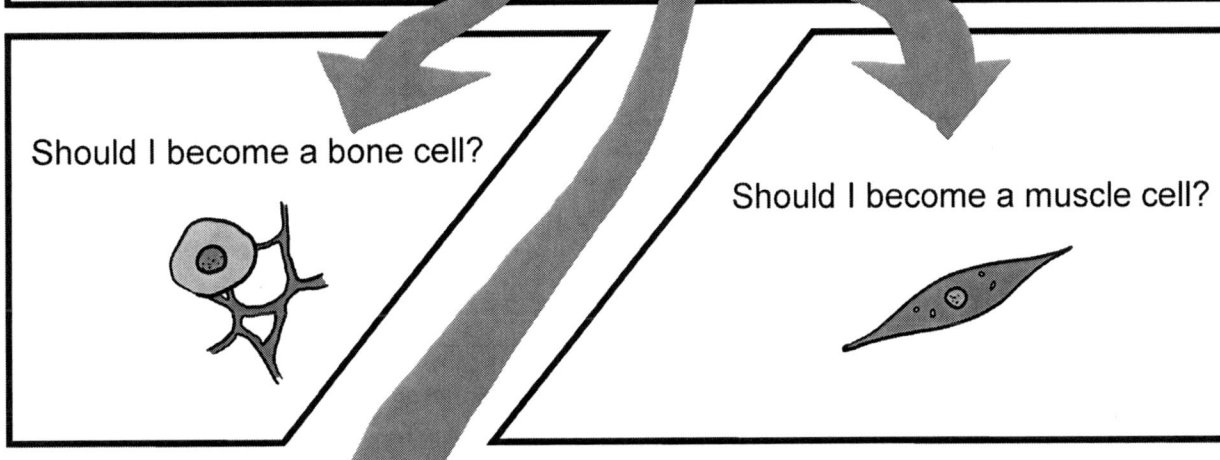

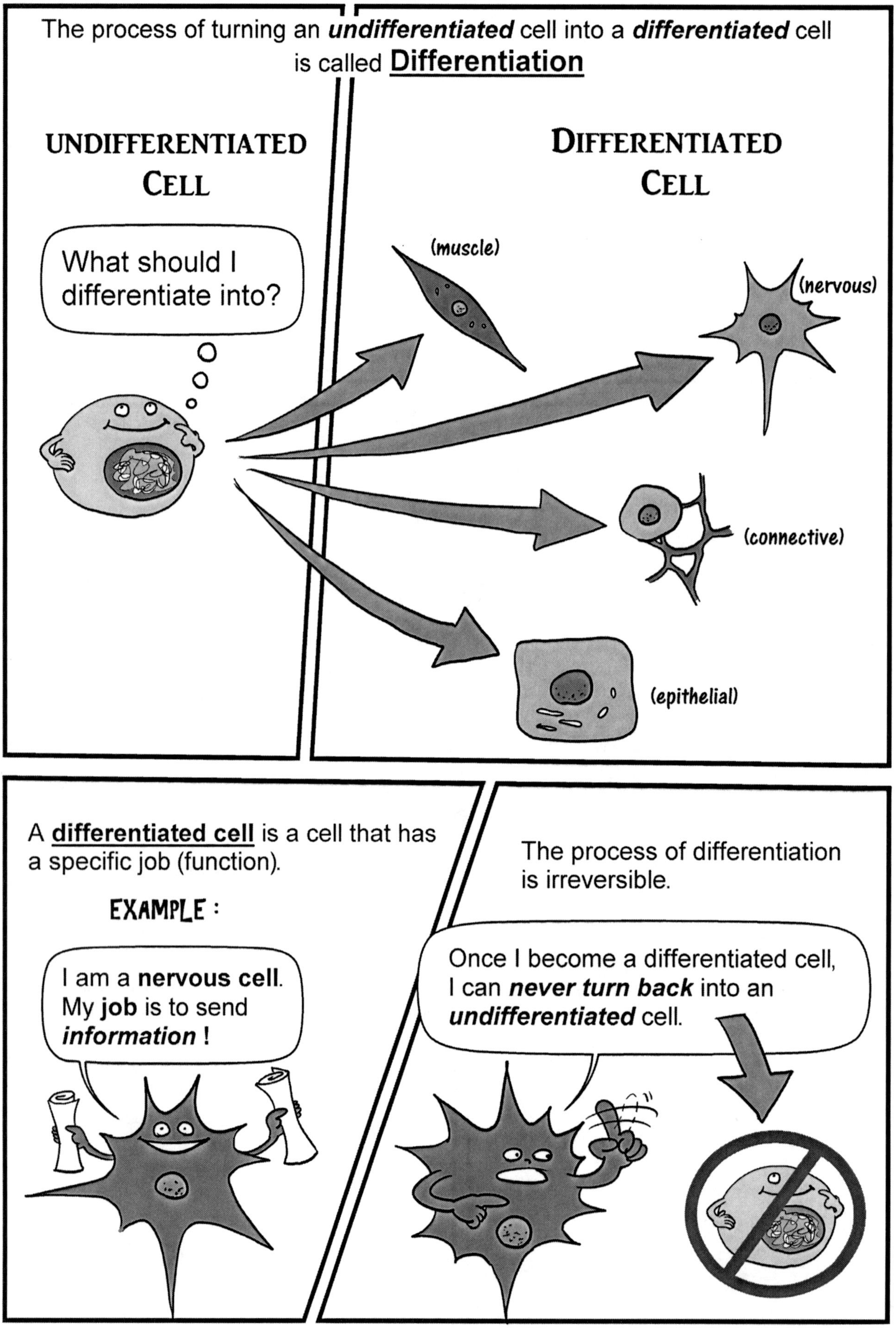

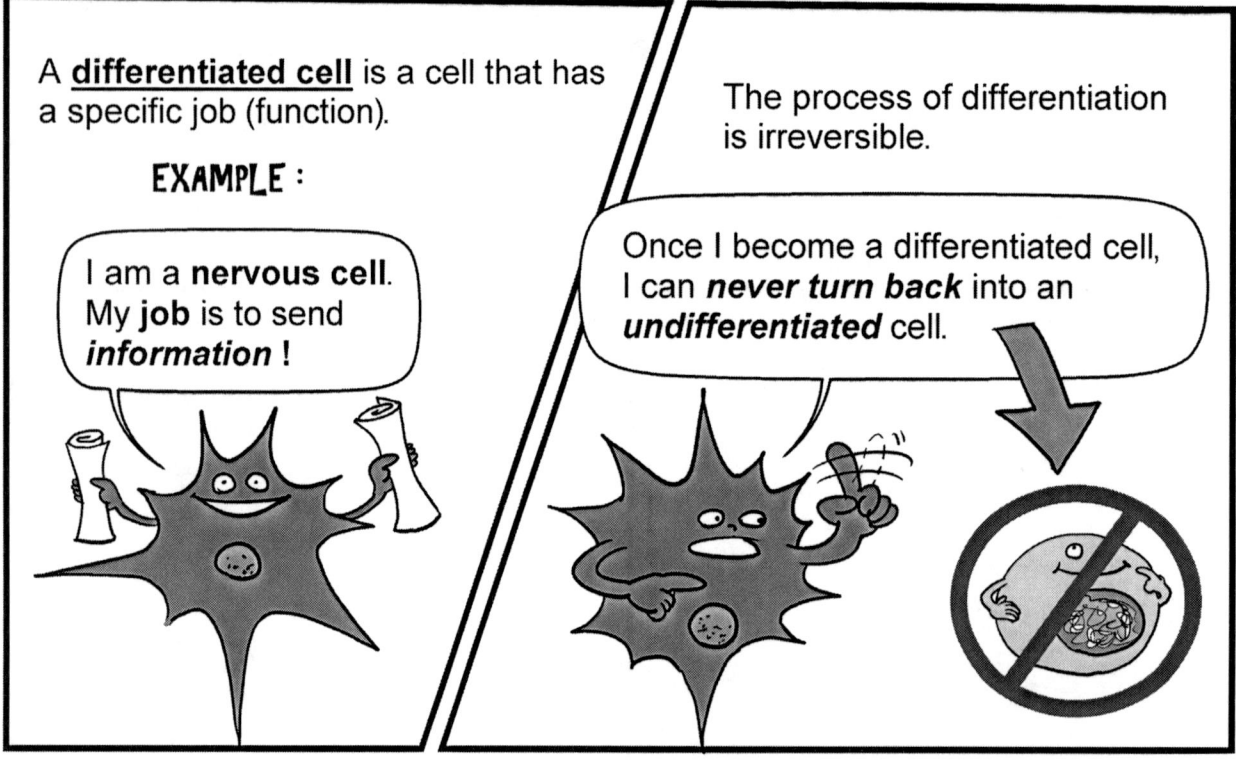

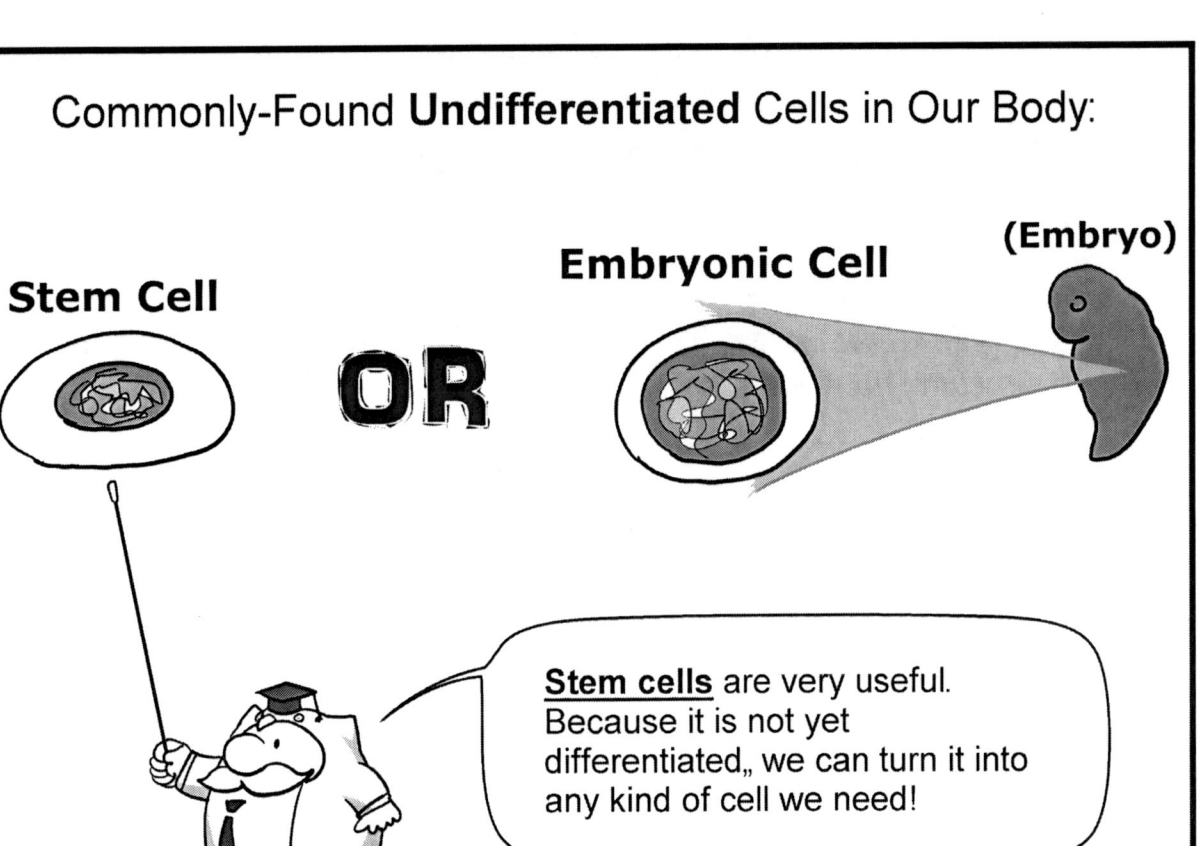
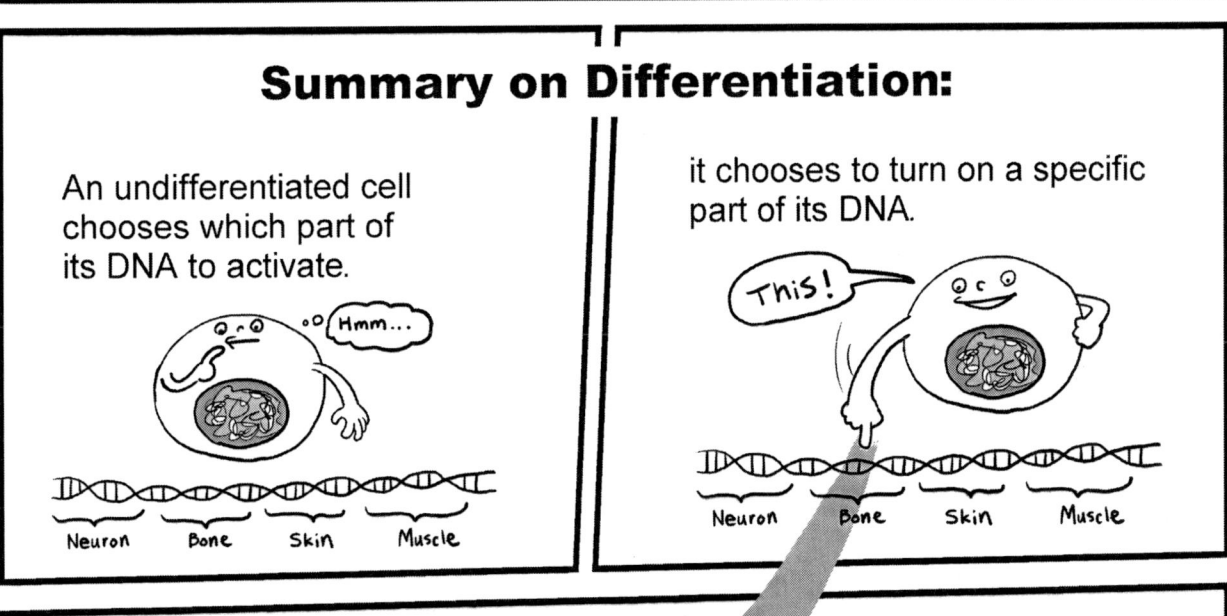
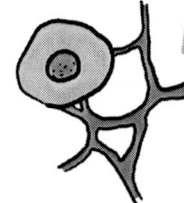

Chapter 33: Bones and Muscles

Science Standard: 7.5.c
Students know how bones and muscles work together to provide a structural framework for movement.

NEW VOCABS

* **Ball and socket joint:** A type of joint that can move in a very wide range of directions. Example: Shoulder joint.

* **Cardiac Muscle Tissue:** A type of muscle that makes up the heart.

* **Cartilage:** a flexible connective tissue that covers bones to protect and cushion the bone against shocks.

* **Hinge joint:** A type of joint that can only move in 1 direction. Example: knee joint.

* **Joint:** A joint is the location between 2 or more bones and allows movement.

* **Ligament:** A type of connective tissue that connects bone to bone.

* **Pivot Joint:** A type of joint that rotates. Example: Neck joint.

* **Saddle joint:** A type of joint that can slightly move in most directions. Example: Thumb joint.

* **Skeletal Muscle Tissue:** Muscles that are attached to bones via tendons. They are voluntarily controlled by our conscious mind.

* **Smooth Muscle Tissue:** Muscles that are used mostly to construct some of our internal organs. It cannot be voluntarily controlled by our conscious mind.

* **Tendon:** A type of connective tissue that connects muscle to bone.

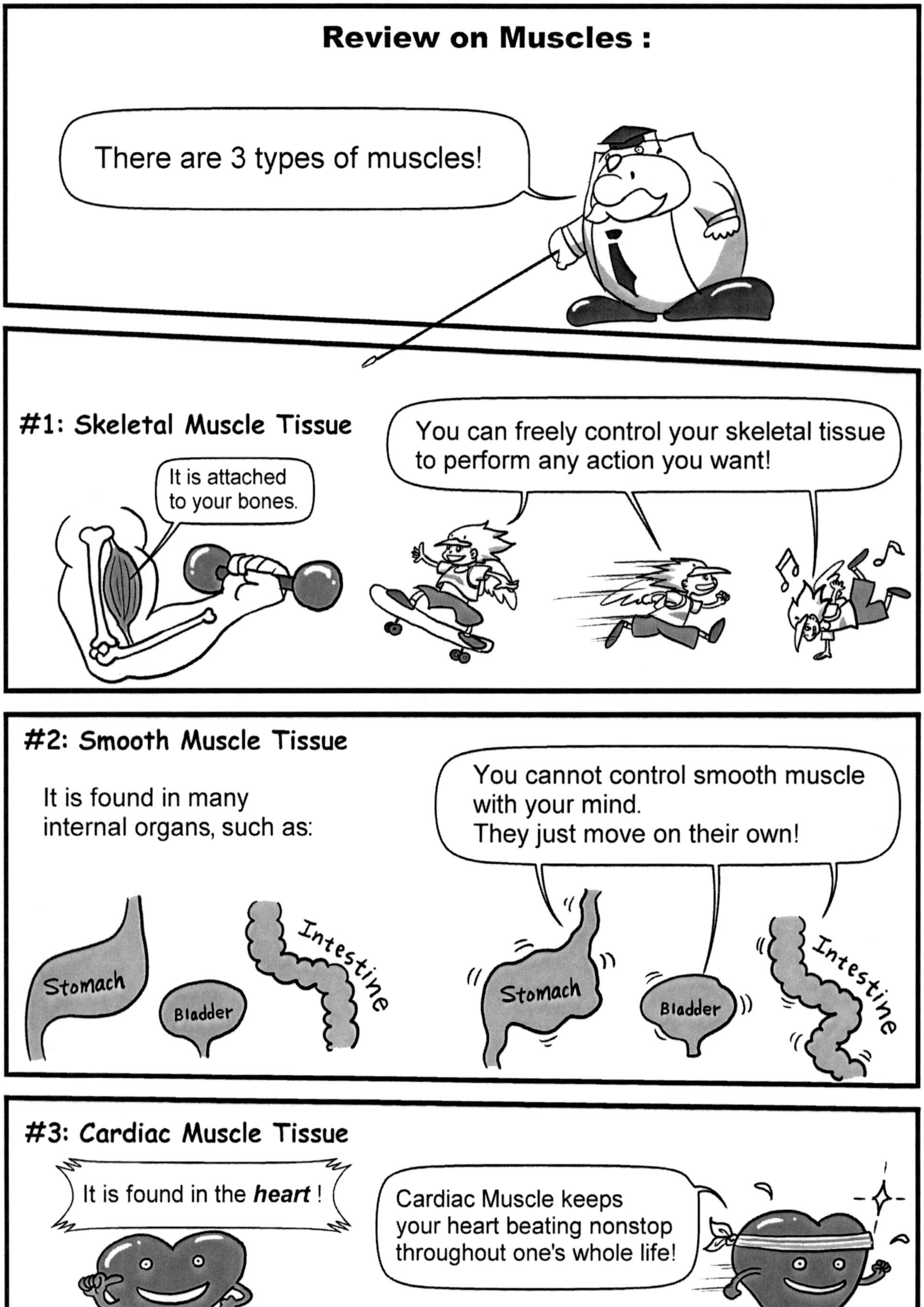

Review on Muscles:

"There are 3 types of muscles!"

#1: Skeletal Muscle Tissue

"It is attached to your bones."

"You can freely control your skeletal tissue to perform any action you want!"

#2: Smooth Muscle Tissue

It is found in many internal organs, such as:

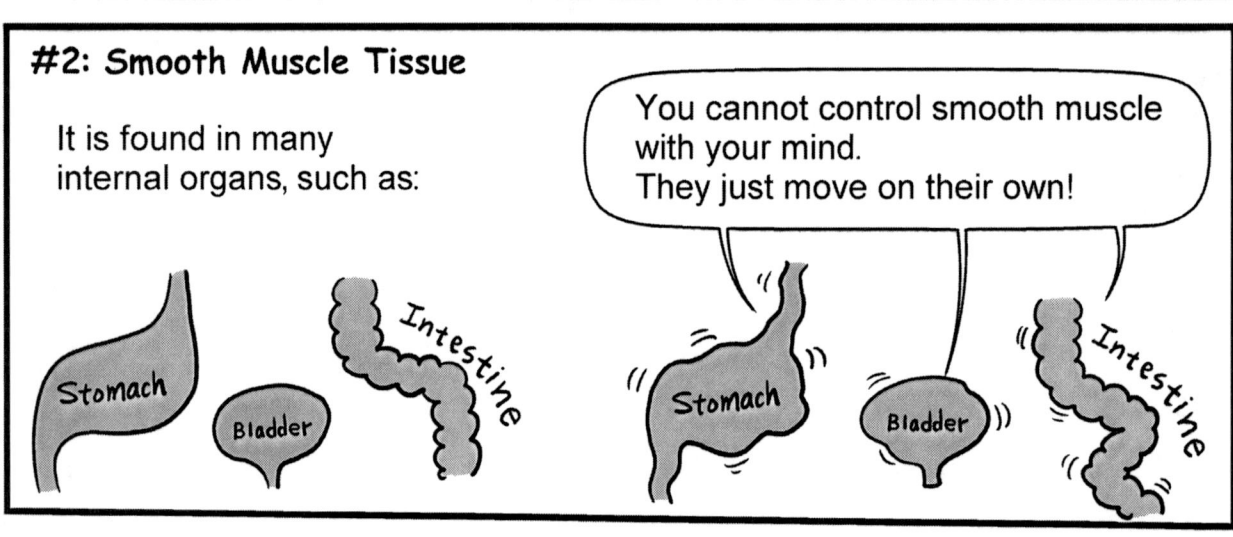

"You cannot control smooth muscle with your mind. They just move on their own!"

#3: Cardiac Muscle Tissue

"It is found in the *heart*!"

"Cardiac Muscle keeps your heart beating nonstop throughout one's whole life!"

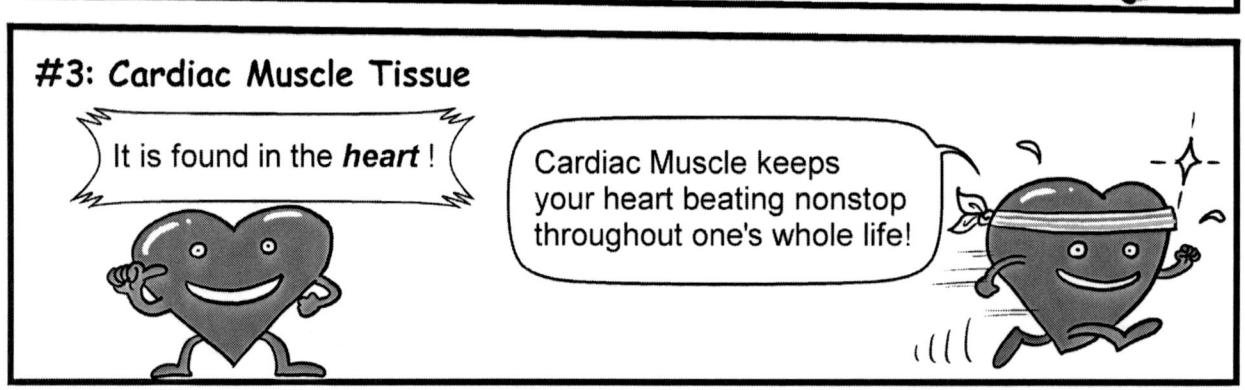

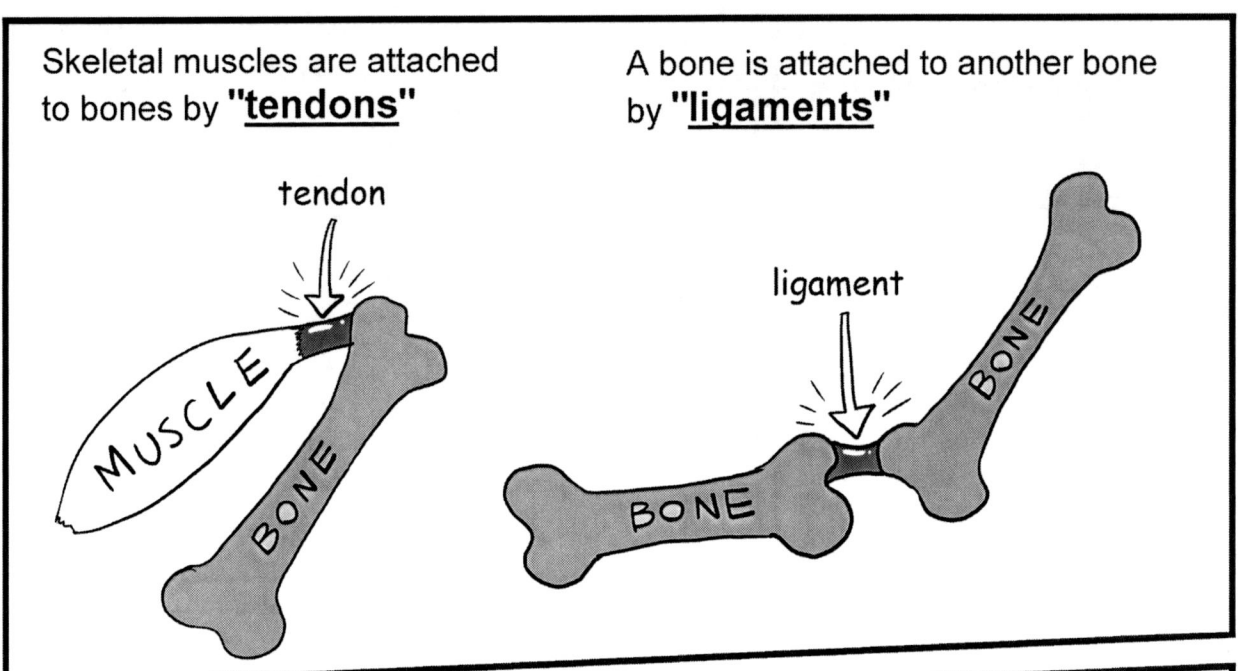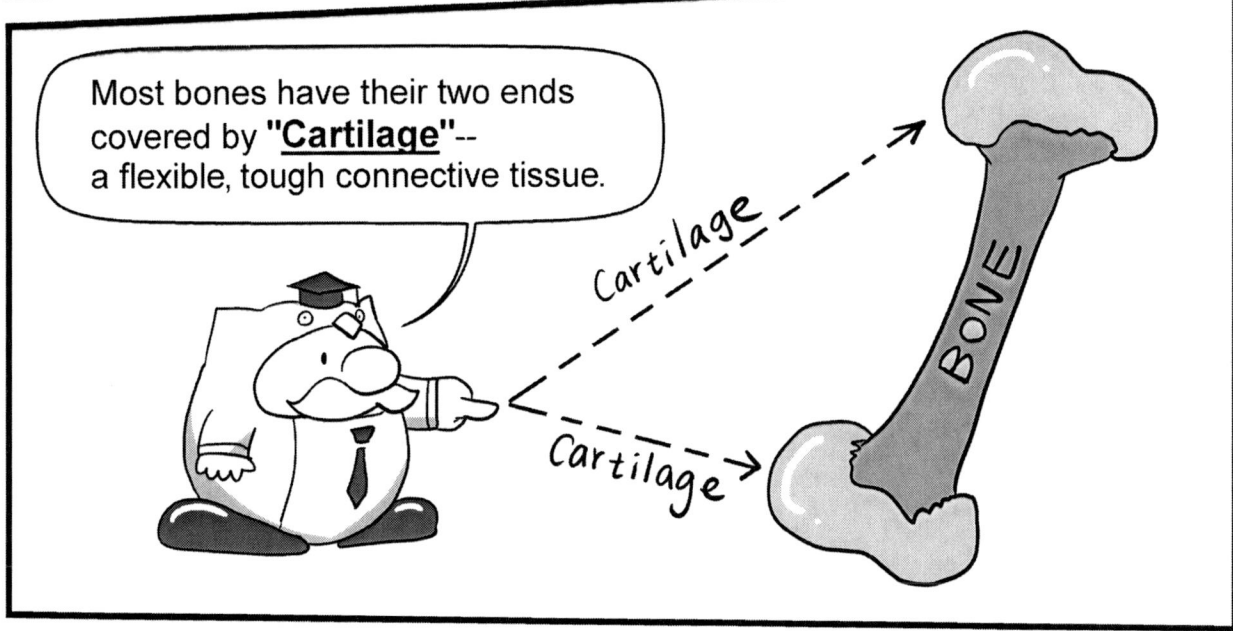

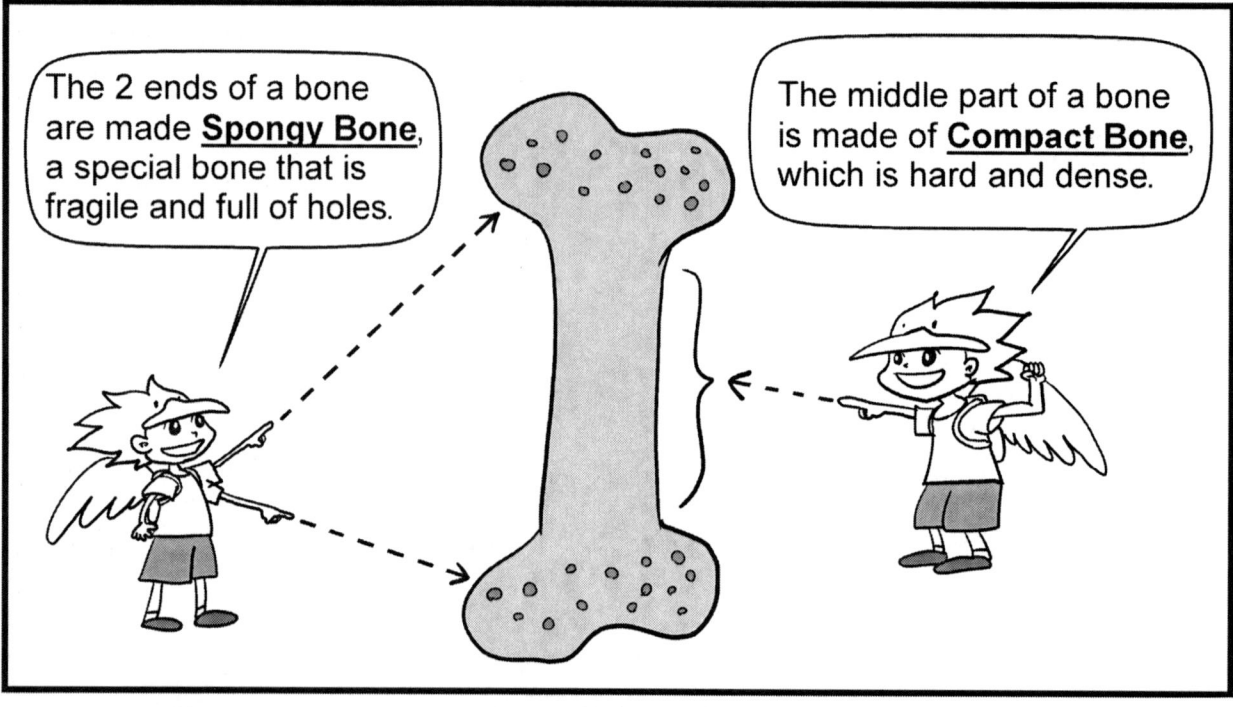
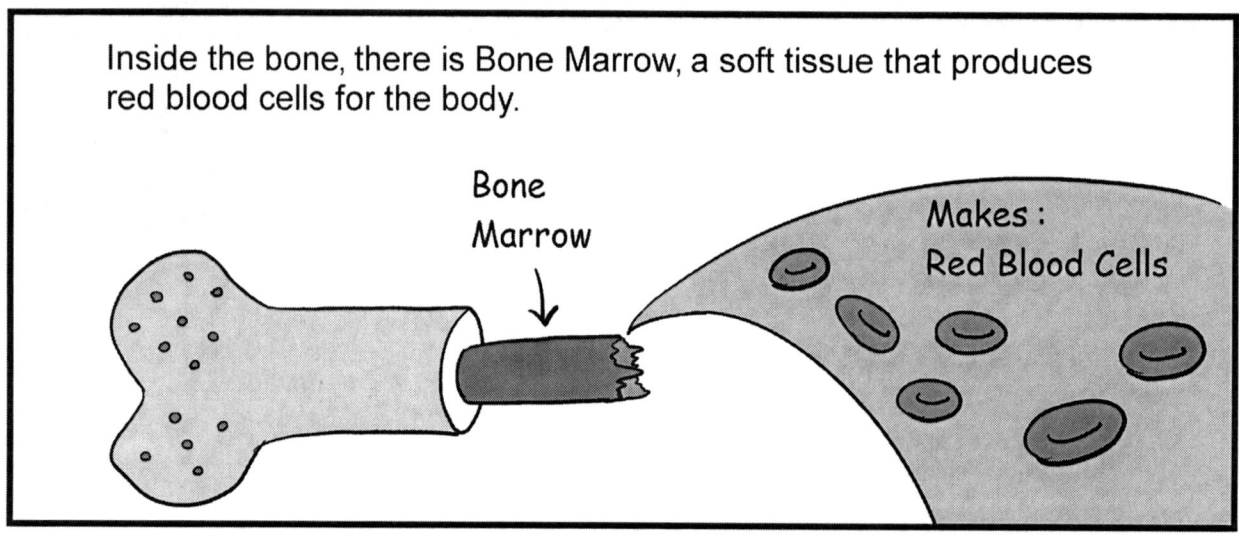

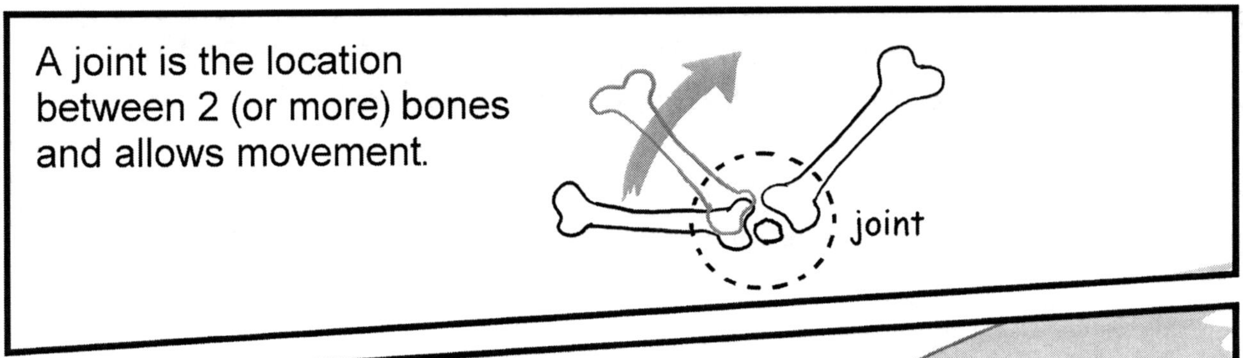

A joint is the location between 2 (or more) bones and allows movement.

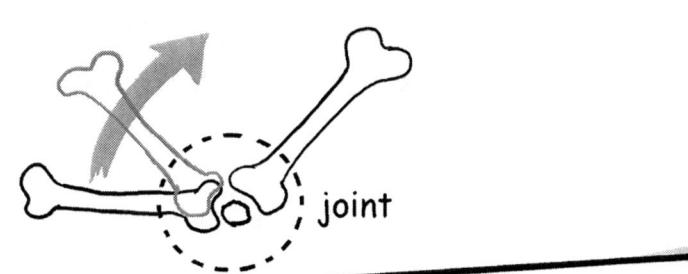

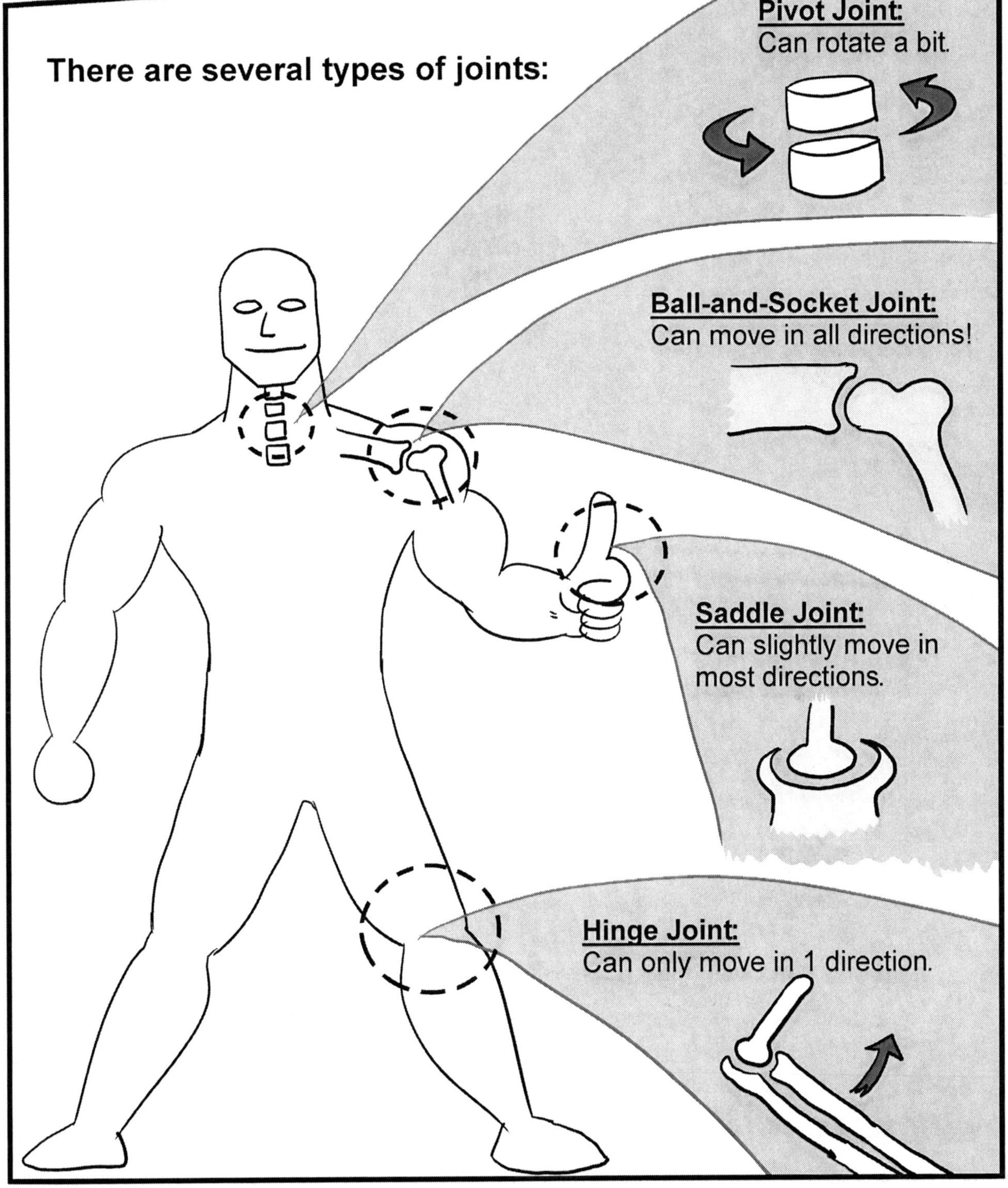

There are several types of joints:

Pivot Joint: Can rotate a bit.

Ball-and-Socket Joint: Can move in all directions!

Saddle Joint: Can slightly move in most directions.

Hinge Joint: Can only move in 1 direction.

Here is how skeletal muscles and bones work together to create movement:

When muscle is relaxed, it lengthens.

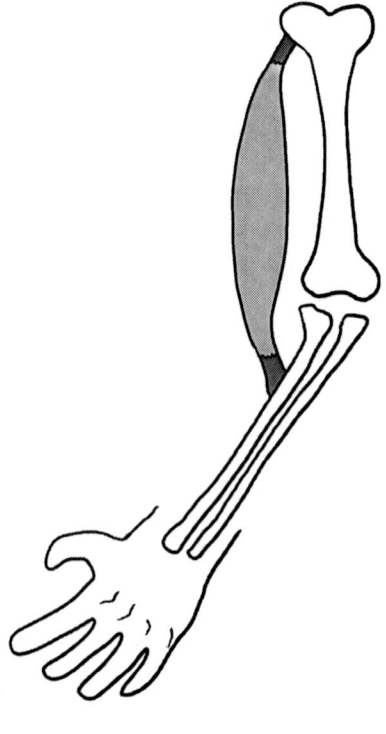

When muscle flexes, it shortens, which pulls on the bone!

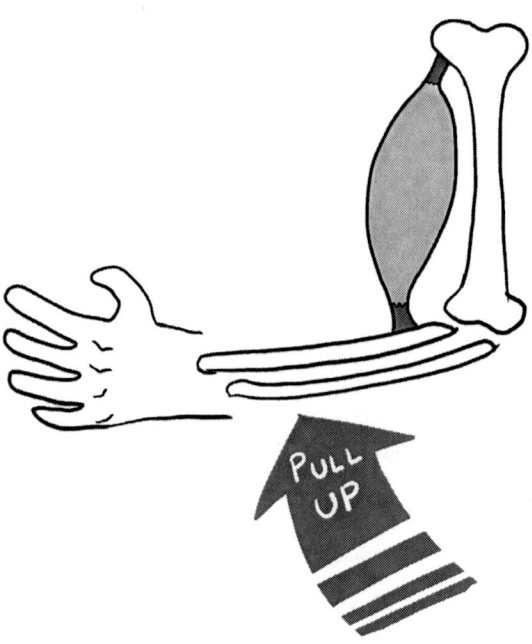

Notice that if the tendon breaks, the muscle will not be able to move the bone!

Chapter 34: Blood Circulation

Science Standard: 7.6.j
Students know that contractions of the heart generate blood pressure and that heart valves prevent backflow of blood in the circulatory system.

NEW VOCABS

* **Artery:** The blood vessels that lead blood OUT OF the heart.

* **Atrium:** (plural: atria) The chamber of the heart that receives the blood flowing into the heart through veins.

* **Blood Vessels:** The tubes in which blood travels through —and throughout the body.

* **Capillaries:** Tiny blood vessels that allow blood to exchange gases, food, and wastes…etc. with various parts of the body.

* **Contraction of the Heart:** The action during which the heart squeezes the blood to keep it flowing.

* **Diastolic Pressure:** The blood pressure generated when the left ventricle relaxes.

* **Systolic Pressure:** The blood pressure generated when the left ventricle's squeezes (contracts).

* **Vein:** The blood vessels that lead blood INTO the heart.

* **Ventricle:** The chamber of the heart that pushes the blood out of the heart.

page 311
Unit 5: Chapter 34

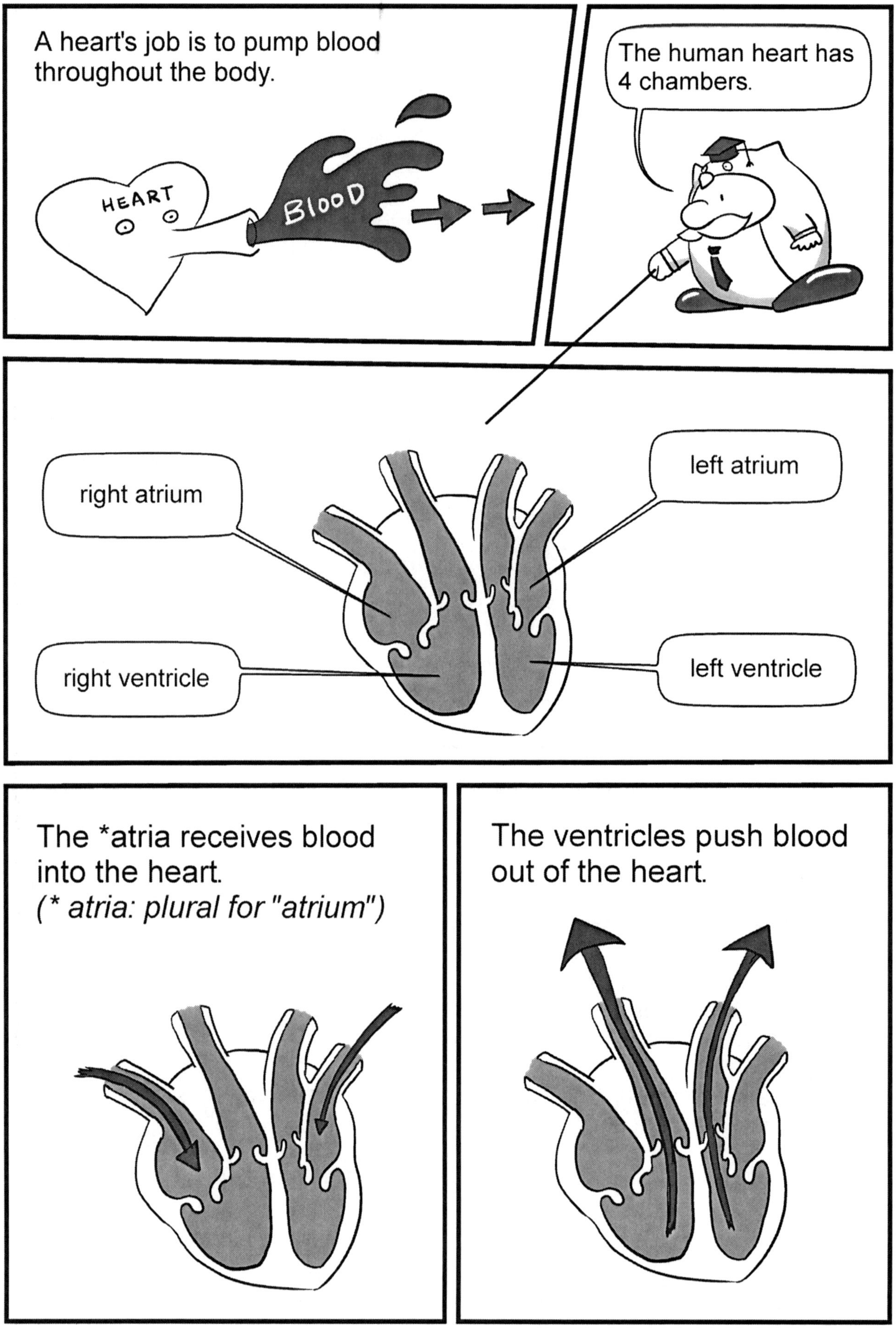

A heart's job is to pump blood throughout the body.

The human heart has 4 chambers.

right atrium

left atrium

right ventricle

left ventricle

The *atria receives blood into the heart.
(* atria: plural for "atrium")

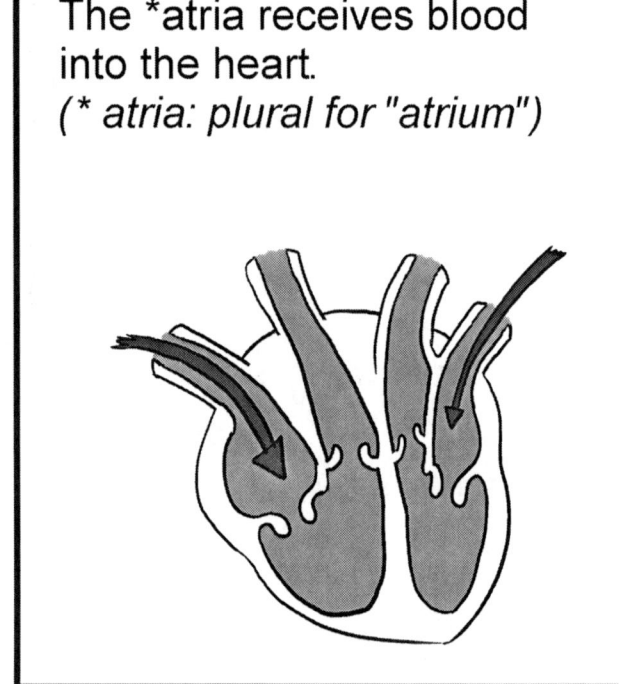

The ventricles push blood out of the heart.

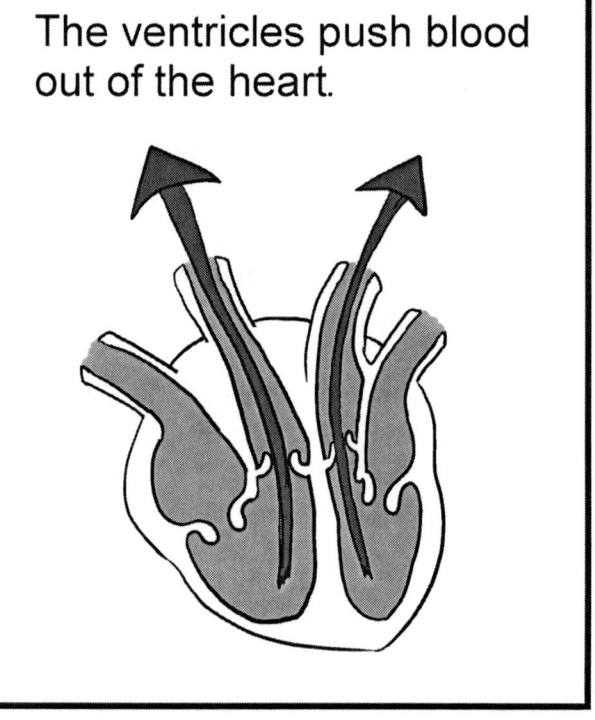

The heart uses a "squeezing" action, which is called a "**contraction**.", to keep the blood flowing.
The force of this blood-flow is called **"Blood Pressure."**

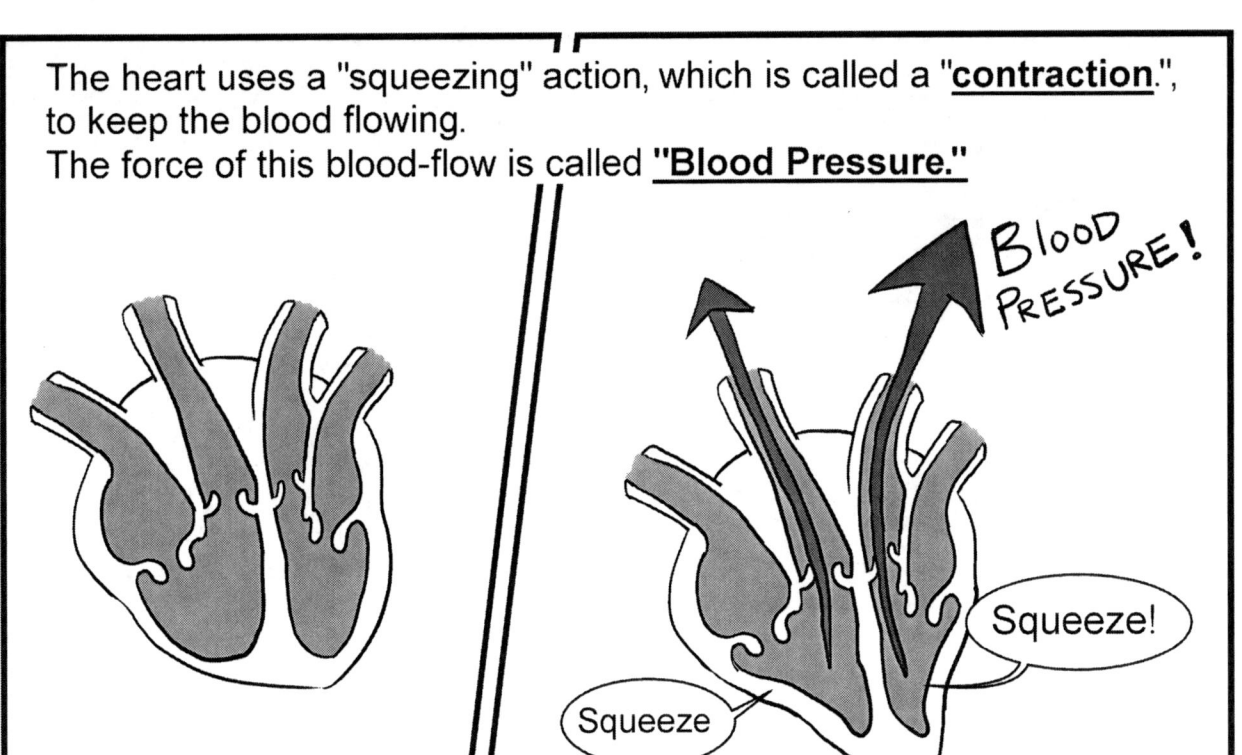

When the left ventricle contracts, it produces a strong pressure of blood-flow called the **"Systolic Pressure."**

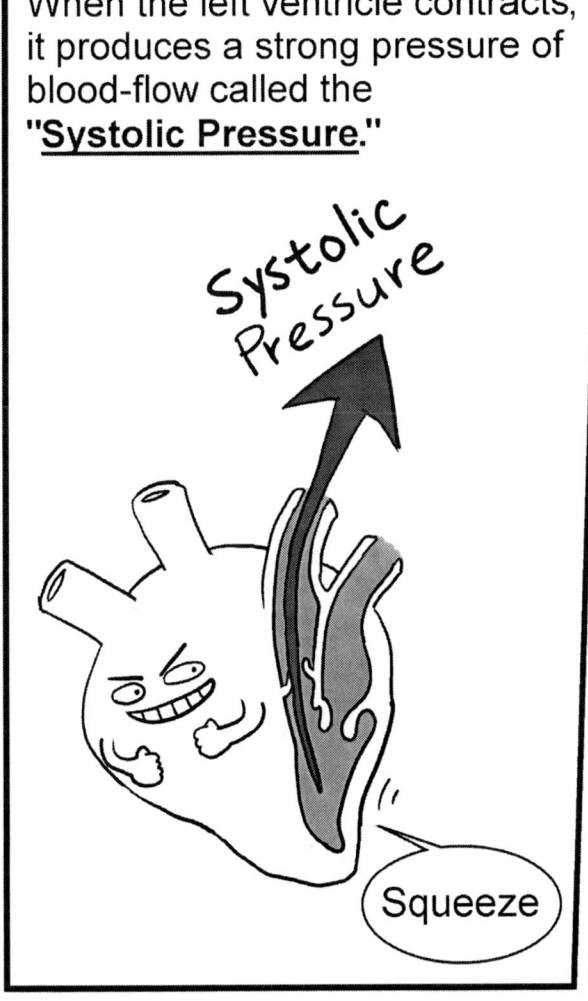

When the left ventricle relaxes, it produces a weak pressure called **"Diastolic Pressure."**

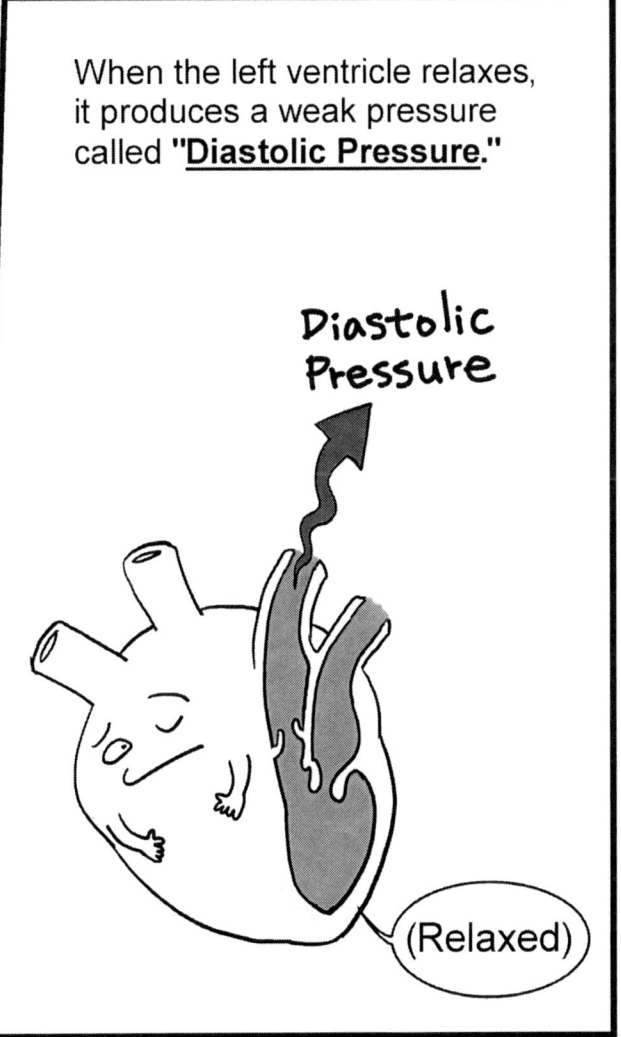

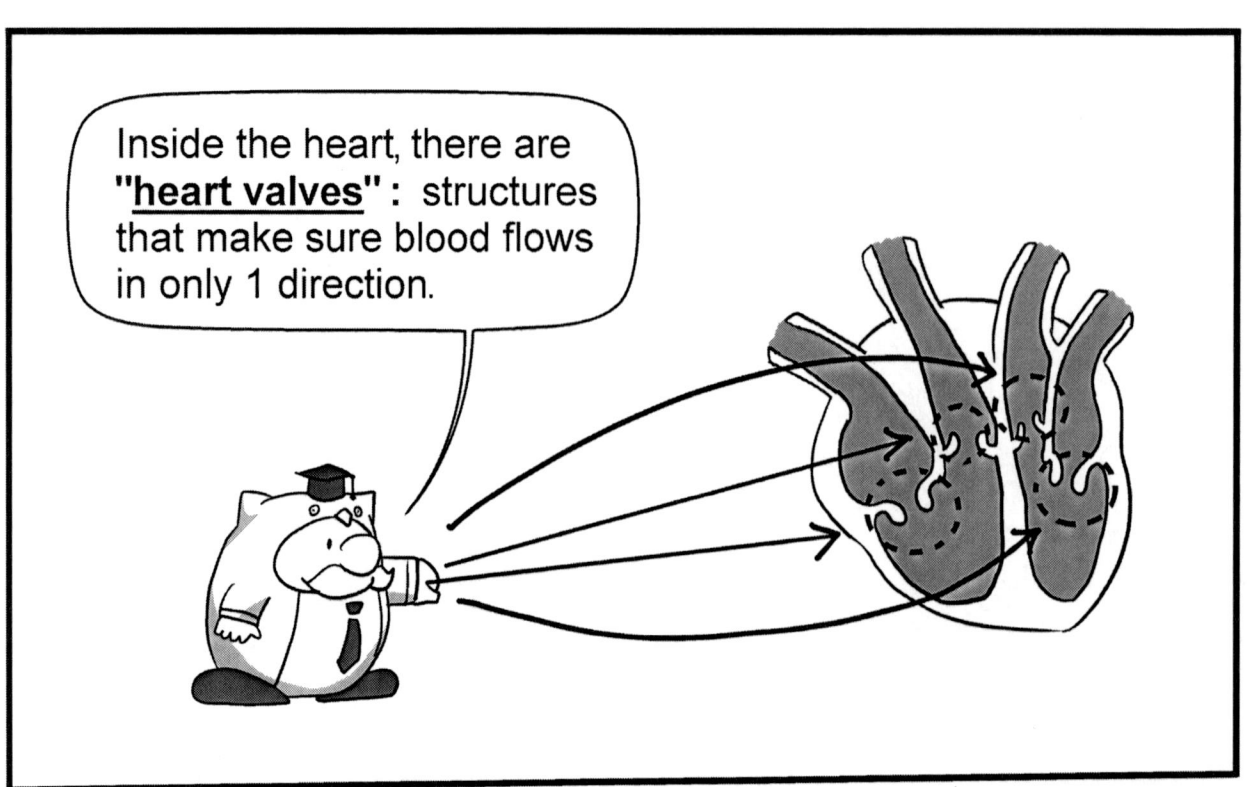

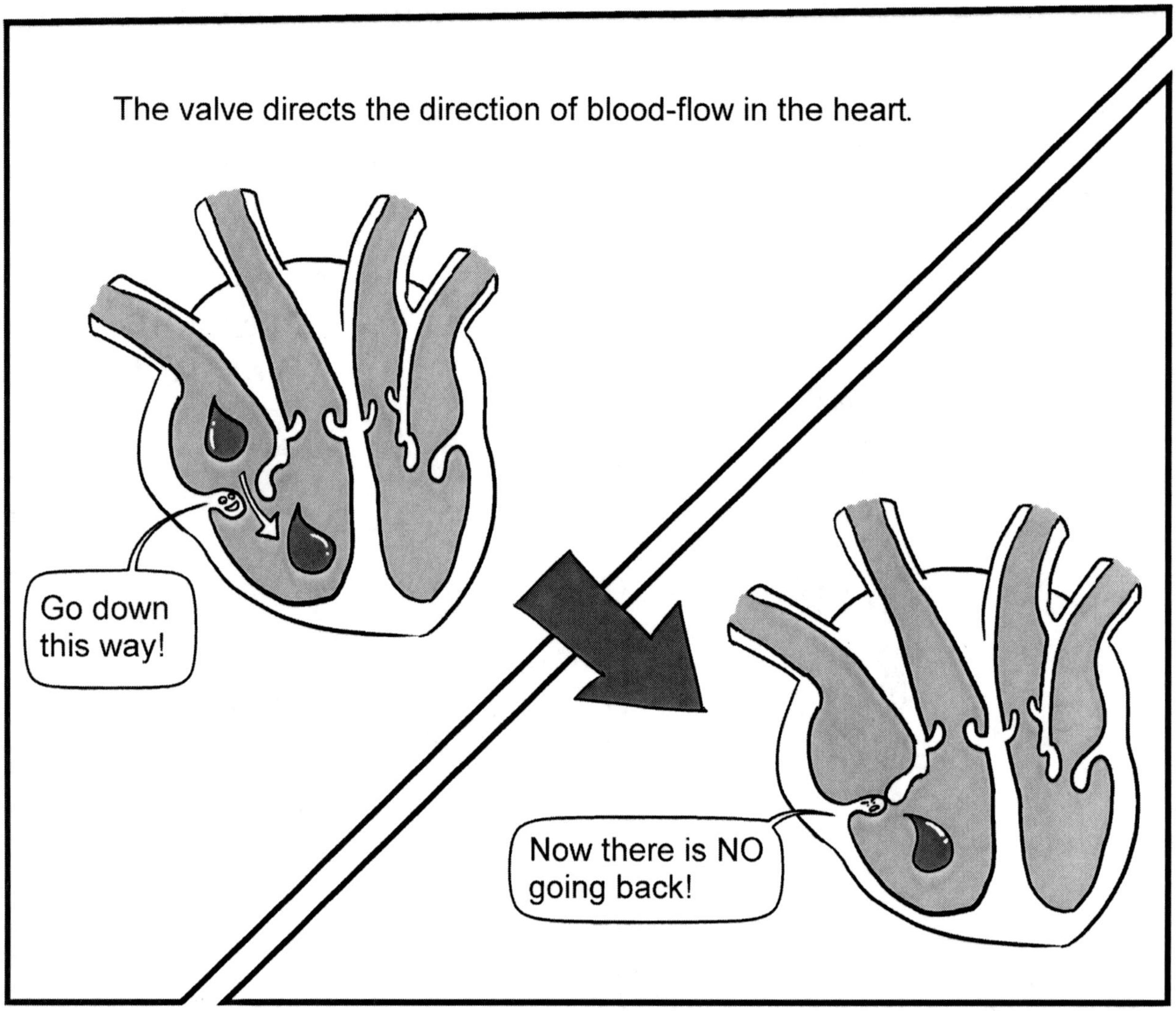

6-Steps of Blood's Journey Around the Body

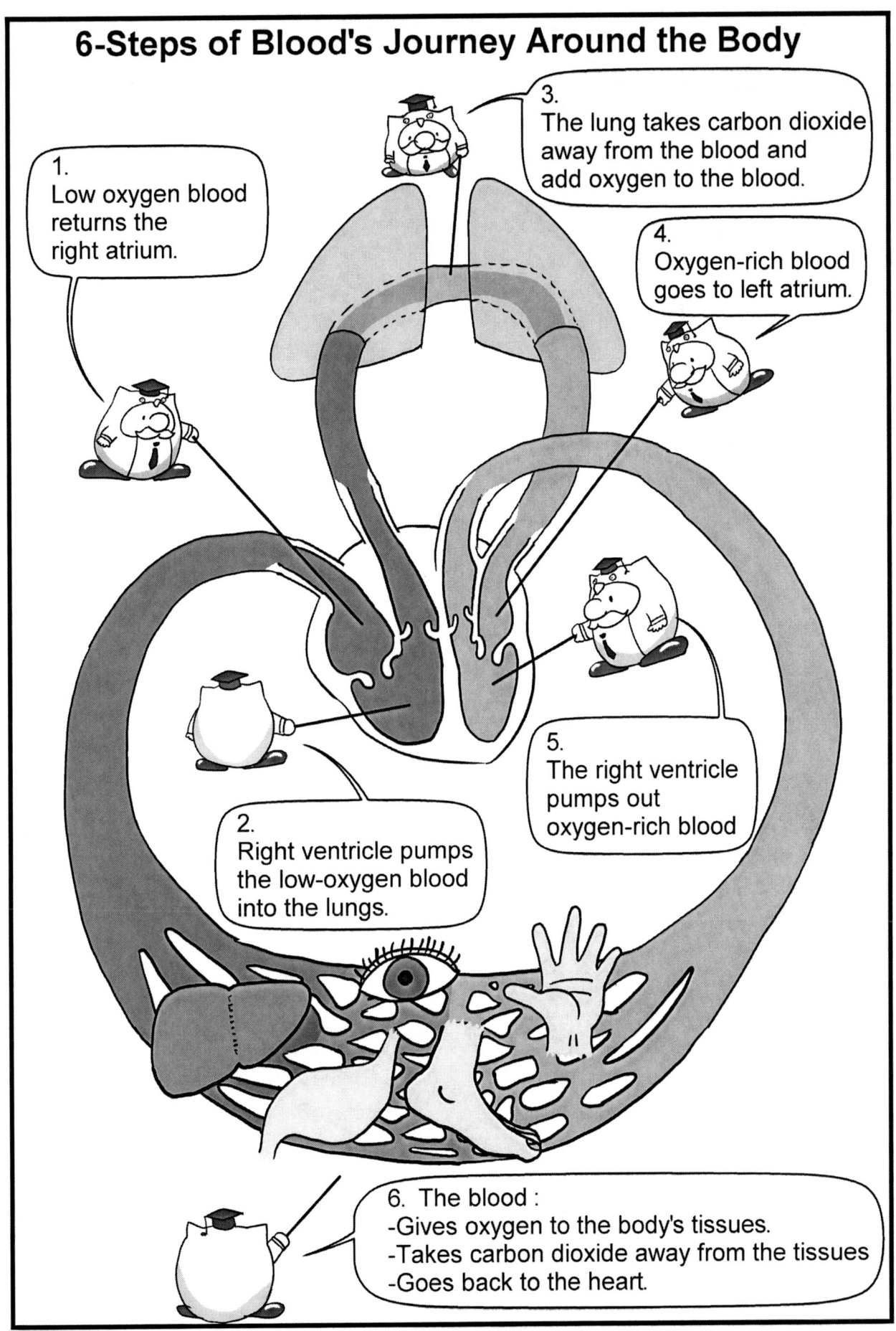

page 315
Unit 5: Chapter 34

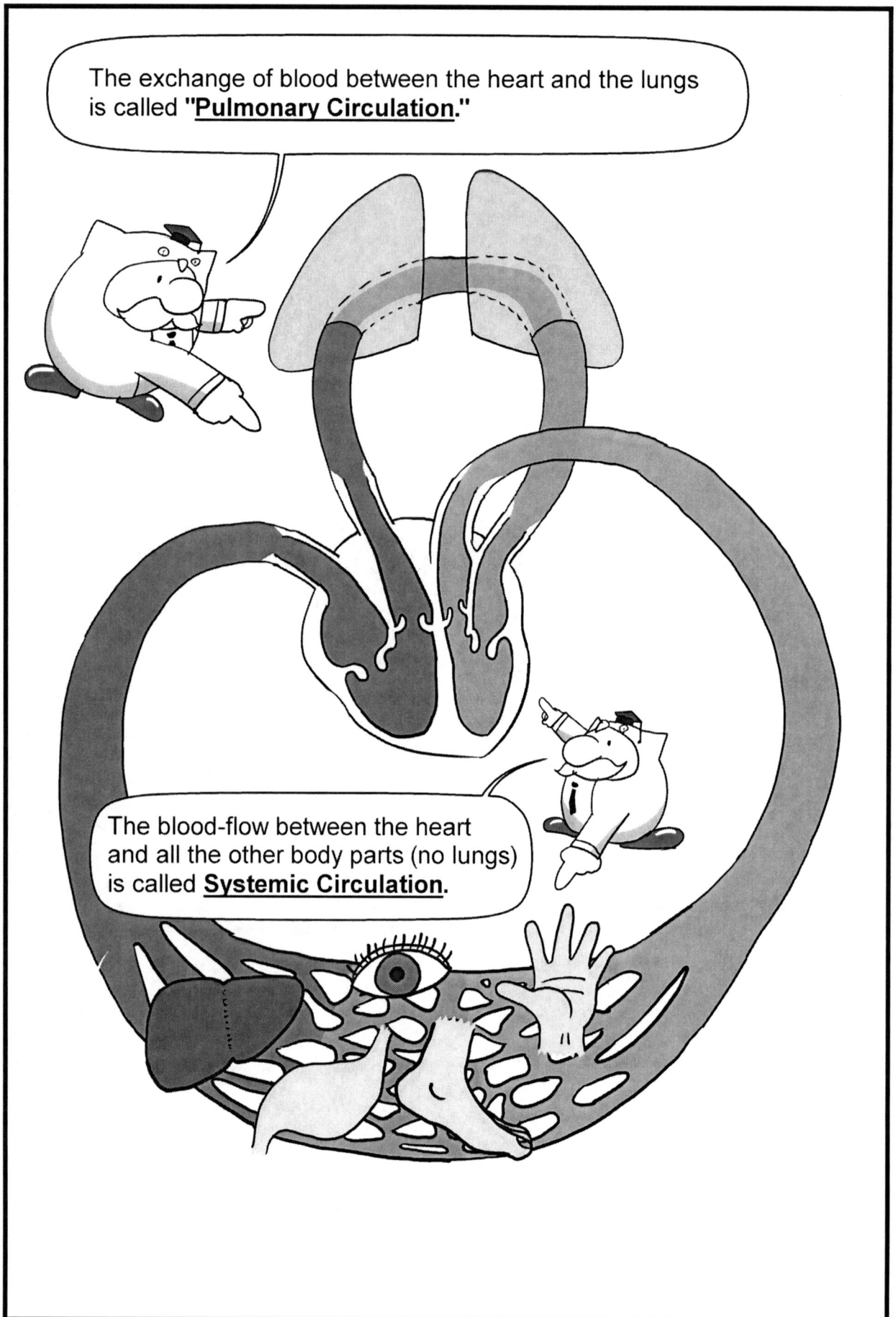

Pulmonary Circulation

During Pulmonary Circulation, the heart pumps **blood** to the **lungs.**

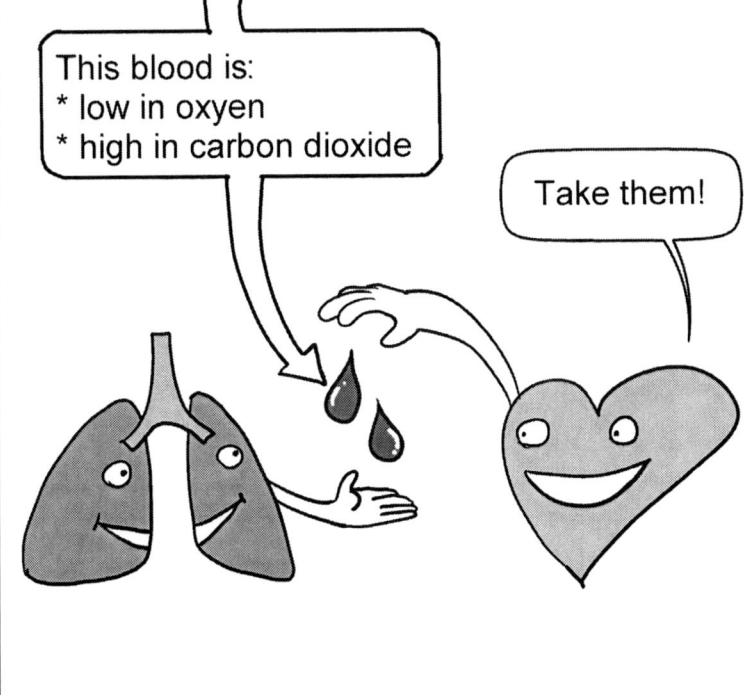

This blood is:
* low in oxyen
* high in carbon dioxide

Take them!

The lungs take away the blood's carbon dioxide, and add more oxygen to it.

Add oxygen… Remove carbon dioxide…

Now, the oxygen-rich blood returns back to the heart.

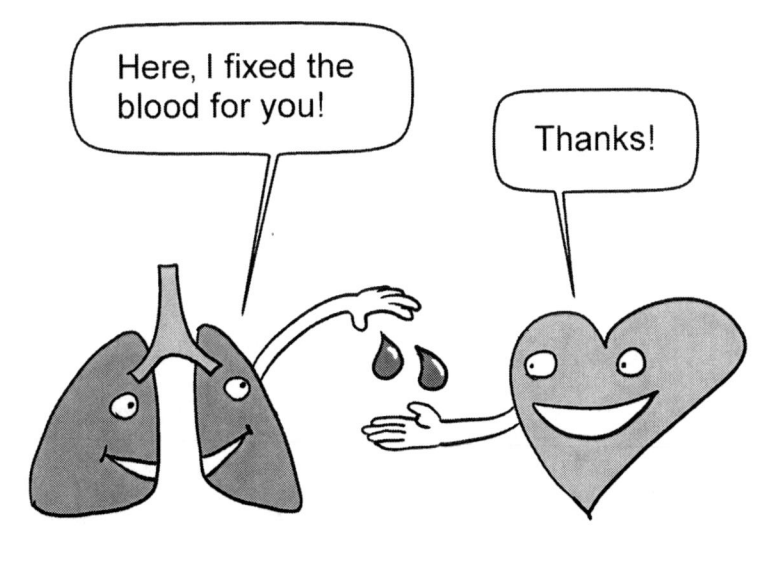

Here, I fixed the blood for you!

Thanks!

So What happens next?

AFTER THAT, SYSTEMIC CIRCULATION TAKES PLACE

The heart sends out the oxygen-rich (thanks to the lungs) blood to the rest of the body.

Your body tissues take the oxygen from the blood.

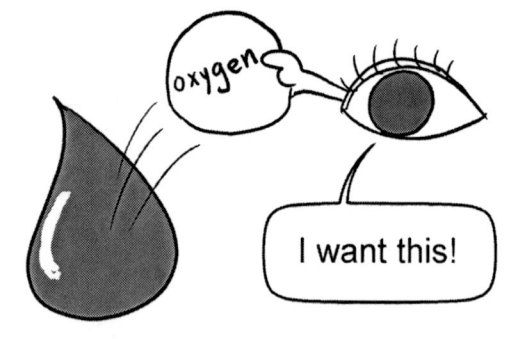

Dump the carbon dioxide into the blood.

The blood then returns the carbon dioxide back to the heart.

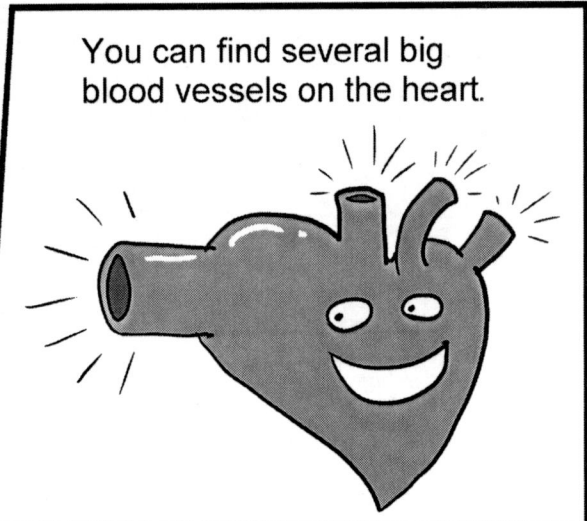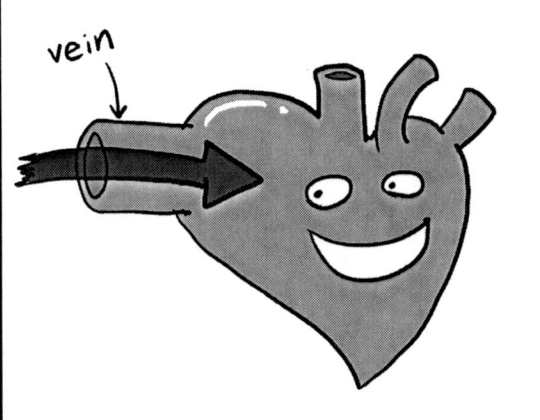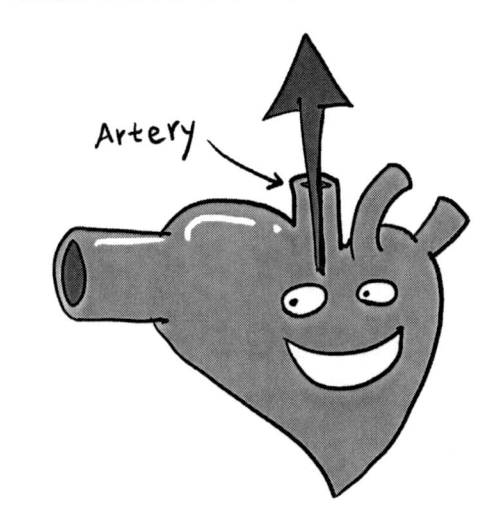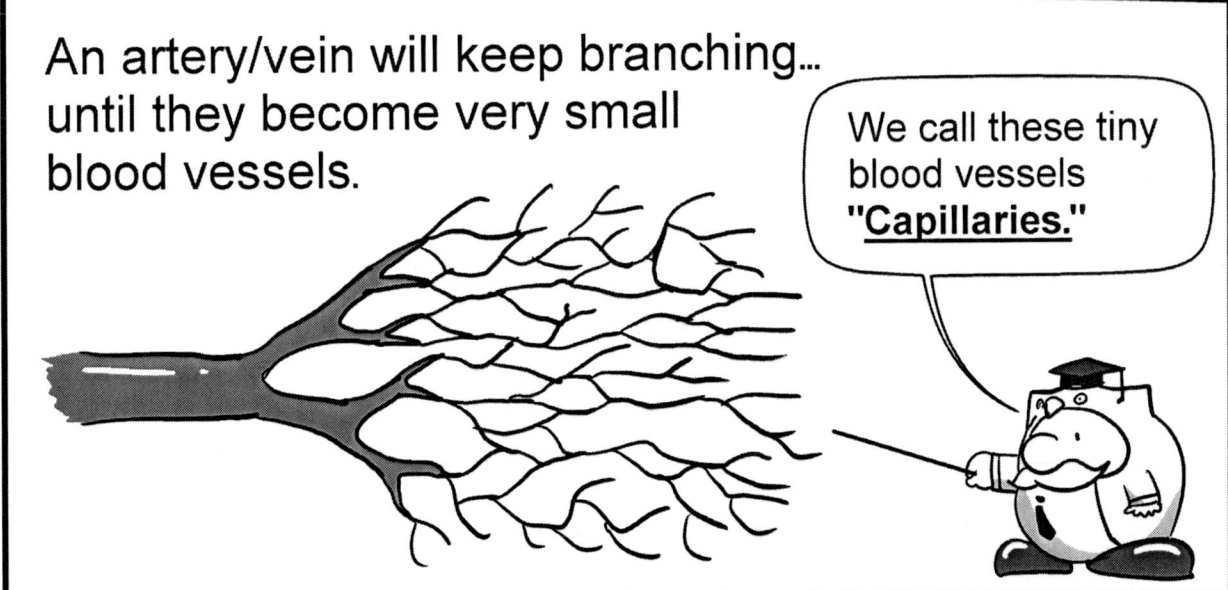

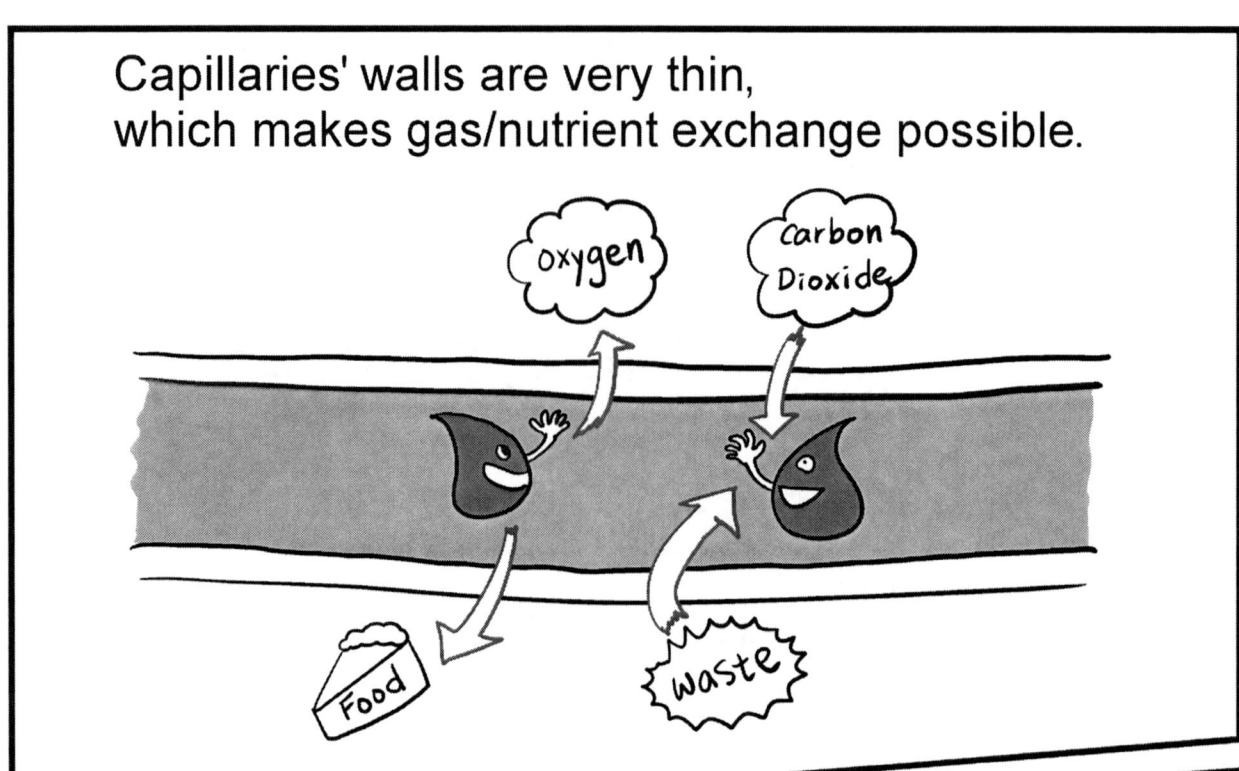

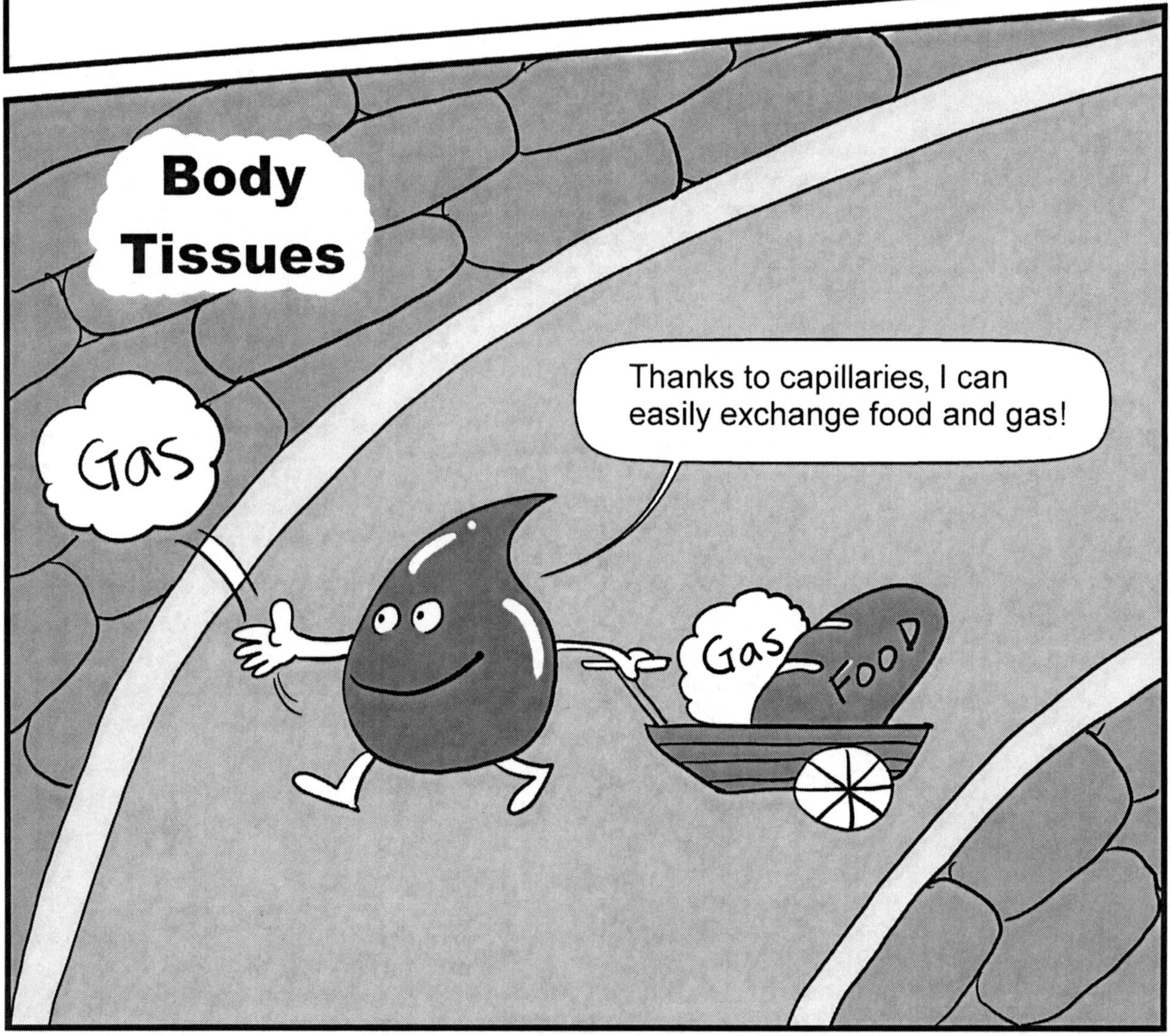

Chapter 35: Material Exchange by Body Systems

Science Standard Bio/LS . 9 . a :
Students know how the complementary activity of major body systems provides cells with oxygen and nutrients and removes toxic waste products such as carbon dioxide.

NEW VOCABS

* **Absorption:** When the food's nutrients go from the food into your blood.

* **Alveolus:** (plural: alveoli) Small, balloon-like air-sacs that make up the lungs. They help with the exchange of oxygen and carbon dioxide between the air and the blood.

* **Bronchus:** (plural: bronchi) At the base of the trachea, it branches into 2 tubes named "bronchi." The 2 bronchi leads into the left and right lungs.

* **Diaphragm:** A layer of muscle located right beneath the lungs. It controls the action of breathing.

* **Digestion:** The act of breaking food down into smaller bits and pieces of molecules of nutrients.

* **Epiglottis:** A flap (made of mostly muscle and cartilage) located right next to the trachea. Its job is to cover the trachea when you are swallowing food, so food does not accidentally fall into the trachea.

* **Esophagus:** The canal (tube) by which food goes down into the stomach.

* **Feces:** Solid wastes (commonly known as "poop") that is to be expelled from the body.

* **Gall bladder:** It is a structured located at the liver and stores the bile juice, so it is ready to be used in times of need.

* **Hemoglobin:** A special iron-rich protein found inside the red blood cells. Hemoglobin gives red blood cells the ability to carry oxygen and carbon dioxide. Hemoglobin is red, which is also why your red blood cells (and your blood) is red.

* **Kidneys:** An organ that takes the toxic wastes out of the blood. It then sends the wastes through 2 tubes called "ureters" to the urinary bladder for storage.

* **Large Intestine:** An organ that absorbs water, minerals, and vitamins from food. As the food passes through the large intestine, it slowly becomes feces (solid wastes).

* **Liver:** It is an organ that secretes "bile," a liquid that helps break down lipids (fats). Bile is secreted into the small intestine.

* **Pancreas:** It is an organ that releases the "pancreatic juice" into the small intestine to help digest food.

* **Plasma:** The liquid part of the blood. Nutrients and toxic wastes are dissolved into the plasma to be carried throughout the body.

* **Rectum:** An organ that expels feces (wastes) out of the body.

* **Red Blood Cells:** Special cells found inside the blood responsible for carrying oxygen and carbon dioxide.

* **Small Intestine:** An organ where food is digested and most of the nutrients are absorbed.

* **Stomach:** It is an organ that digests food with acid and enzymes. It also "churns" the food by pressing/pushing on it.

* **Trachea:** The canal (tube) by which the air goes in and out of the lungs.

* **Urinary Bladder:** An organ where urine is stored, so it can be released later.

* **Urine:** The liquid wastes from your body. It is commonly known as "pee" and is made of mostly the chemicals "urea" and "uric acid."

* **Villus:** (plural: villi) Small, finger-like projections found inside the small intestine. Villus helps the small intestine to absorb nutrients from food.

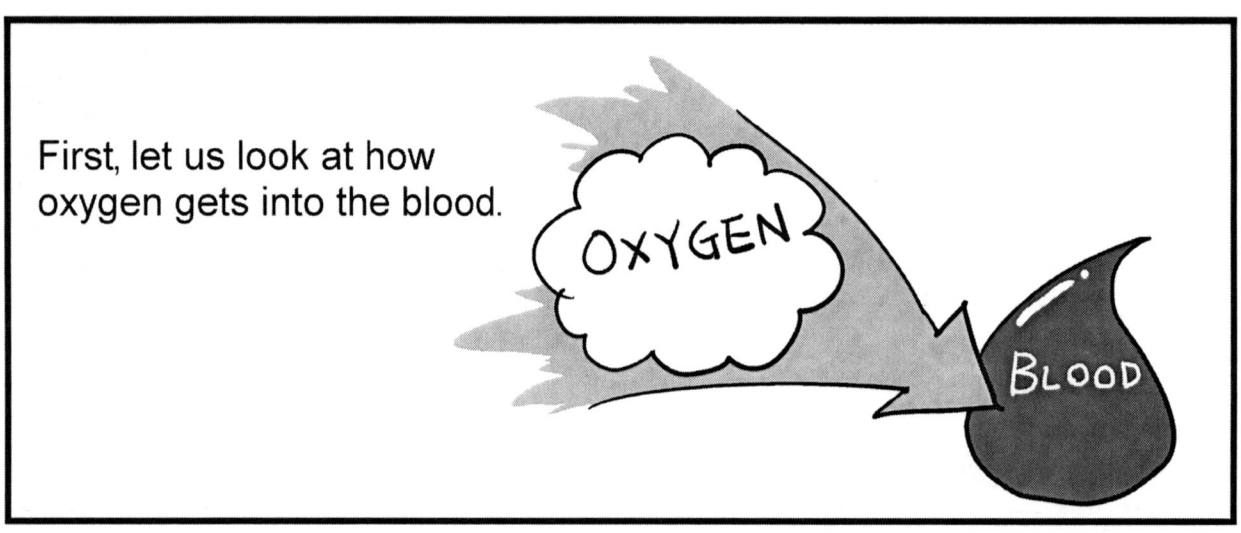

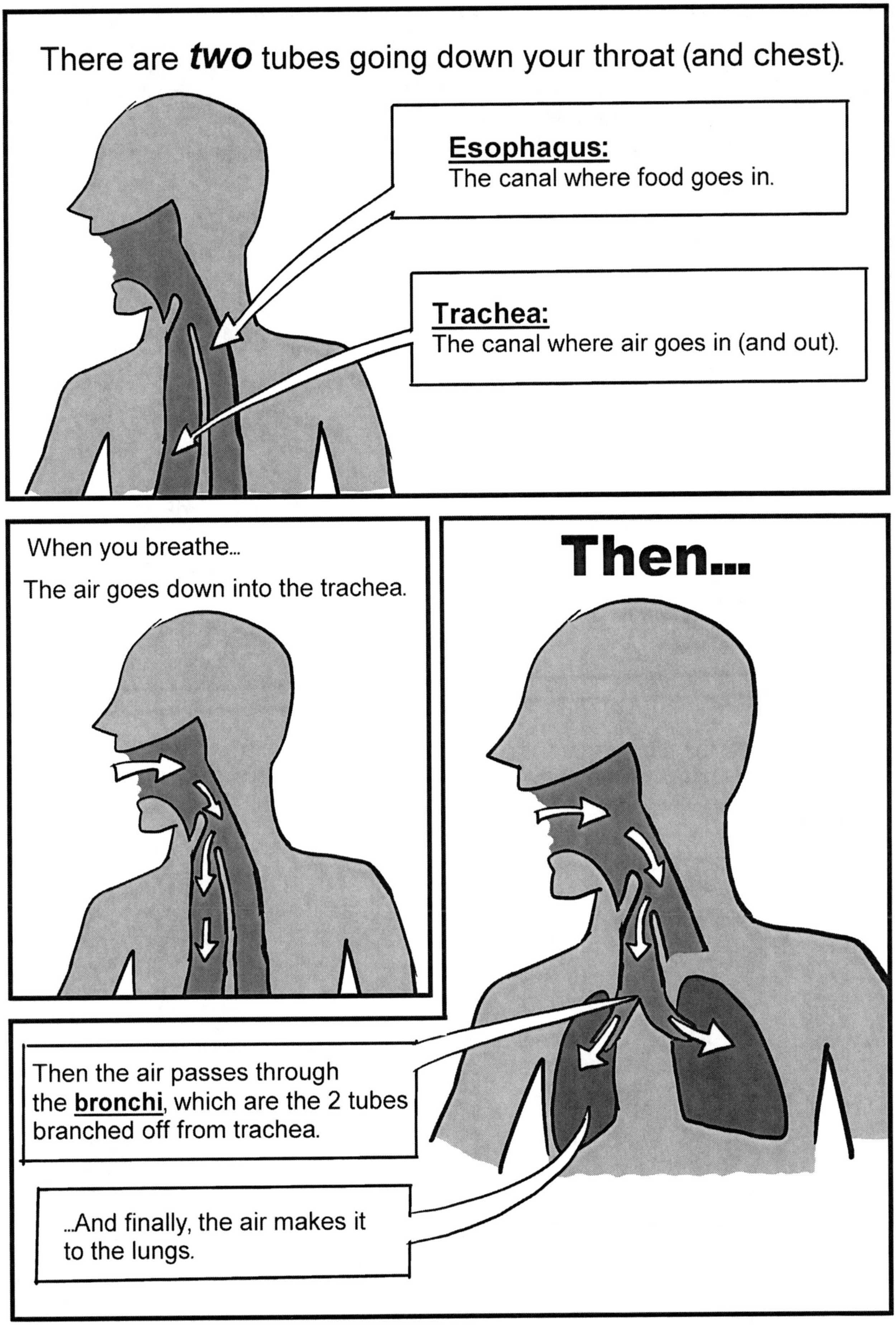

There are *two* tubes going down your throat (and chest).

Esophagus:
The canal where food goes in.

Trachea:
The canal where air goes in (and out).

When you breathe…
The air goes down into the trachea.

Then…

Then the air passes through the **bronchi**, which are the 2 tubes branched off from trachea.

…And finally, the air makes it to the lungs.

page 325

Unit 5: Chapter 35

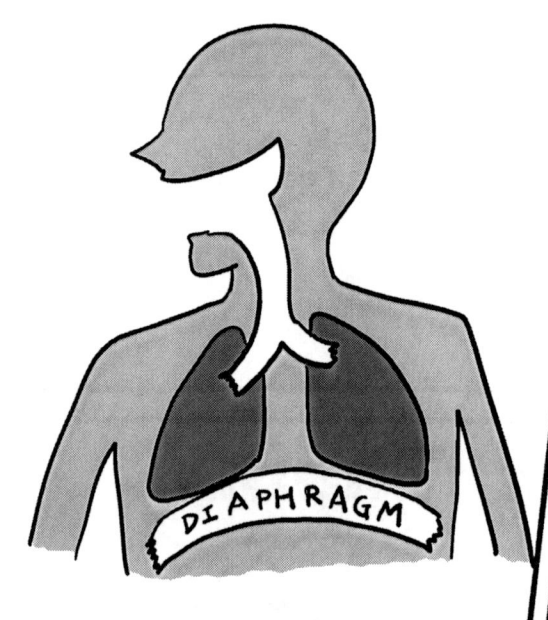
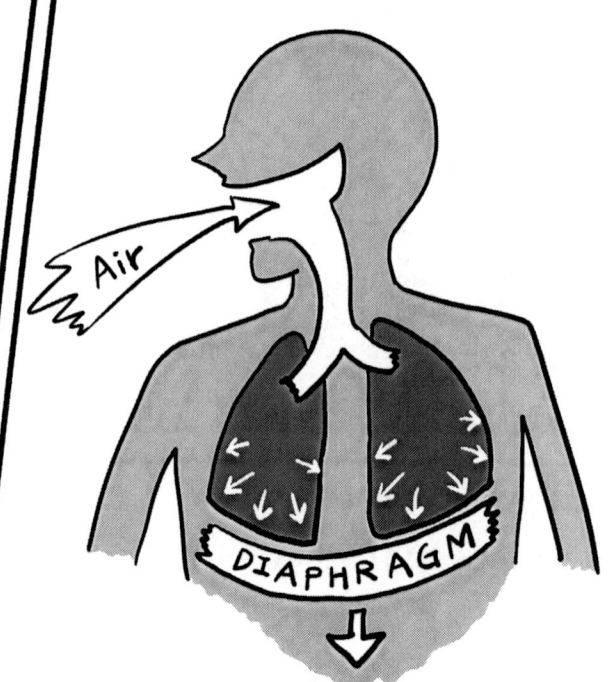
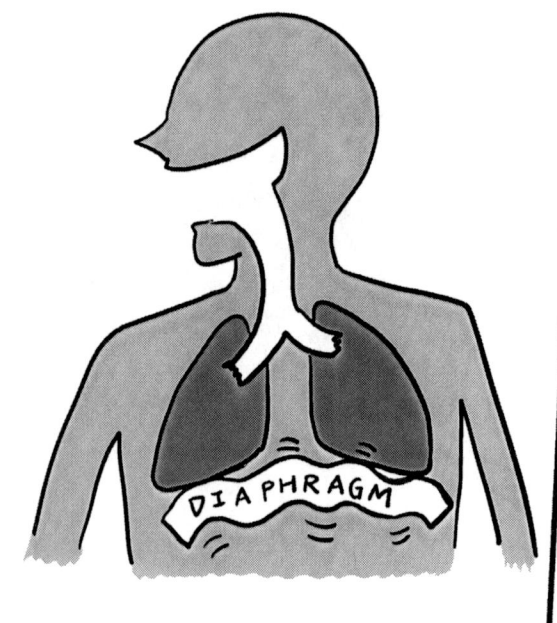
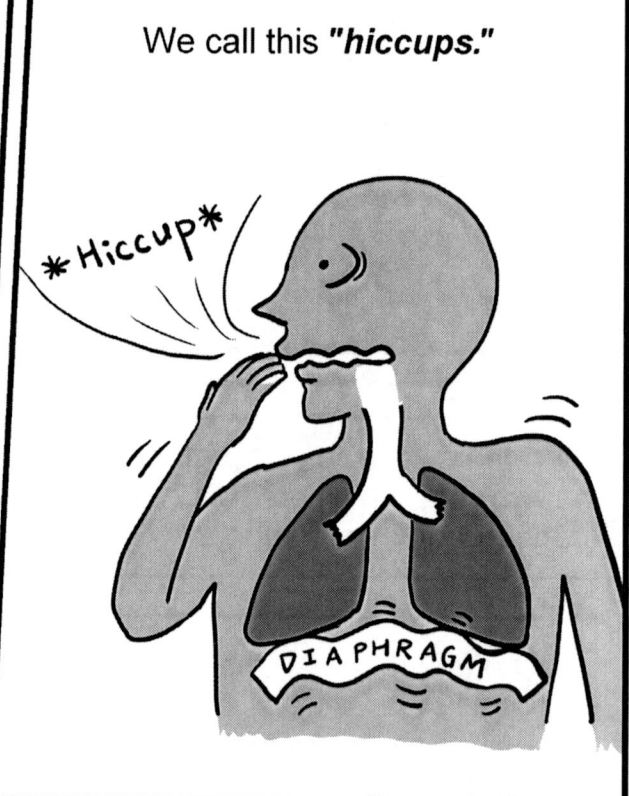

The lungs are made of structures called ***Alveoli**, which are small balloon-like sacs.
(* "alveoli" is the plural form of "alveolus")

When you breathe, each alveolus is filled with fresh air.

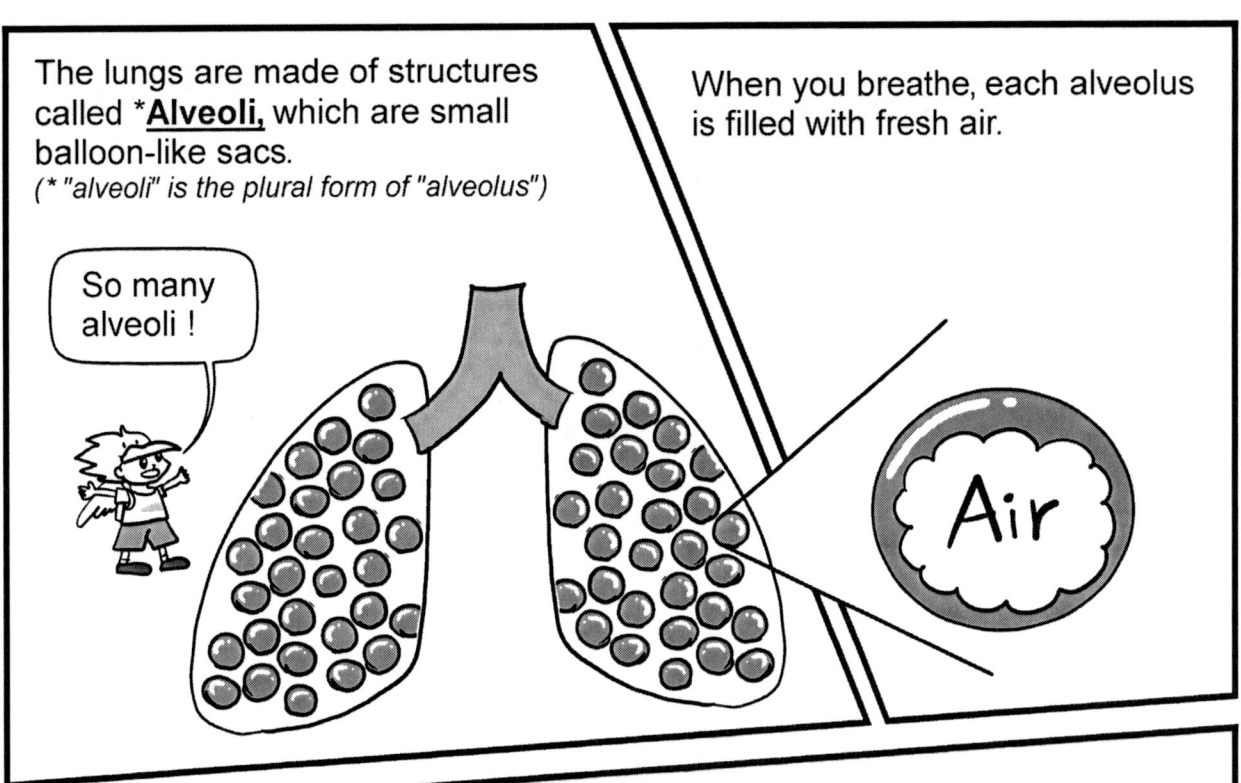

It is important to know that every alveolus is covered with capillaries (tiny blood vessels).

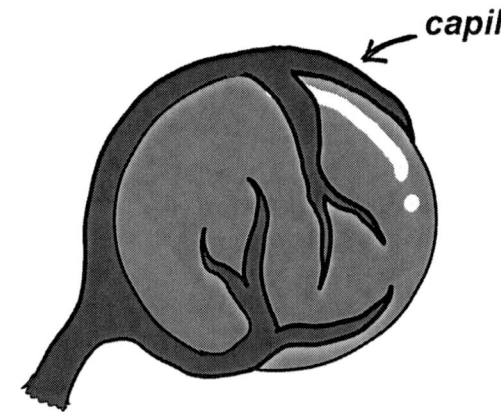

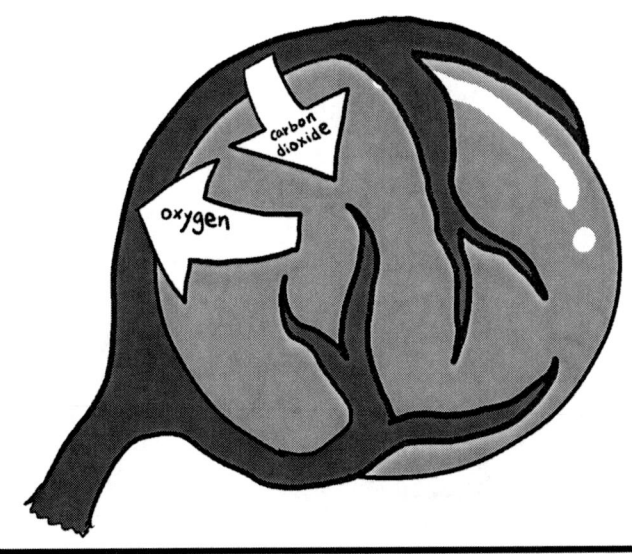

As the blood flows past an alveolus, the blood **gains** oxygen and loses carbon dioxide.

Unit 5: Chapter 35

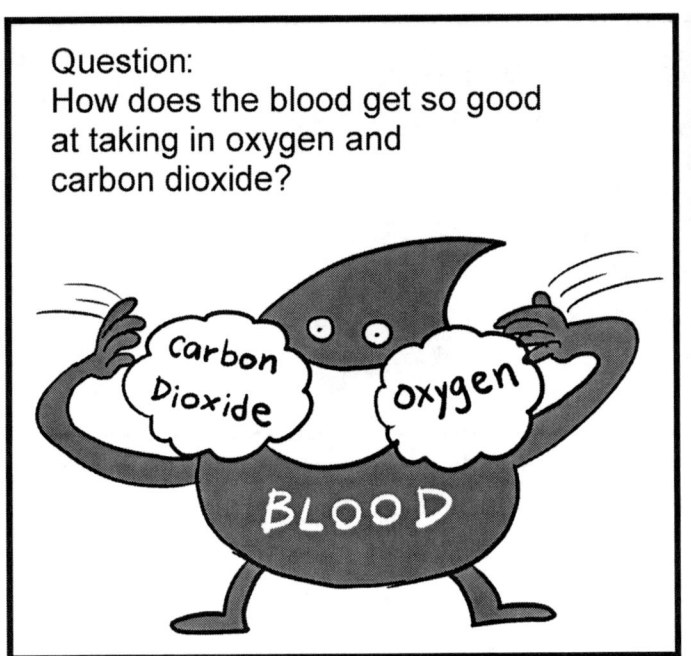

Question:
How does the blood get so good at taking in oxygen and carbon dioxide?

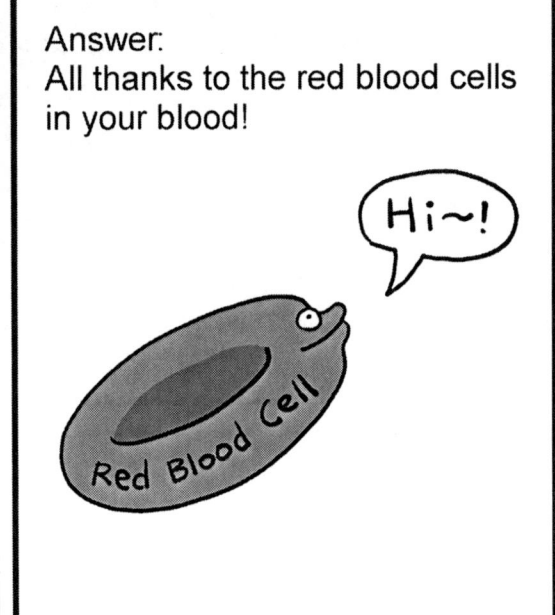

Answer:
All thanks to the red blood cells in your blood!

A **red blood cell** can carry oxygen and carbon dioxide because it contains a lot of **hemoglobin**, a special iron-rich protein.

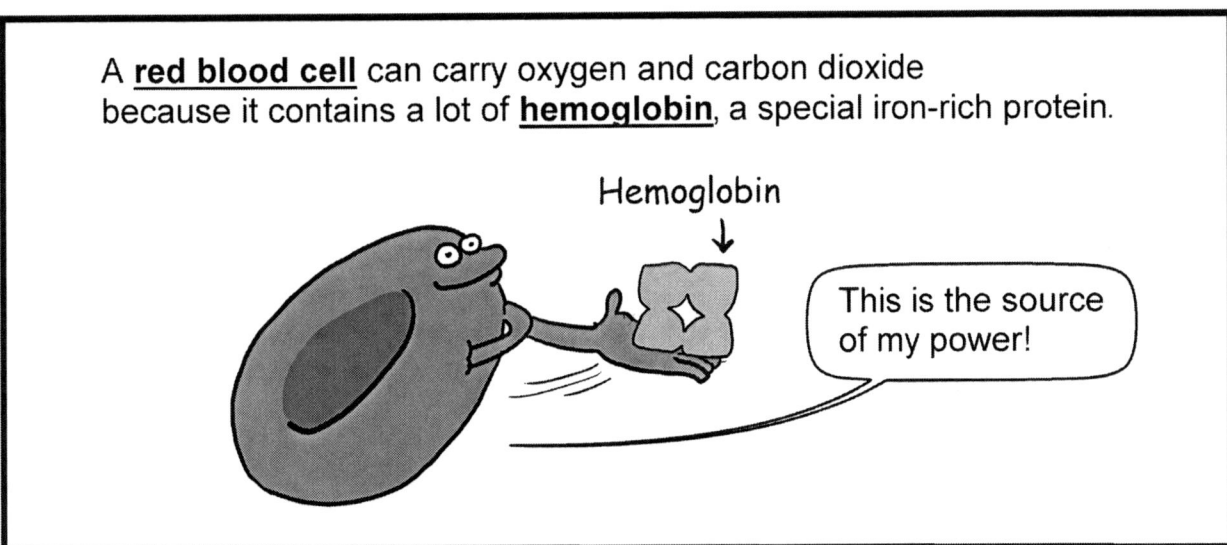

Hemoglobin is what gives your blood a red color.

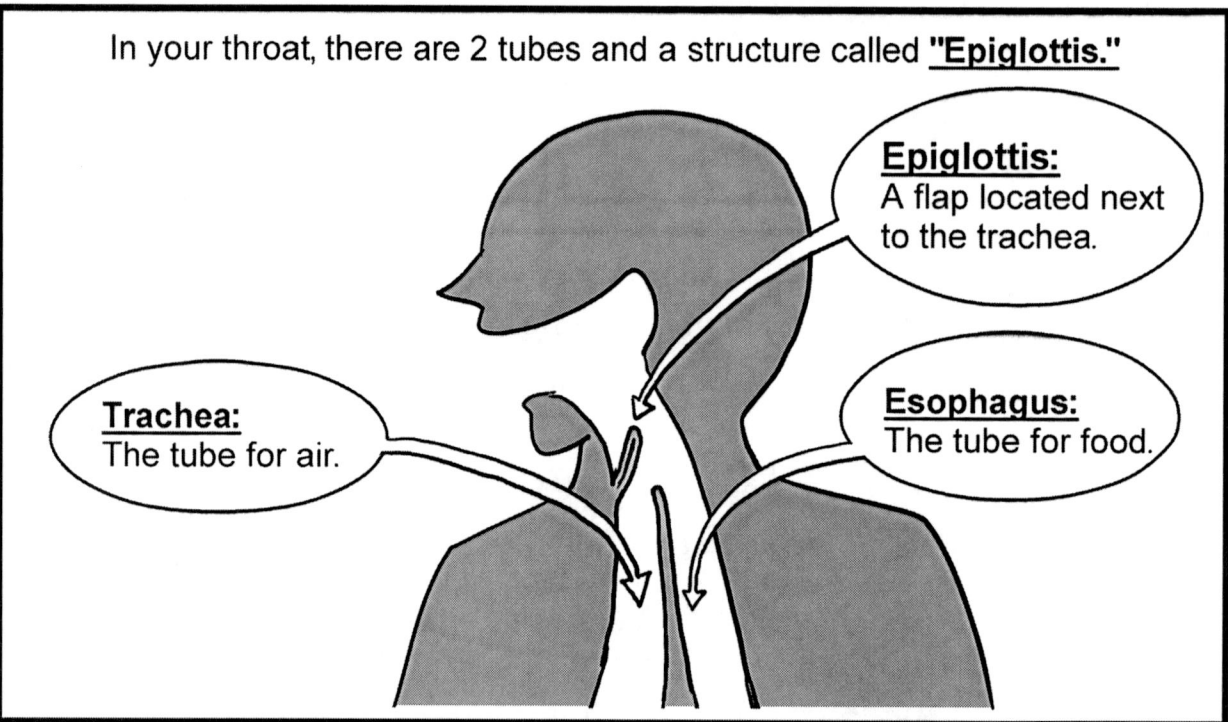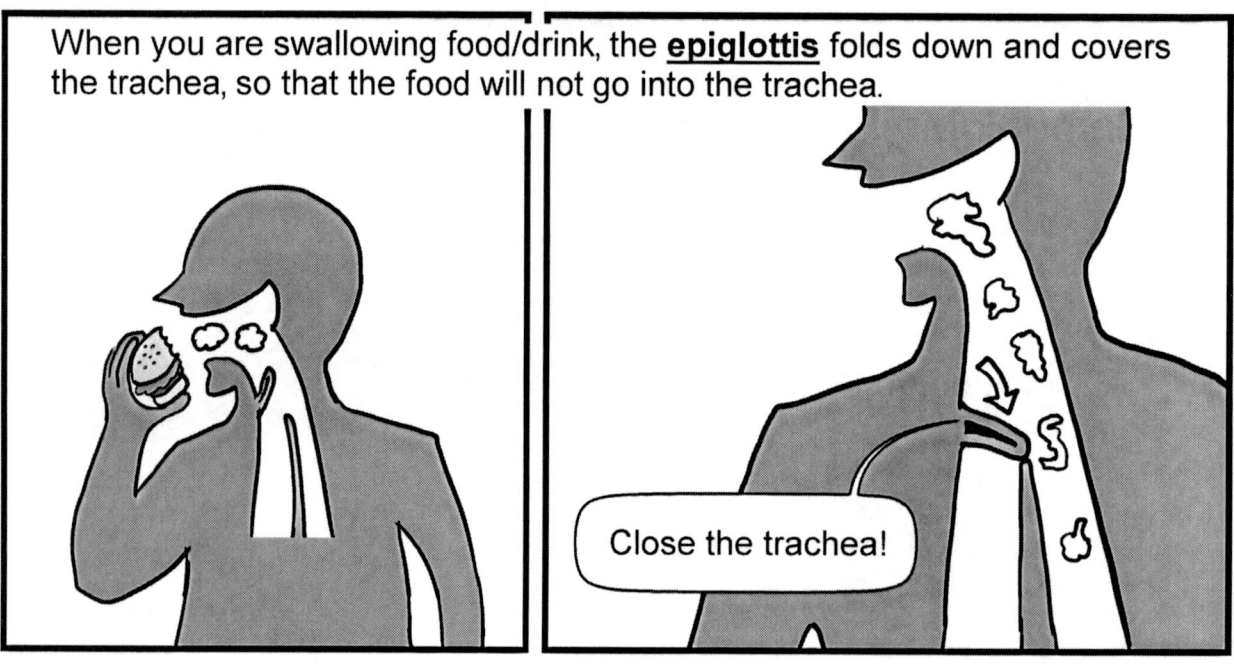

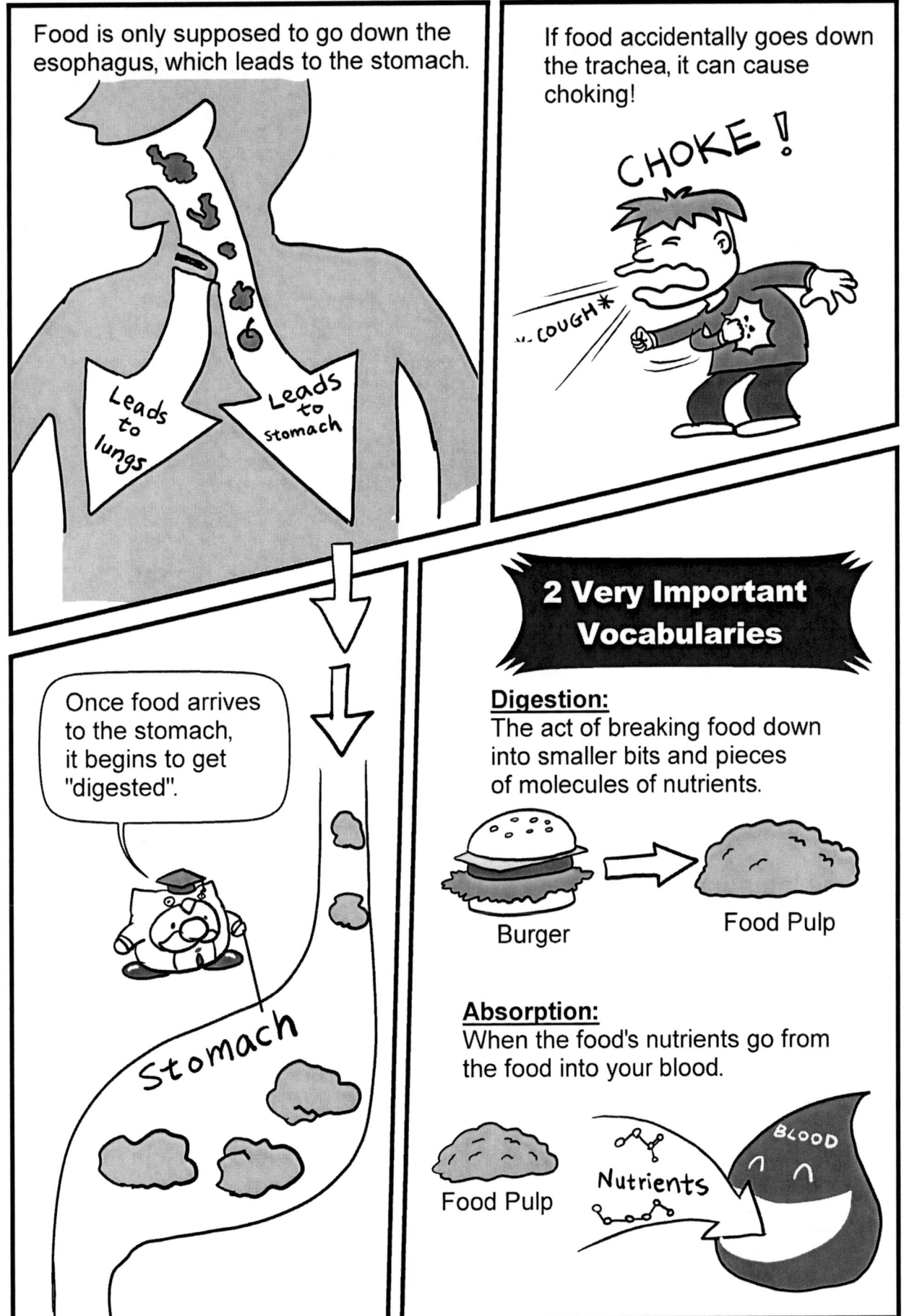

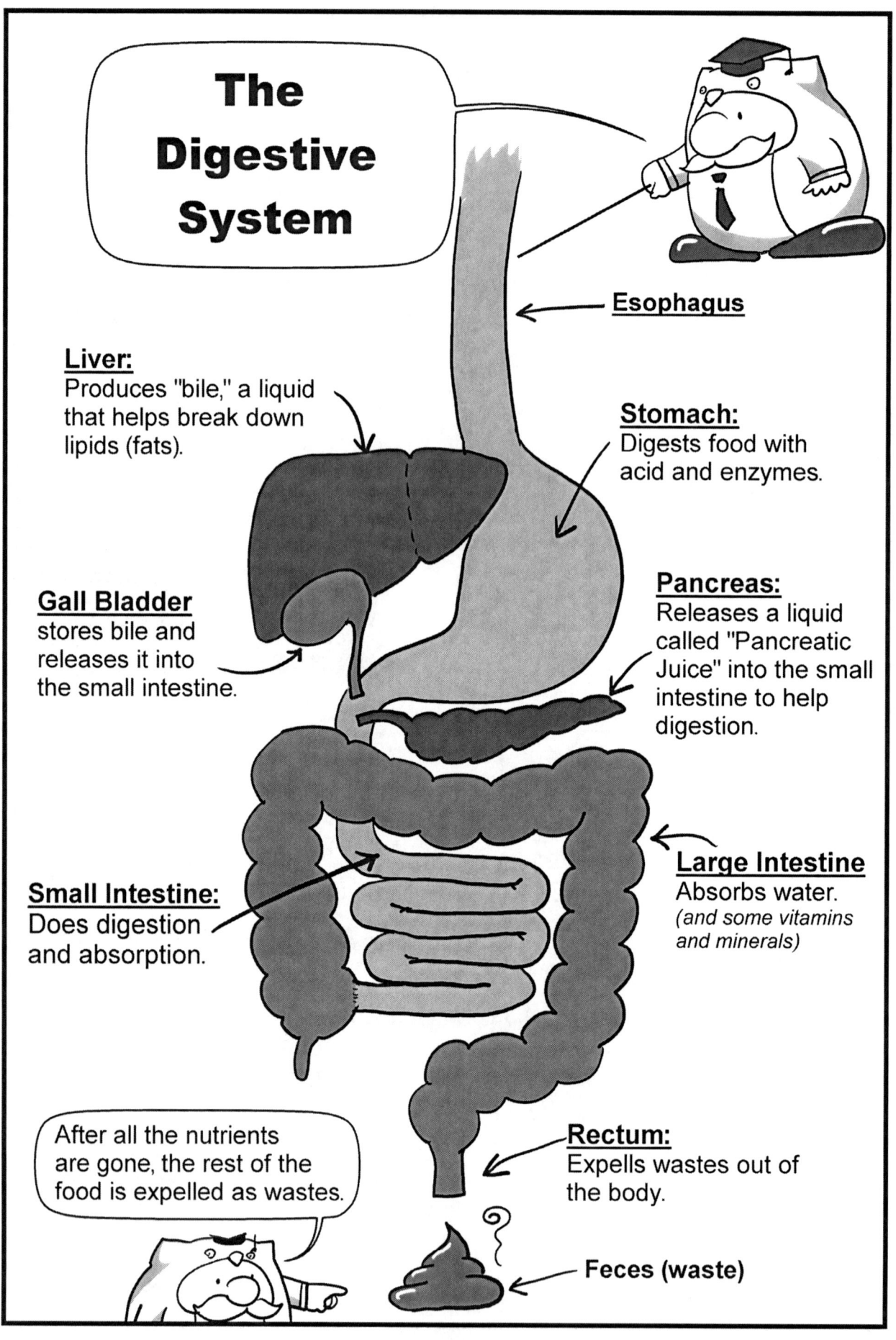

The Process of Digestion in Detail

Step 1: Food goes down the esophagus

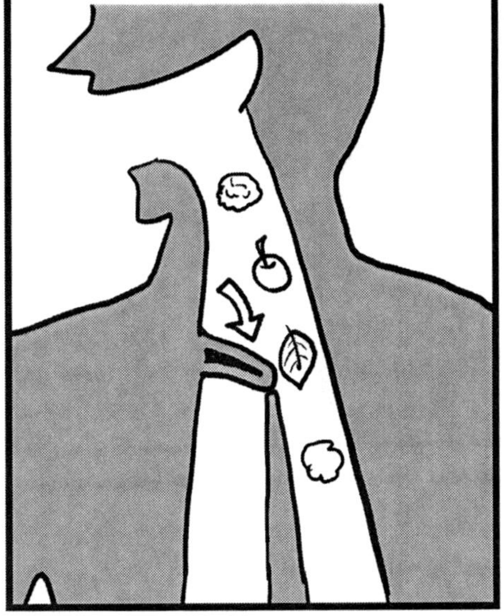

Step 2: In the stomach, digestion happens, but there is no absorption.

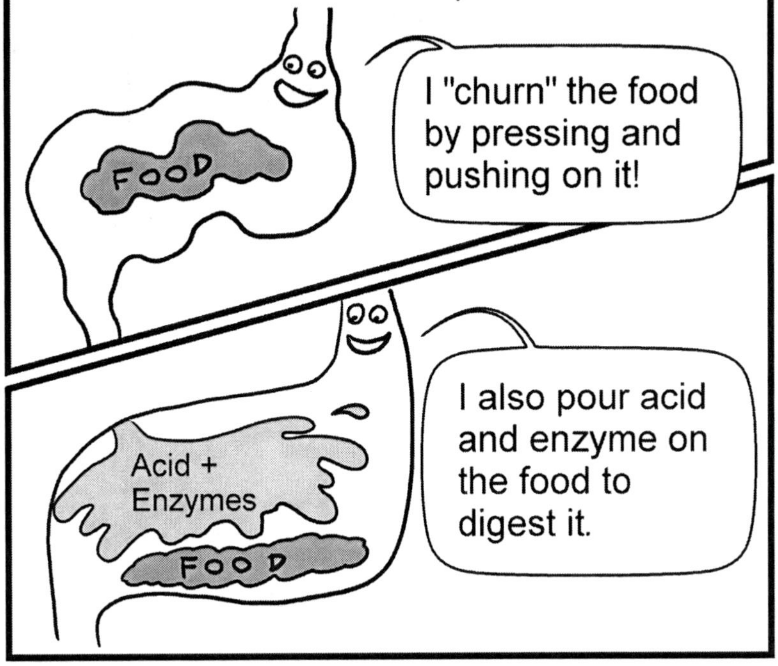

I "churn" the food by pressing and pushing on it!

I also pour acid and enzyme on the food to digest it.

Step 3: Food arrives at the Small Intestine.
Liver secretes *bile juice* unto the food.
Pancreas secretes *pancreatic juice* unto the food.
Both juices help **digest** the food.
Then the small intestine **absorbs** nutrients from the food.

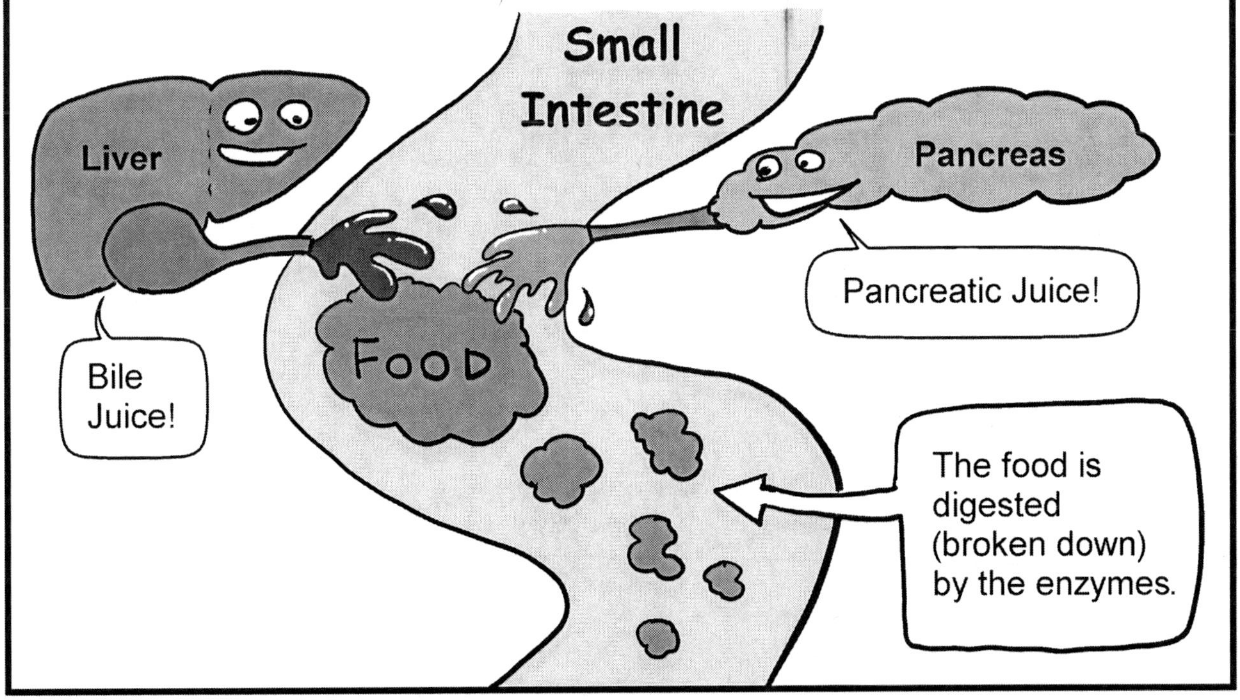

Liver — Bile Juice!

Pancreas — Pancreatic Juice!

The food is digested (broken down) by the enzymes.

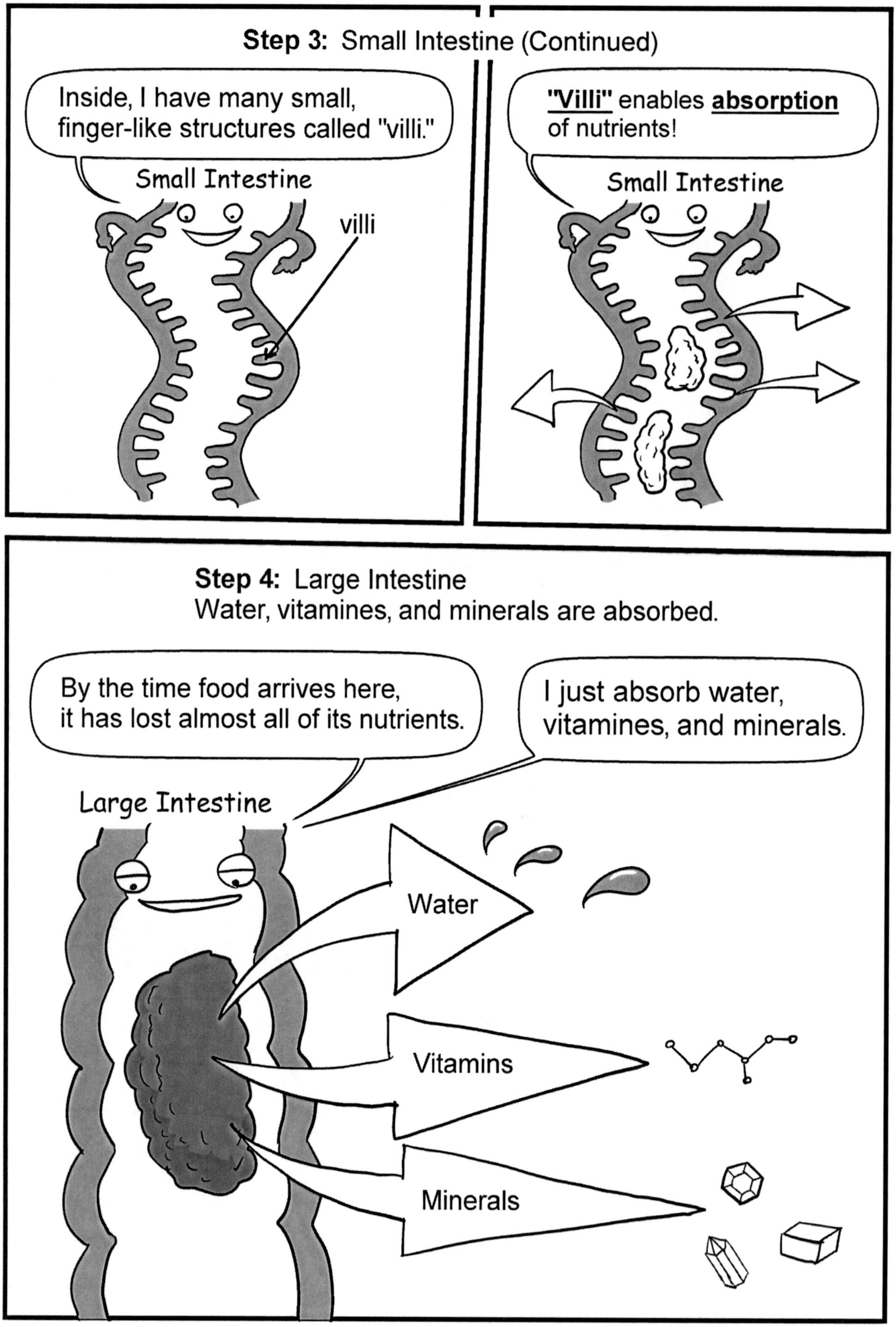

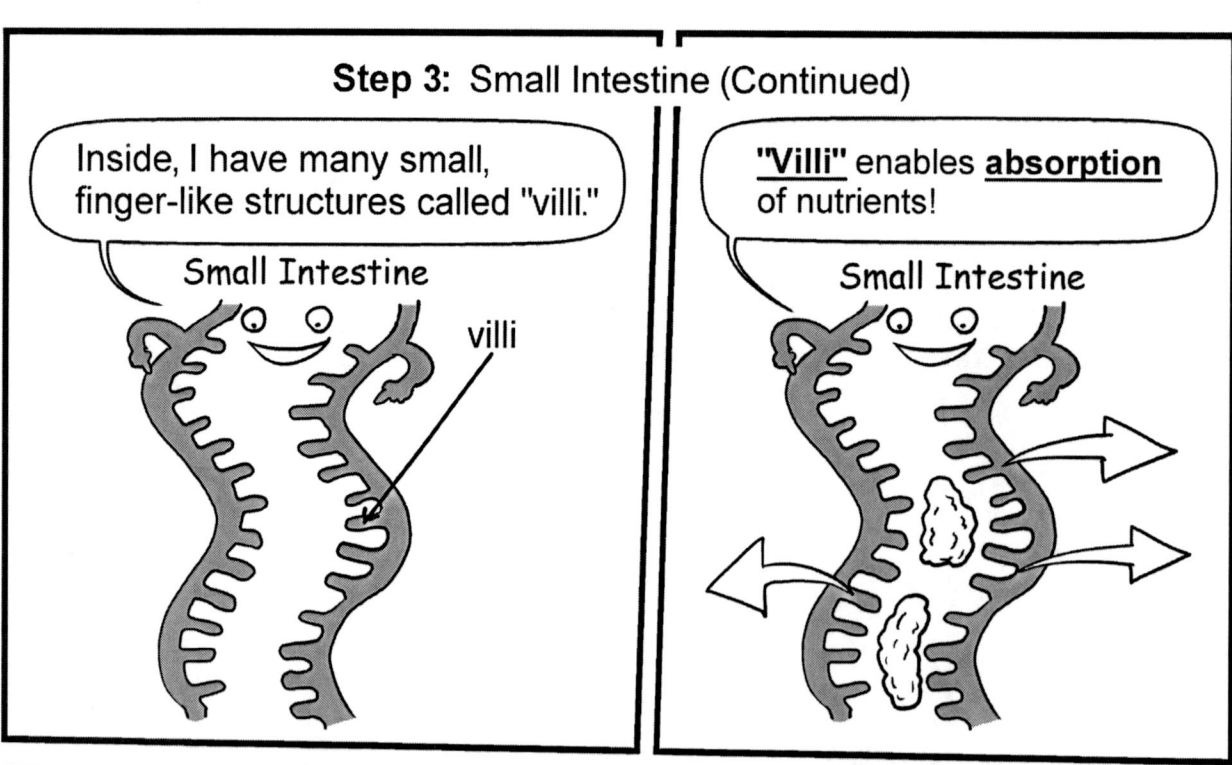

Side Note:
Where did the toxic wastes come from?

After cell takes in oxygen and nutrients...

...It releases urea and uric acid as wastes.

These things are toxic (poisonous) to living things.

The blood picks up the toxic wastes (they dissolve in blood's plasma)

"I pick up the toxic wastes!"

The wastes are dropped off at the kidneys.

"I am here to drop off toxic wastes!"

"Leave it to me!"

page 337

Unit 5: Chapter 35

Chapter 36: The Nervous System

Science Standard Bio/LS . 9 . b :
Students know how the nervous system mediates communication between different parts of the body and the body's interactions with the environment.

NEW VOCABS

* **Central Nervous System:** It consists of the brain and the spinal cord.

* **Interneuron:** A type of neuron. It helps forward important information to different parts of the body.

* **Motor Neuron:** A type of neuron. It controls the actions of your muscles.

* **Nervous System:** The network of neurons all over your body. It sends, receives, and processes information. It is made of the Central Nervous System and the Peripheral Nervous System.

* **Peripheral Nervous System:** It consists of nervous tissues outside of the brain and the spinal cord. The peripheral nervous system is spread throughout the body.

* **Sensory Neuron:** A type of neuron. It detects sensations.

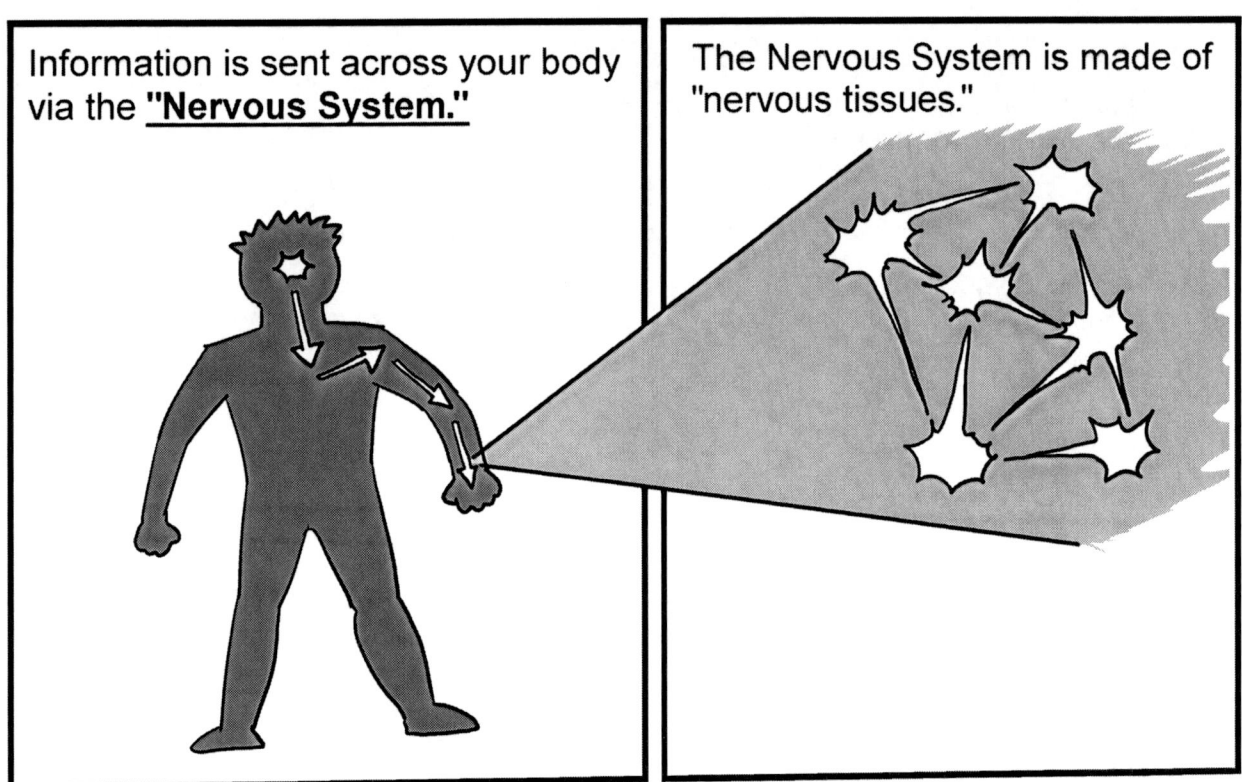

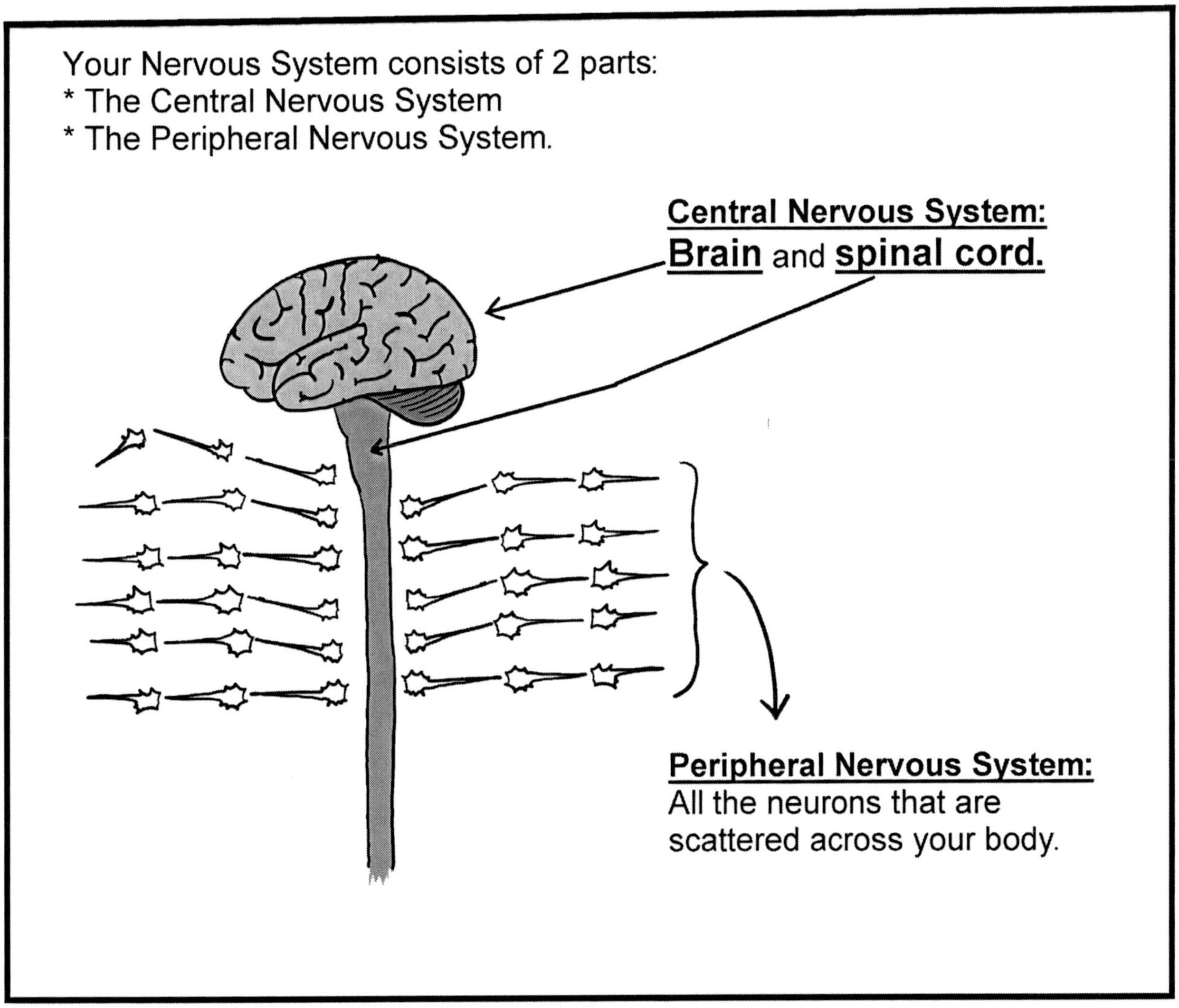

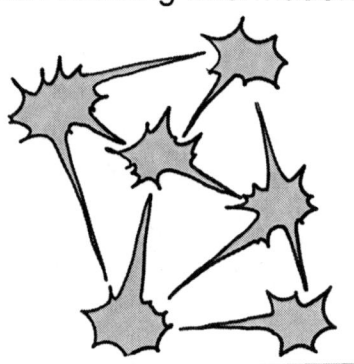

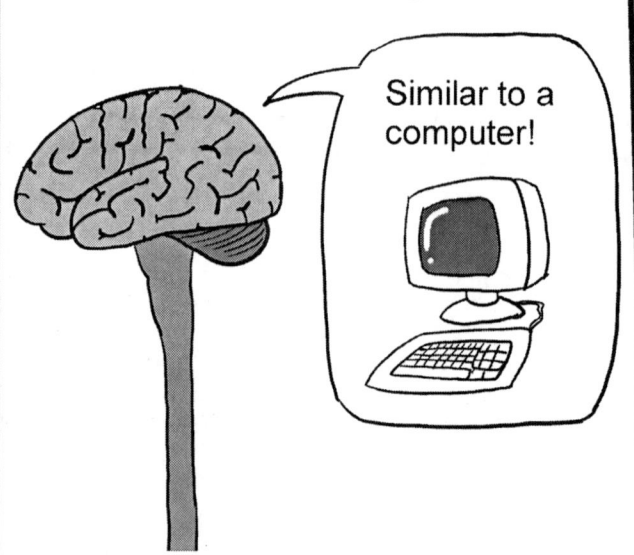

The Peripheral Nervous System is responsible for sending and receiving information.

From 1 neuron to another... Similar to a relay race!

The Central Nervous System is responsible for processing information.

Similar to a computer!

The brain is better at processing information than the spinal cord.

Brain: More powerful.

Spinal Cord: So-so.

There are 3 main types of neurons:

Sensory Neuron:
It feels sensations.
(pain, warmth, itch...etc.)

Interneuron:
It passes on information.

Motor Neuron:
It controls muscle movements.

Detect!

Special Delivery!

I move the muscles!

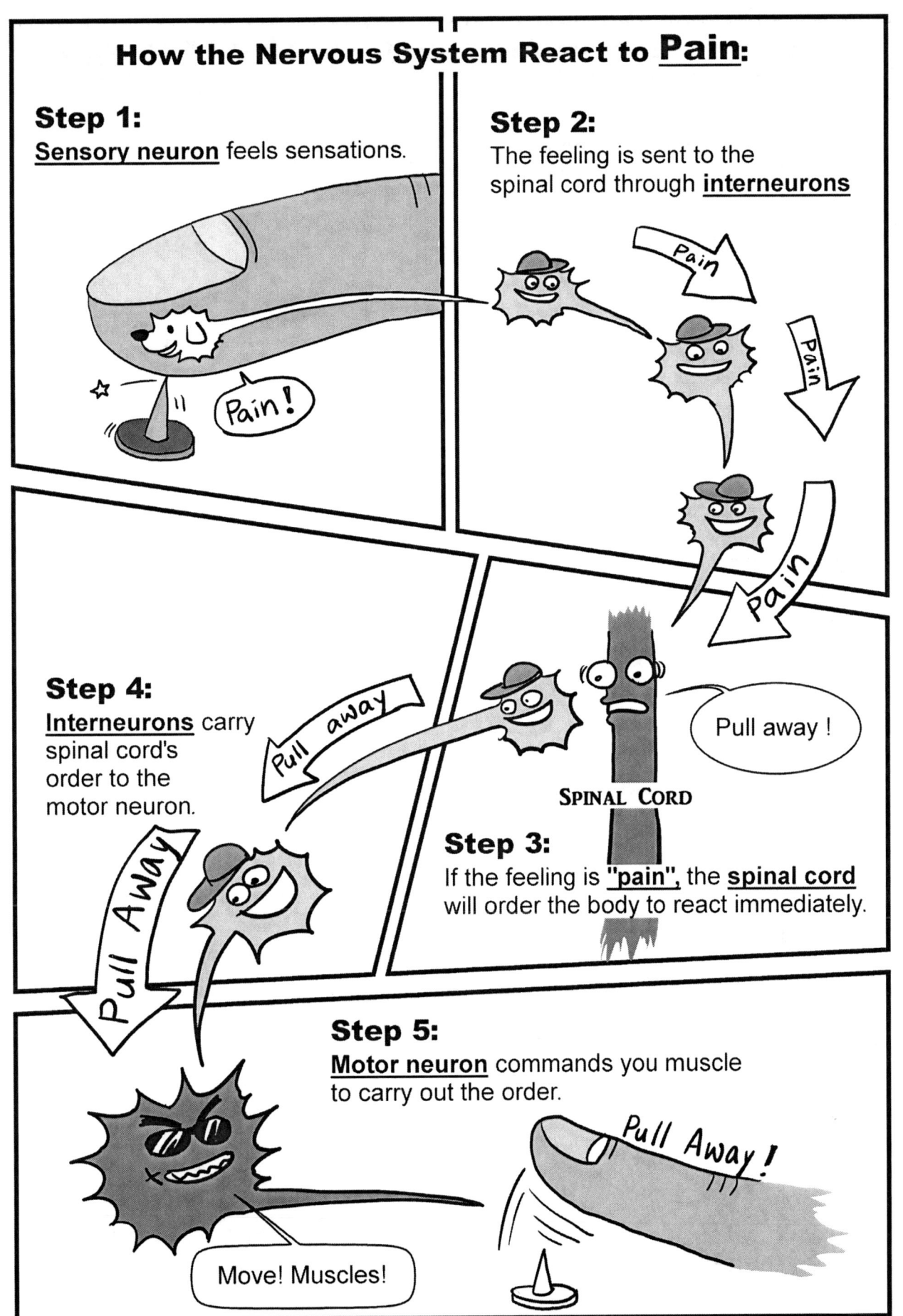

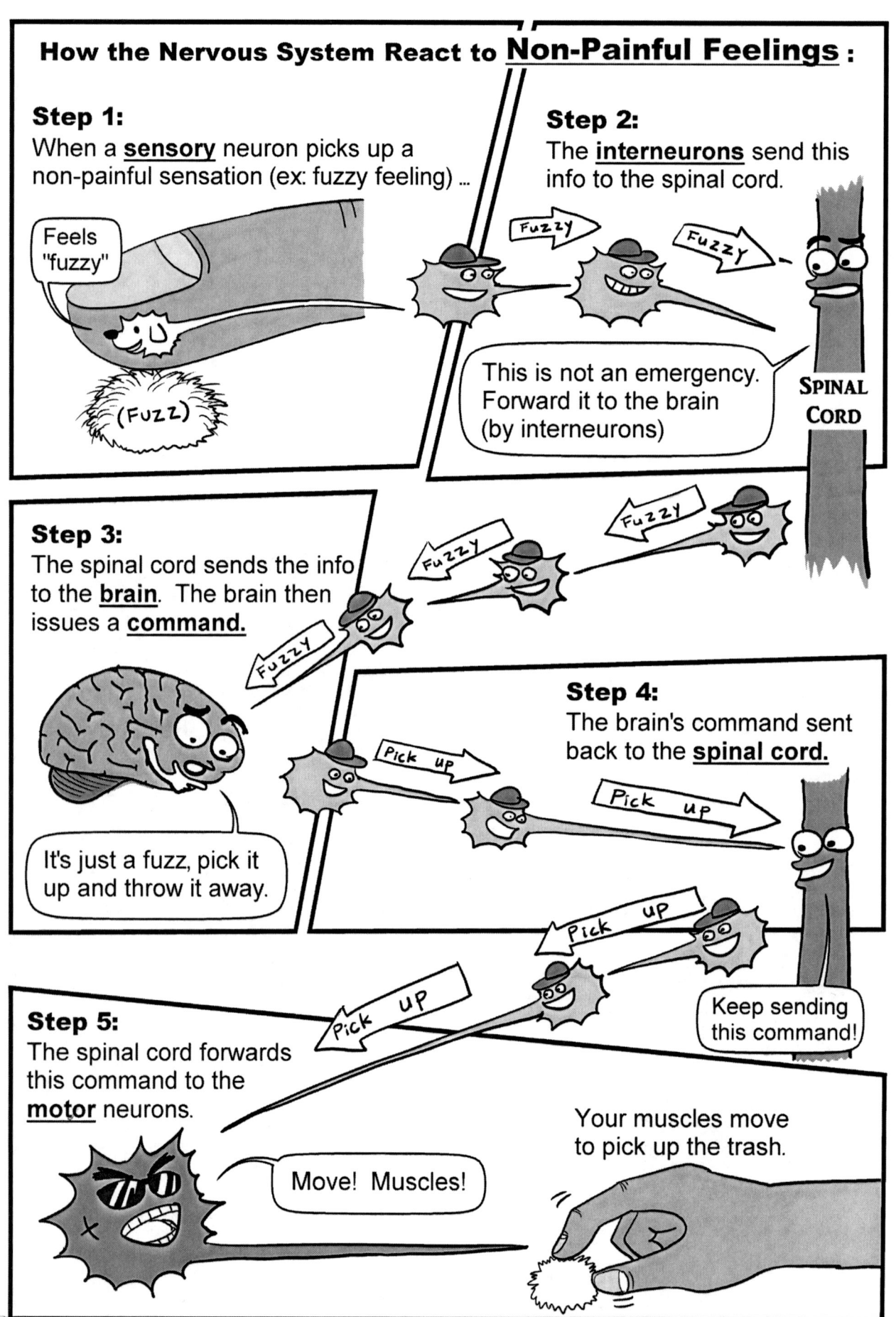

Chapter 37: The Immune Response

Science Standard Bio/LS . 10 . b :
Students know the role of antibodies in the body's response to infection.

NEW VOCABS

* **Antibody:** A substance produced by your white blood cells. It can attach to the antigens on the surface of the pathogen to weaken the pathogen, make it less powerful, and make it more vulnerable to the white blood cells.

* **Antigens:** A substance (usually made of protein or carbohydrate) that covers the surface of pathogens, and it can cause an immunological response.

* **Immune System:** A system responsible for defending an organism's body against foreign intruders. White blood cells are an important part of the immune system.

* **Immunological Response:** A process that involves your body going into "alert mode" to deal with foreign intruders.
This process is usually triggered by a white blood cell finding an antigen.

* **Inflammatory Response:** A strong-than-usual immunological response, characterized by: redness, pain, itch, swelling, and/or increasing in temperature of the affected areas of the body.
This response, while uncomfortable, can give white blood cells faster access to the affected areas.

* **Pathogen:** A micro-organism that can cause sickness and is contagious. (Common examples are viruses, bacteria, and fungus.)

* **White Blood Cells:** A type of blood cells that is responsible for defending your body against foreign intruders (Such as bacteria/viruses).

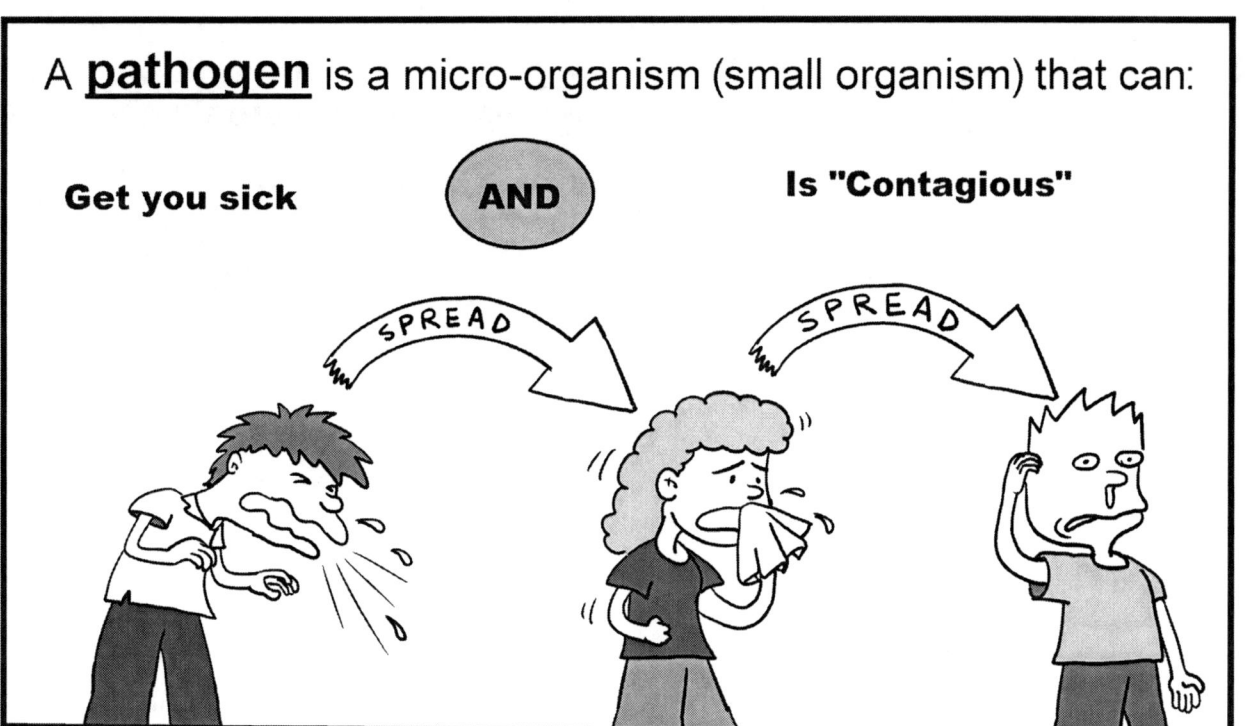

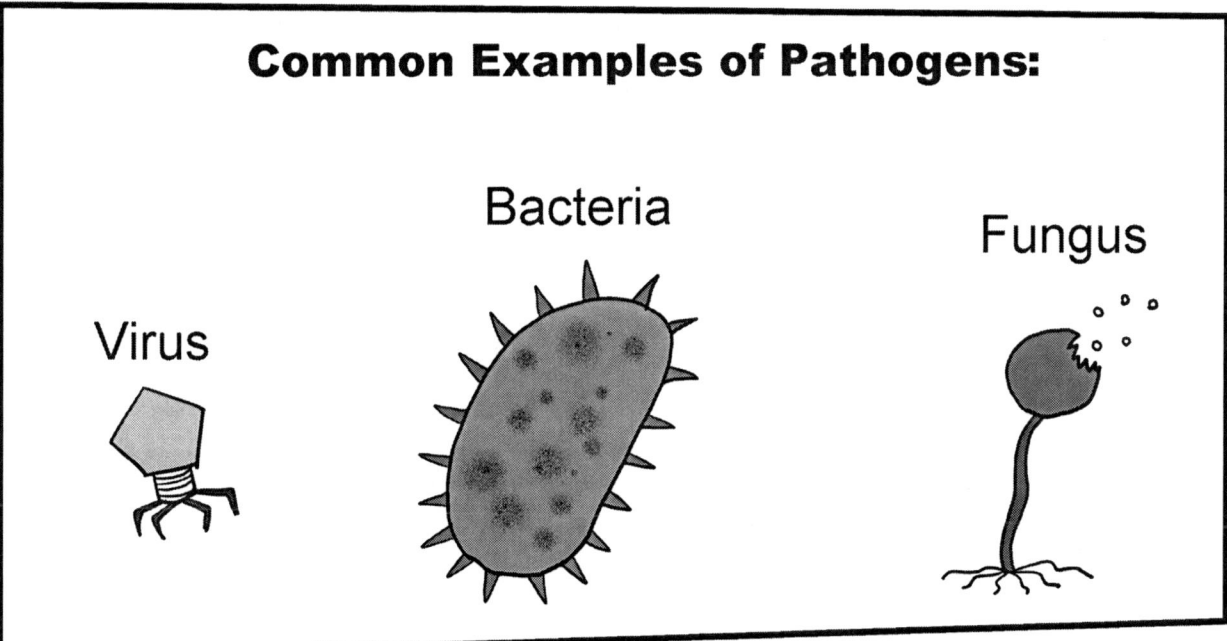

A pathogens body is covered in substances called **"antigens"**

"This stuff!"

Antigens are usually made of either **proteins** or **carbohydrates**.

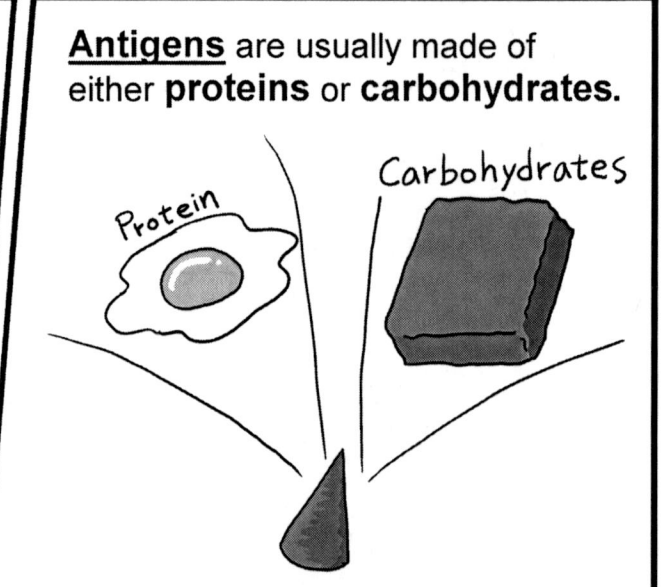

The antigens fall off from time to time...

The **white blood cells** in your body are part of your **"Immune System"**, a system responsible for defending your health..

"We *look for* antigens!"

When the white blood cells picks up the antigen, it *alerts* the other white blood cells. This alerting process is called **immunological response.**

"What!?"

"Hey guys, We have a problem!"

page 346
Unit 5: Chapter 37

When the white blood cells are alerted, they start making **"antibodies"**: A substance that attacks pathogens.

Antibodies can attach to the surface of pathogens and *weaken* them.

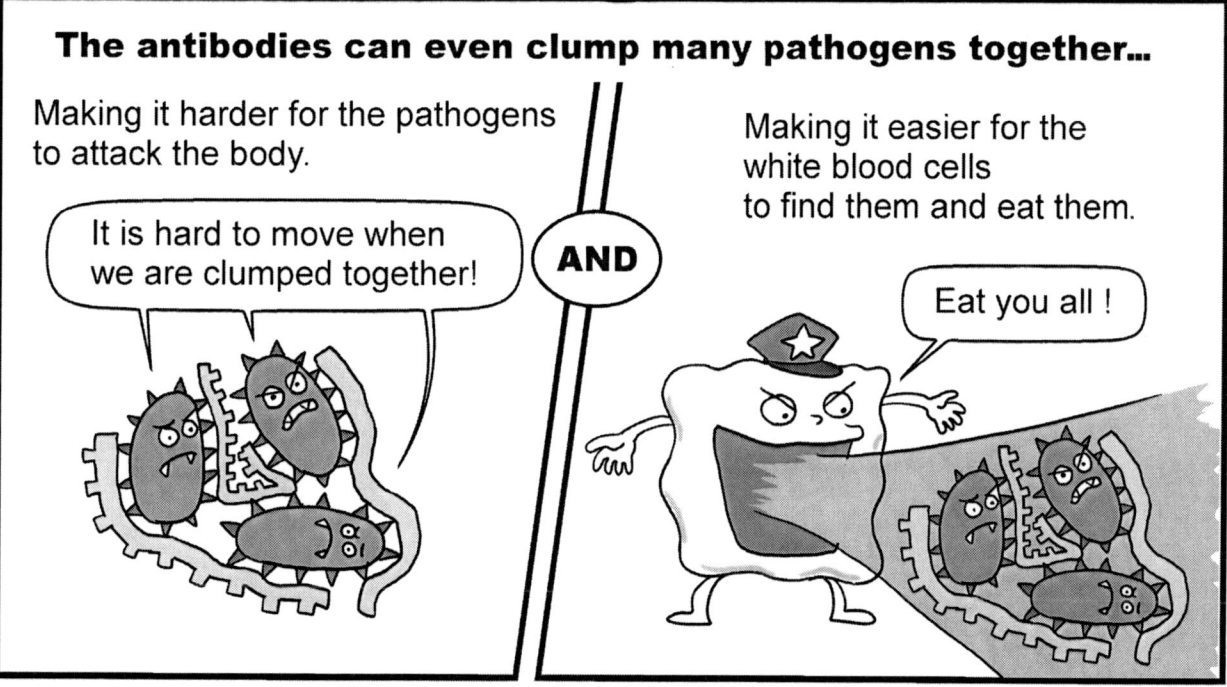

Different types of antibodies are used to attack different pathogens.

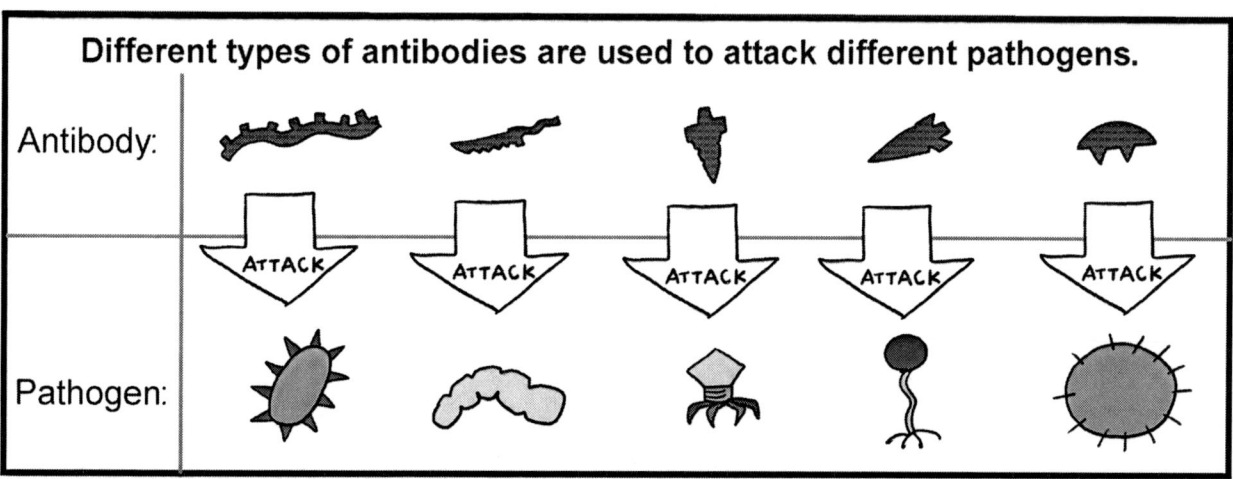

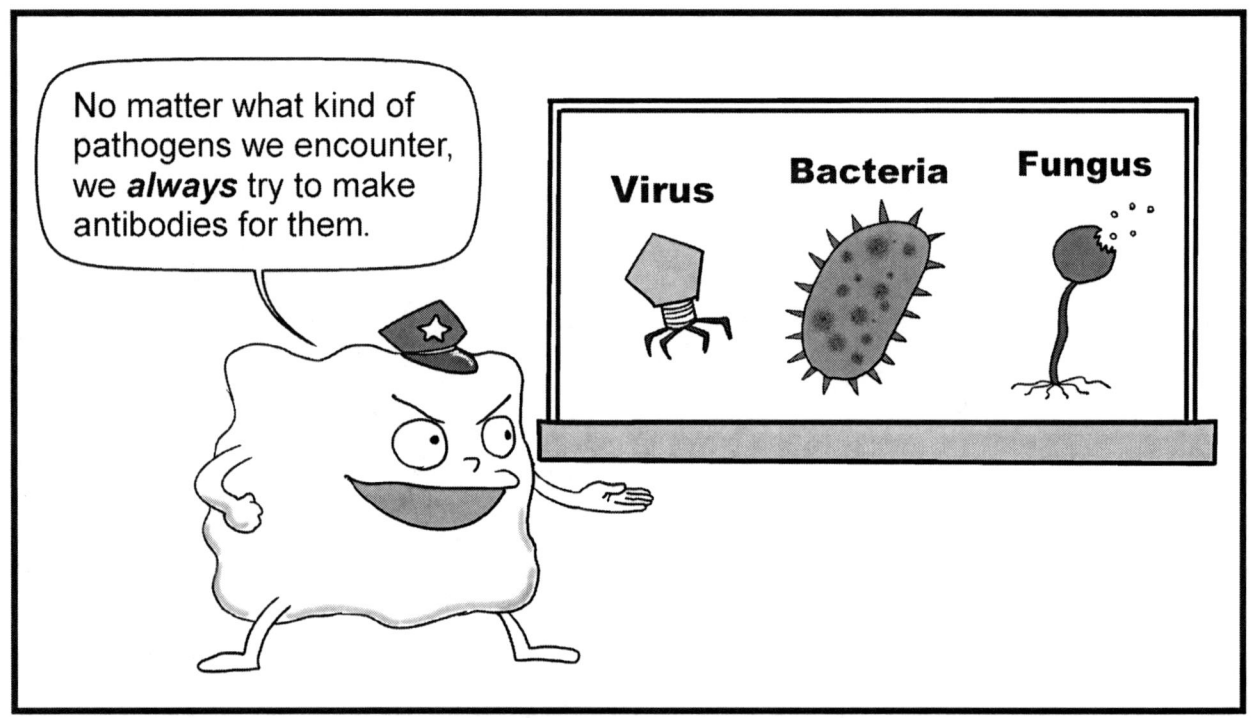

Aside from pathogens, Inflammatory Response can also be caused by...

Burns

Certain Chemicals

Pollens

And many other reasons!

ETC.

Chapter 38: Vaccines

Science Standard Bio/LS . 10 . c :
Students know how vaccination protects an individual from infectious diseases..

NEW VOCABS

* **Antibody:** A substance produced by your white blood cells. It can attach to the antigens on the surface of the pathogen to weaken the pathogen, make it less powerful, and make it more vulnerable to the white blood cells.

* **Antigens:** A substance (usually made of protein or carbohydrate) that covers the surface of pathogens, and it can cause an immunological response.

* **Pathogen:** A micro-organism that can cause sickness and is contagious. (Common examples are viruses, bacteria, and fungus.)

* **Vaccine:** A substance that increases your immunity to a particular disease. It does so by educating your immune system with information on a particular pathogen.

* **White Blood Cells:** A type of blood cells that is responsible for defending your body against foreign intruders (Such as bacteria/viruses).

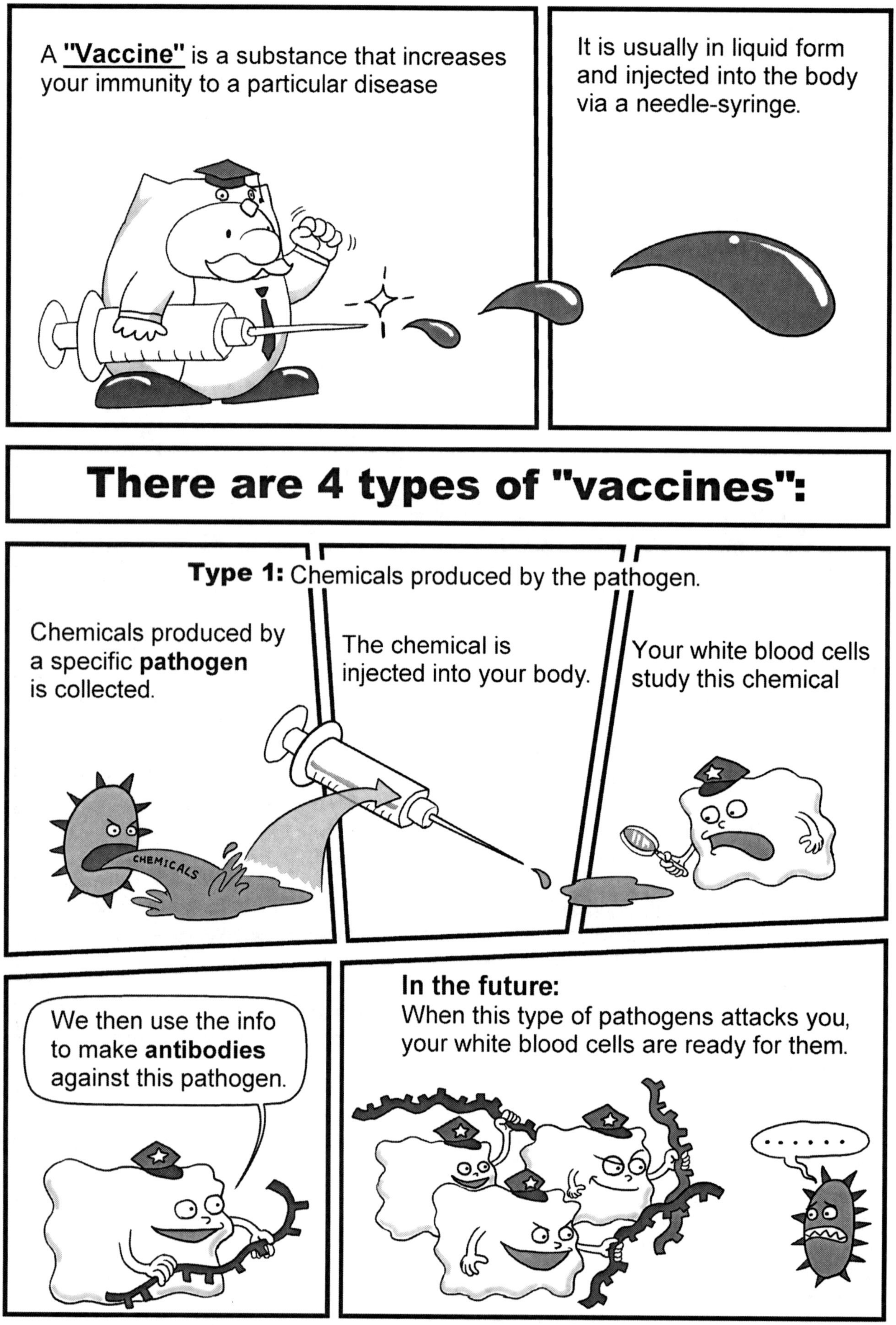

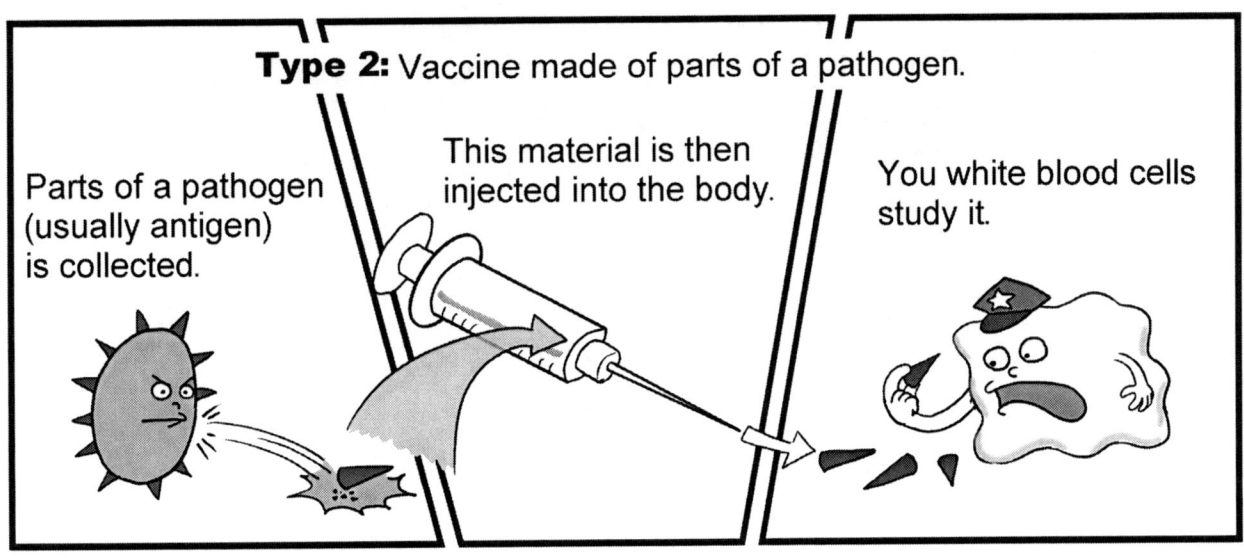
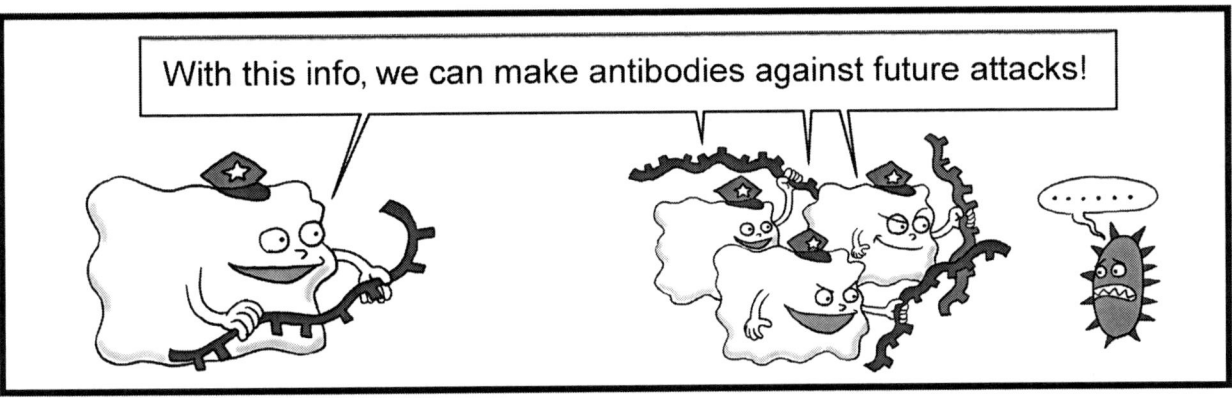
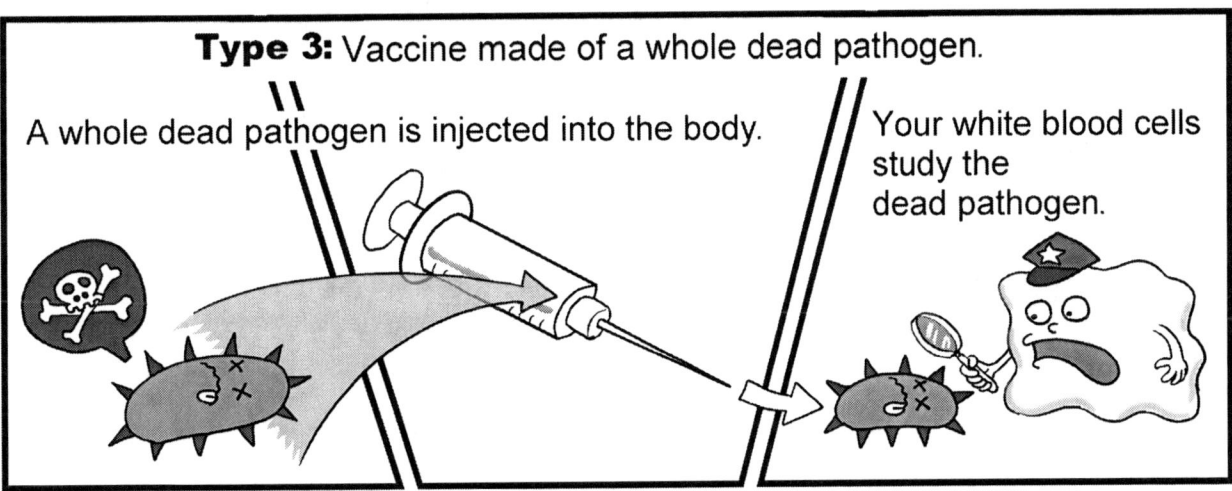
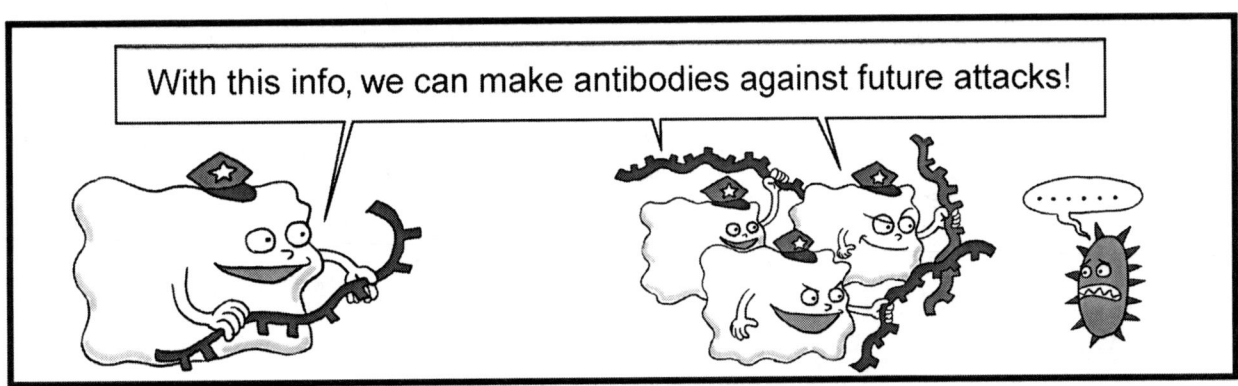

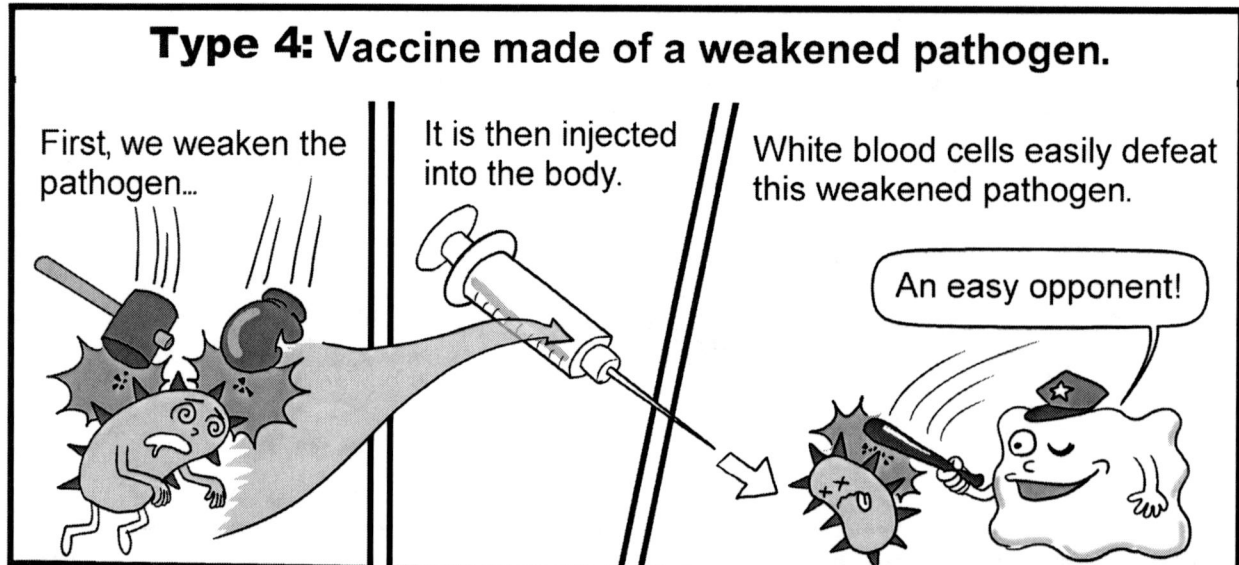

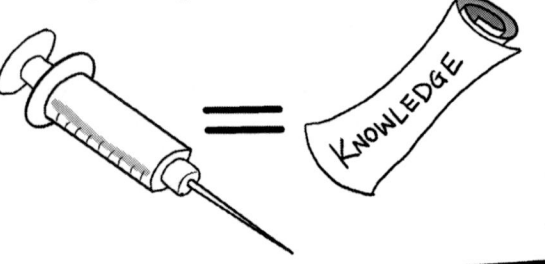

Chapter 39: The War against Pathogens

Science Standard Bio/LS . 10 . d :
Students know there are important differences between bacteria and viruses with respect to their requirements for growth and replication, the body's primary defenses against bacterial and viral infections, and effective treatments of these infections..

NEW VOCABS

* **Antibiotics:** A medicine made from collecting the chemicals secreted by certain types of fungi. Antibiotics are used to kill bacteria.

* **Antifungal Medication:** A medicine used to kill fungi and treat fungal infections.

* **Antiviral Medication:** Medications used to treat viral infections. Antiviral medications cannot kill viruses. However, they can slow down the viruses' activities and/or kill host cells that are infected by viruses.

* **Interferon:** A special chemical secreted by the host cell when it is attacked by pathogens. Interferon can attack the pathogens (especially viruses), destroy the virus-infected cells (along with the viruses inside the cell), alert the white blood cells, and help start an inflammatory response.

* **Spores (fungal):** Tiny reproductive units dispersed by fungus in order to reproduce and make more offspring.

We have already learned that there are 3 main types of *pathogens:
(*pathogens: Infectious agents that can make a person sick)

Bacteria   Viruses   Fungus

All pathogens have **antigens** all over their *surface*. The antigens fall off from time to time.

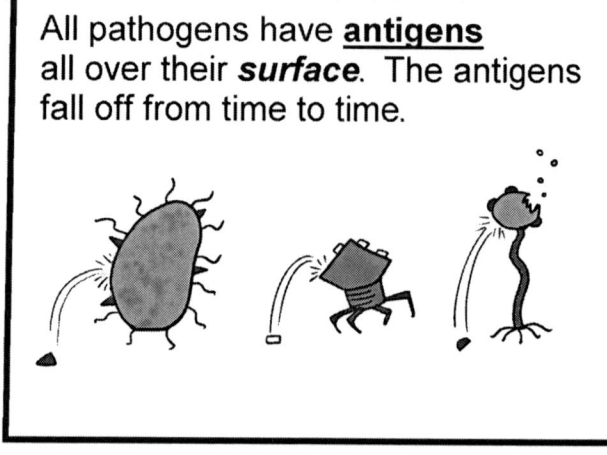

Let us look at how the 3 types of pathogens reproduce:

Fungus (plural: fungi)

"**Spores**" are tiny reproductive units spread out by fungus to reproduce..

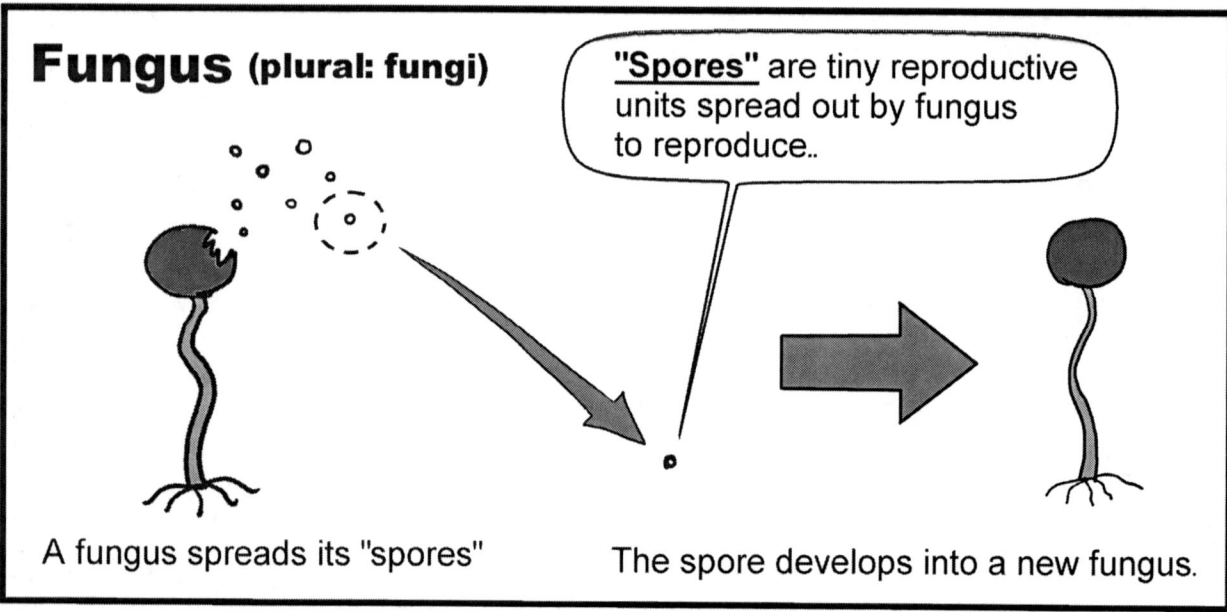

A fungus spreads its "spores"

The spore develops into a new fungus.

Bacterium (plural: bacteria)

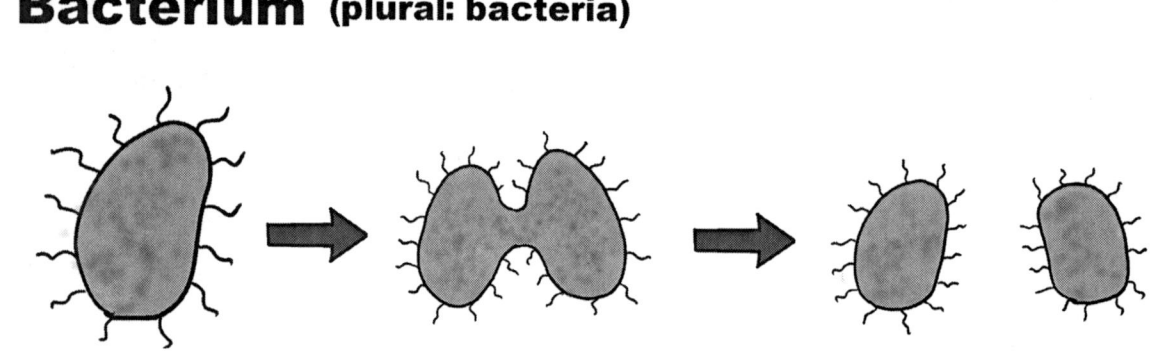

A bacterium goes through mitosis cell division

Now we have 2 bacteria.

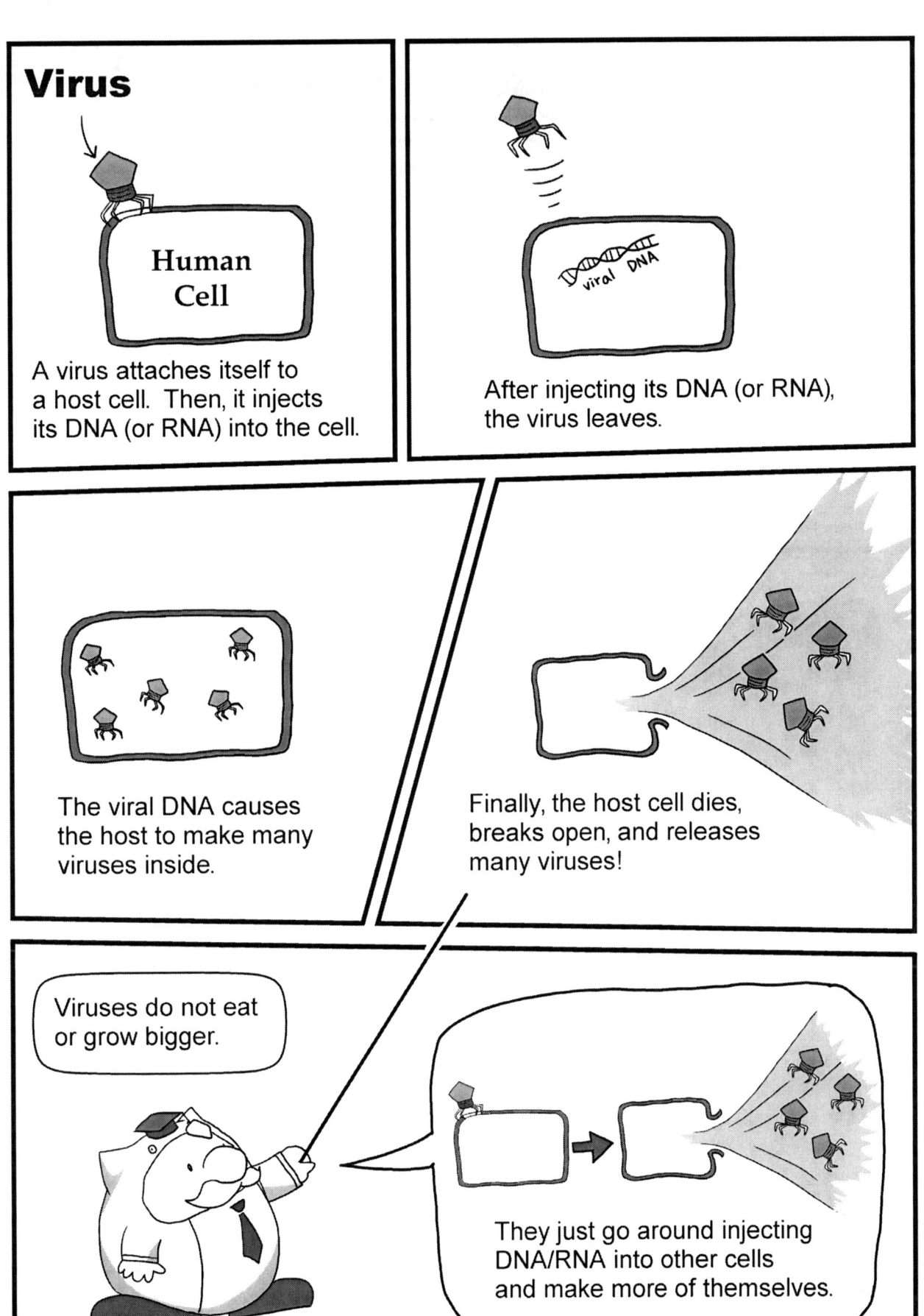

TREATMENT FOR : BACTERIA, FUNGI, AND VIRUSES.

You can kill bacteria with **"Antibiotics"**, which is a chemical we collected from certain fungi. It can effectively kill bacteria.

You can kill fungi with **anti-fungal** medications.

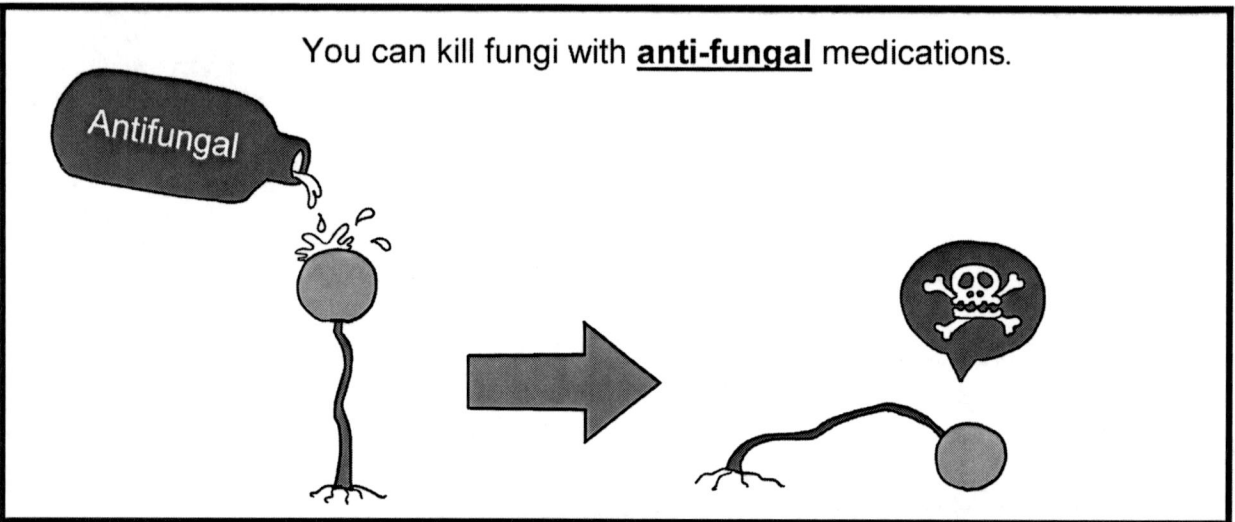

There is **no** medication that can really **"kill"** a virus.

But certain **"anti-viral medicine"** can slow down viruses' activities, causing them to do less damage to you.

How does our body fight against pathogens?

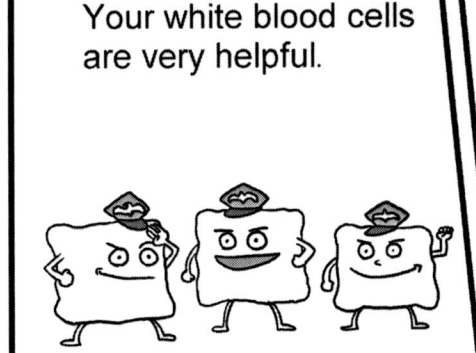

Your white blood cells are very helpful.

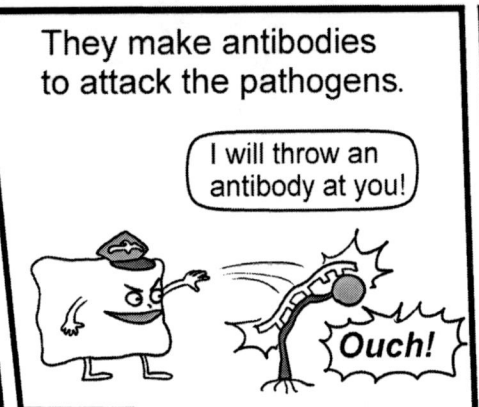

They make antibodies to attack the pathogens.

"I will throw an antibody at you!"

Ouch!

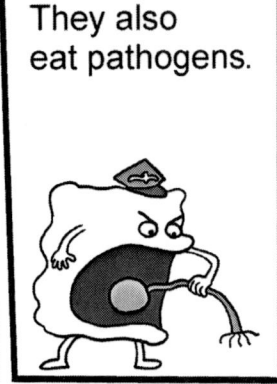

They also eat pathogens.

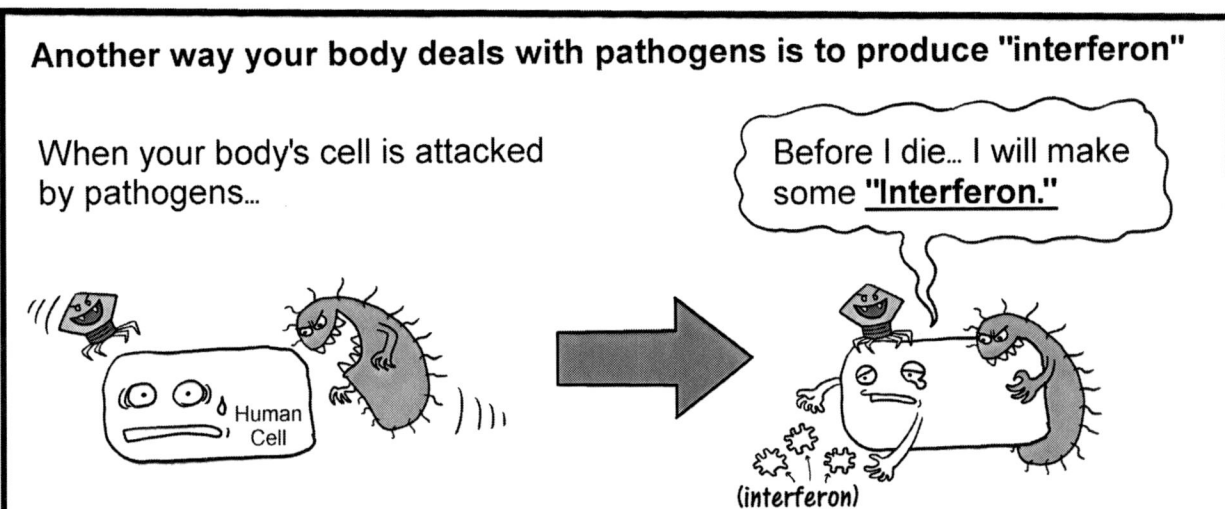

Another way your body deals with pathogens is to produce "interferon"

When your body's cell is attacked by pathogens...

Before I die... I will make some **"Interferon."**

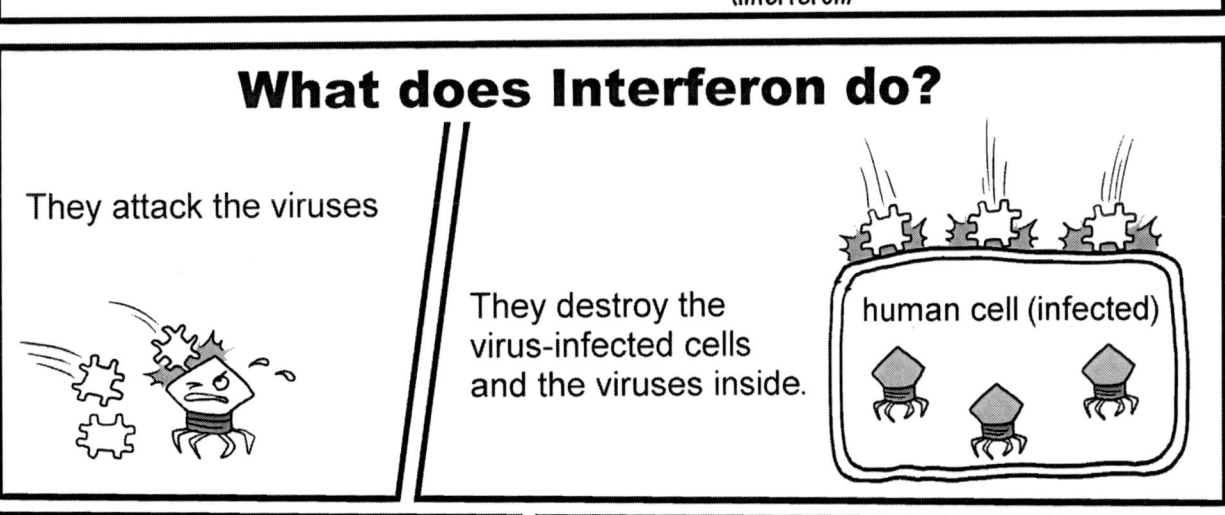

What does Interferon do?

They attack the viruses

They destroy the virus-infected cells and the viruses inside.

human cell (infected)

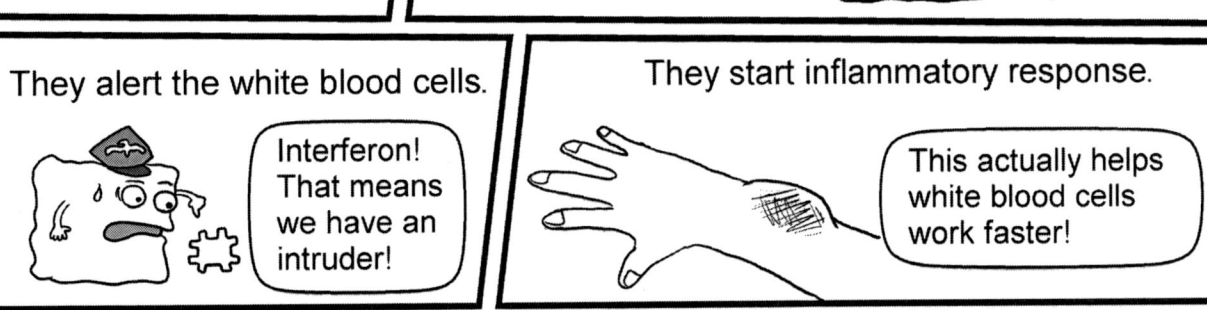

They alert the white blood cells.

Interferon! That means we have an intruder!

They start inflammatory response.

This actually helps white blood cells work faster!

CPSIA information can be obtained at www.ICGtesting.com
Printed in the USA
LVOW09s0028141015

458183LV00019B/240/P